DELAWARE AVIATION HISTORY

George J. Frebert, AUTHOR

Debbie Haskell, EDITOR

DOVER LITHO PRINTING CO., Dover, DE

Cover design
Sonja Frey

Electronic Pre-press Page Design
Sandi Irons

Electronic Halftone Scanning
Gretchen Wagner

ISNB: 0-924117-09-5

First Printing August 1998

Library of Congress Catalogue Number: 98-073148

Printed By Dover Litho Printing Company
1211 N. DuPont Highway
Dover, Delaware 19901
U.S.A.

e-mail: dovlitho@dmv.com

Dedication

To my wife, Barbara, who never complained about the countless hours I have spent at the computer, nor the mountains of research material spread all over the house.

About The Author

It sounded like distant thunder. It grew louder and louder. I ran from under the big oak tree and looked up into the sky and saw this gigantic Zeppelin. I yelled for my mother, who came running from the house. We both stared in silence as it flew overhead no higher than five hundred feet. We could plainly see the people waving from the gondola and the row of mammoth Maybach diesel engines mounted to the sides. I wondered how the earth looked to the passengers. The airship was so big it seemed out of place in the sky as it lumbered on its course and eventually out of sight.

The year was 1937, and I was a mere eight years old. Was it the *Hindenburg*? I have often wondered if it was. That event created a love for flying that has stayed with me all my life. I built model airplanes by the score. Later, I soloed from College Park Airport in Maryland, on my sixteenth birthday, August 15, 1945, which was also V-J Day. I rode my bicycle to the airport where I had worked as a line boy after school and weekends in trade for flight instruction and flying time. George Brinkerhoff was the Fixed Base Operator (FBO) and my instructor in a yellow J-3 *Cub*.

Fifty-one years later I still enjoy flying. I prefer the freedom and intimacy of the private airplane rather than airliners. A private pilot in his airplane can pick the time and the place for a flight to remember. An early sunrise, an evening sunset, a climb-out above the cumulus clouds, a day when the visibility is severe-clear, or a full moonlit night. These are some of my special times for flying.

FOREWORD

In writing this book, it was never the intention of the author to list everyone in Delaware who was connected with aviation. Countless people dedicated to aviation, regardless of race, color, creed, or gender, contributed much to Delaware aviation history. To research and document each contribution would take a dedicated staff, and years of research which were not available to the author. The objective was to locate the public-use airfields and airports (grass landing facilities are defined as "airfields," hard-surfaced facilities are defined as "airports" in this publication) throughout Delaware and document how they came to be. Delaware airfields and airports did not just happen. It took people who, for a number of reasons, committed their lives and energies to aviation beyond just flying an airplane. They created the opportunity for others to fly and aviation to expand in Delaware.

Many of the early Delaware airfields, which are now covered with buildings or closed permanently, hold an interesting history of those pioneers and the airplanes they flew. Pilots flew from airfields during all four seasons, in all kinds of weather. Navigation in the early days of flying, for the most part, was IFR (I Follow Roads, Railroads and Rivers). Some pilots flew for fun, others flew to make a living, and some died flying. They all flew for the same reason. They loved flying more than anything else.

Material Sources

Research material was obtained from the following sources:

Spirit of St. Louis, Charles Lindbergh

Record Flights, Clarence Chamberlin

Delaware's Role in World War II, Leon deValinger, Jr.

Airway to Everywhere, Lewis & Timble

"The All American World," *All American Newsletters,* National Air & Space Museum

"Bellanca News Bulletin," Company Newsletters

My Friend G.M., Elinor Smith

Giuseppe Bellanca, William M. Leary

Aviation History, Bimonthly

Air Classics, Bimonthly

Flying Minute Men, Robert E. Neprud

Coastal Air Patrol, Ray Lacharite

Airmindness, John C. Rumm

History of the Columbia Aviation, 1927

Delaware Birdmen, Delaware History, Allen Hawk

The Literary Digest, June 25, 1927

From Bonfires to Beacons, Nick A. Komons

Famous Flights That Changed History, Lowell Thomas & Lowell Thomas, Jr.

Bellanca CF, Jay P. Spenser

Ryan Broughams, William Wagner

Aircraft Piston Engines, Herschel Smith

Delaware Aeronautical Company. Hagley Museum & Library

Term paper, *Richard C. duPont, 1933-1943,* Philip B. Weymouth, III

Pioneer Pilot, Jo Cooper

History of New Castle County Airport, Delaware River & Bay Authority files

Jane's All the World of Aircraft, Jane's Information Group, Ltd.

ACKNOWLEDGEMENTS

Had this book been written fifty years ago, there would have been more early aviation people alive to help it along. But it wasn't. And thus, putting the story together was much like working on a large jigsaw puzzle. In some cases I had no information whatsoever about an airfield. Its memory was only a fleeting suggestion. In others there was conflicting information. But, with the help of many people and the tiresome poring over of maps and photographs and faded newspaper print, the story here unfolds.

The most accurate information came from early aviation publications. Original photos were very difficult to find, especially ones between 1910 and the 1940s. However, when least expected, original source materials seem to have fallen into my hands.

To *Debbie Haskell,* Editor. My inspiration. Whose encouragement, suggestions and help were invaluable when the going got tough.

To *Marguerite Sheridan,* Delaware Heritage Commission, whose proofreading expertise was essential.

To *August Bellanca,* who loaned me historic information and photos of his father's business.

To *George H. Jones, Esq.* who entrusted me with priceless photos and negatives from his grandfather's collection.

To *Frances Rees Hayford,* secretary to Giuseppe Bellanca, who loaned me her treasured photo collection of Bellanca airplanes and people.

To *Johnny Ward, Eddie Edwards, John Hite,* and *John Guidias* who lived through the era of Golden Aviation.

To *Barbara D. Hall,* Archival Specialist, Pictorial Collection, Hagley Museum and Library, who always kept her eyes and ears open for fragments of important material she knew would be useful to me.

To *Caroline du Pont Prickett,* President, Summit Aviation. A very special thanks Caroline, you opened doors to which I did not have keys.

To *Hugh Horning,* who furnished contacts and gave continued support and encouragement to the finish.

To *Bill McDaniel,* Kent County pilot and photographer, who lent me early photos from his vast collection.

To *Mike Leister and the staff of the AMC Museum,* whose help was invaluable to reconstruct the DAFB history.

To three avid supporters, and long-time friends, G*retchen Wagner, Allen Hedgecock* and *General Fred W. Vetter, Jr.,* U.S.A.F. ret.

To *Harry Van Den Huevel,* State Aeronautics Commission, who found treasured photos and material he knew I could use.

To *Eleanor Wallin,* widow of Hal Wallin, who lent me her family photo album after our interview on Rehoboth Airfield.

To *Helga Jackson,* widow of Willard Jackson, who lent me family photos and gave me a history of Milford Airfield.

To *Joe Jenkins,* whose interview covered his entire life in aviation.

To *Tom Murray, Jr.* whose interview and historic photos of the family business were invaluable contributions.

To *Reynolds Jones,* pilot, airplane owner, instructor, airfield owner, and farmer, who served Smyrna with a pilots' paradise since 1946.

To *Floyd Durham,* long-time friend and fellow pilot, who gave me anything and everything he had on Delaware Airpark and other memories of early Delaware aviation.

To *Daniel Koble,* who collected photos and material on Bellanca for years before his untimely death in 1996. His beloved wife Marjorie donated his collection to use in this book.

To *Frank Shahan,* who entrusted me with historical photos and documents from the New Castle County Airport which were treasures that were not available anywhere else.

These, and many more, are the real people who made this book possible. Many photographs, which had been stored for years, were family treasures turned over to me with misty eyes. To these individuals go my most heartfelt thanks. You have loaned more than your photos, you have given your hearts. You have also played an important part in making it possible for others to experience some of the excitement that a young industry felt as it created many aviation "firsts" in the small "first state" of Delaware.

CONTENTS

CHAPTER I

The Impossible Dream

An Aeronaut of the 17th Century.

Imagine this. The lone hawk spiraled effortlessly in the warm summer sky, taking advantage of every air current. It scarcely used its energy to flap its wings and stay aloft. On the ground a caveman, hunched in the shadows of a tree, became obsessed with the flying of the bird. He could not help but wonder how easy it might be to spot his prey from the bird's vantage point. Suddenly, the hawk's wings folded and the bird dived toward the ground at great speed, snatched his prey and flew off in the distance. The hungry caveman marveled at the ease with which the bird could obtain food without being exposed to danger. He started running across the open field as fast as he could, flapping his arms frantically in a hopeless effort to become airborne. Gasping for breath, he stumbled and fell to the ground frustrated and exhausted.

Of all the animals on earth, birds are the only ones able to fly. From the very beginning, human beings were envious because they were earthbound. Even gods and angels have been perceived enviously as having wings in the sky. Over the years the envy gave way to challenge, and the impossible dream eventually became a reality.

The Chinese flew tethered kites, the first form of aircraft, around 350 B.C. As in the drawing on this page, jumping from high places might have given the impression of flight, but it was dangerous and it didn't work. Prolonged flight had to be accomplished at all costs, and many who tried paid the ultimate price for their gallant efforts.

These early attempts at flying started with leaping from a high place with some sort of contraption that resembled a bird. Most early innovations were fashioned after observing birds in flight and, more often than not, the attempts ended in disaster. One such attempt was made in 1503 by an Italian mathematician, Giovanni Danti, when he jumped from his parish church in Perugia. He sailed about 300 paces then plunged to the earth and died.

In the 16th century, Leonardo da Vinci contributed greatly to aviation progress by inventing the airscrew, or propeller, and the parachute. He also created three different types of ornithopters, which were manpowered aircraft designed with flapping wings. He also attempted the development of hand-held gliders. In the end he broke his back testing one of them.

As early as the eighteenth century serious studies were made of eagles, with the idea that they could be harnessed to carry a man aloft. Although the idea is recorded, tests are not documented, and it can be assumed the idea was dropped. However, Sir George Cayley (1773-1857), a British inventor, built and flew the first full-size gliders successfully. Although they "flew," directional control was limited to shifting one's body weight. Shifting body weight is still widely used today in hang gliders. However, modern

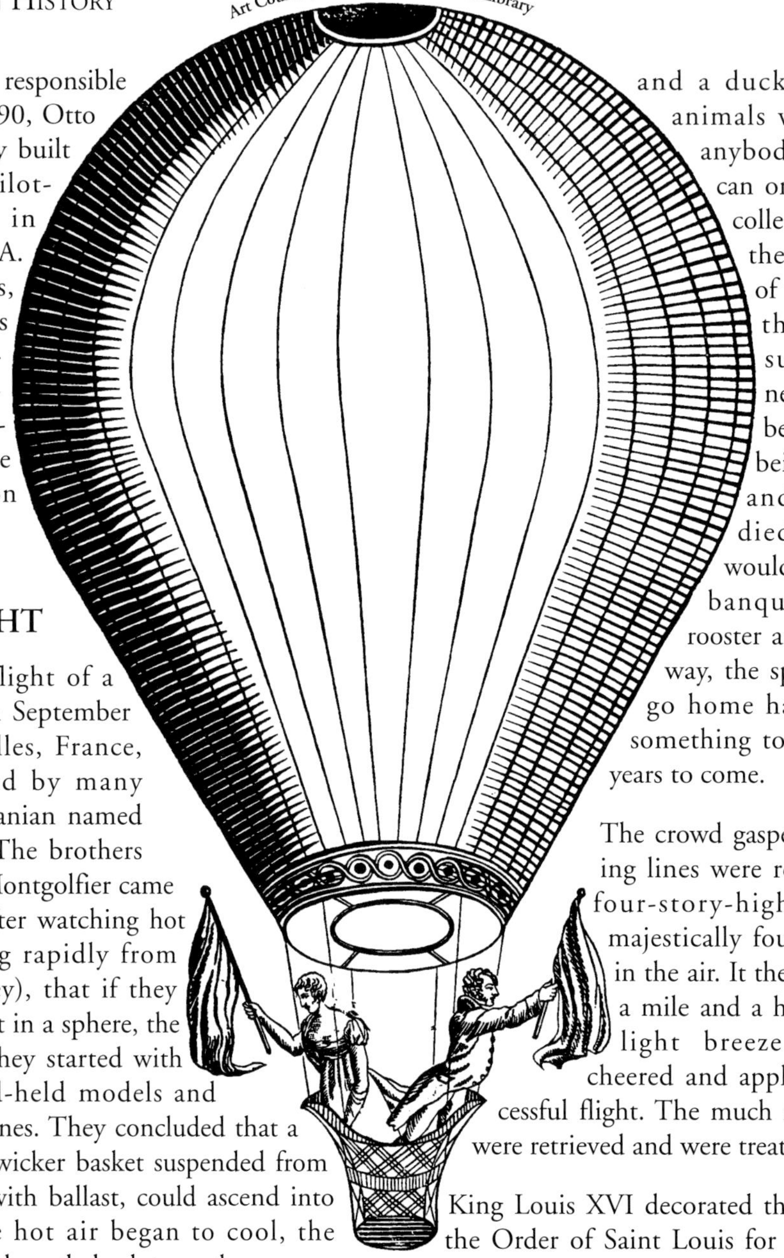
Art Courtesy of Hagley Museum & Library

aeronautical design is responsible for its stability. By 1890, Otto Lilienthal of Germany built and sold the first pilot-controlled glider in Europe and the U.S.A. The Wright brothers, on many occasions seeking advice, corresponded with Lilienthal, who was considered at this time the ultimate authority on aeronautics.

BALLOON FLIGHT

The first recorded flight of a balloon was made on September 17, 1783, in Versailles, France, which was observed by many including a Pennsylvanian named Benjamin Franklin. The brothers Joseph and Etienne Montgolfier came to the conclusion (after watching hot sparks and air rising rapidly from their factory chimney), that if they could contain the heat in a sphere, the sphere would rise. They started with small tethered hand-held models and progressed to larger ones. They concluded that a large balloon with a wicker basket suspended from the bottom, loaded with ballast, could ascend into the heavens. As the hot air began to cool, the balloon would descend gently back to earth.

They proceeded to build a forty-foot diameter balloon which they filled with hot air by building a huge platform with a fire pit below. The large mouth of the silk balloon was placed over a hole in the center of the platform to capture the hot air. As the balloon filled with hot air, the mooring lines became taut.

A large gallery of local peasants and notables including King Louis XVI and the Court of France attended the grand finale. The two brothers loaded three live animals into the wicker basket; a sheep, a rooster and a duck. (How these animals were chosen is anybody's guess. One can only assume they collectively matched the weight of one of the brothers.) If the flight was successful, the next flight would be with a human being. If it failed, and the animals died, then they would simply hold a banquet of sheep, rooster and duck. Either way, the spectators would go home happy and have something to talk about for years to come.

The crowd gasped as the mooring lines were released and the four-story-high balloon rose majestically four hundred feet in the air. It then drifted slowly a mile and a half away with a light breeze. The crowd cheered and applauded the successful flight. The much relieved animals were retrieved and were treated as heroes.

King Louis XVI decorated the brothers with the Order of Saint Louis for their incredible feat. He even offered the brothers two criminals, who had been condemned to death at the gallows, to be dispatched on the next attempt toward heaven. If they died in the attempt it would not be considered a loss to the state since they were condemned to death anyway. The downside of the king's offer was, if the flight was successful, the prisoners would become heroes or make good their escape. The King's offer was turned down when two young Italians, named Pilatre de Rozier and Marquis d'Arlandes, volunteered to accept the challenge. On November 21, 1783, they ascended from the Bois de Boulogne and traveled over five miles at 300 feet in

25 minutes. They seem to have been the first human beings who saw the earth from a manmade flying machine. This flight dispelled all thoughts of those who held steadfast that an airborne man could not survive the altitude.

Word of the event spread like a brush fire through a prairie. Once man got aboard and drifted above the towns and villages, looking down at the waving spectators, it became addictive. Building a balloon was a simple task. After buying the material, usually silk or muslin, the rest was simply a labor of love and the daring to fly it.

The problem of directional control of the balloon was addressed with some interesting innovations. One such idea was to harness trained eagles to the balloon. They would be released from aloft in the belief that they would return home, bringing the balloon in tow. Another was hand-operated air-screws. And still another was compressed-air, steam jets. None of these ideas proved practical. The balloonist finally accepted the fact that he would go wherever the balloon went.

An enterprising forty-year-old Frenchman named Jean-Pierre Blanchard had made forty-four flights over the European continent, including a spectacular flight across the English Channel by 1792. Blanchard tried to obtain directional control by making the gondola in the form of a small boat. The boat was equipped with a rudder and silk-bladed oars. It did nothing to achieve control, but did add to the show that Blanchard was all too eager to provide for a sum of money.

Blanchard then realized that no flights of any kind had been made in America. He was now ready to prove to America that balloon flight was not only real but safe as well. He announced in France that he would soon journey to America and ascend from the city of Philadelphia, the national capital.

When he arrived in America, he found public reaction so roused he felt that the masses of people could threaten the safety of his flight. He finally found an appropriate site, the exercise yard of the Walnut Street Gaol, which was surrounded by very high walls. Once the location was secured, an announcement was made that the flight would take place on January 9, 1793, at ten o'clock in the morning, weather permitting of course. The interior of the compound was reserved for the press and the invited guests who would observe the actual launching. Additional entry was charged at $5.00 per person. Sales of tickets were later reduced to $2.00 since there were many vantage points available on rooftops from which to observe the event.

Notables and politicians, including President George Washington, John Adams, Thomas Jefferson, James Madison, Alexander Hamilton, James Monroe, Edmund Randolph, and the Governor of Pennsylvania Thomas Mifflin, attended the gala affair. They watched in silence as Blanchard operated his hydrogen generator, secured the wicker basket and placed ballast aboard. The ballast included a couple bottles of fine wine and a loaf of French bread donated by the invited guests. After all, how could a self-respecting Frenchman, about to risk his life flying this contraption, depart this earth without an ample supply of wine? The bread was optional.

The balloon began to get bigger and bigger as Blanchard frantically ran around it pulling wrinkles free. With all the fanfare a Frenchman could muster, Blanchard got in the wicker basket with his companion, a little black dog. President Washington came forward, shook his hand, and presented Blanchard with a document that offered safe passage throughout the land of America.

Cannon fire had sounded periodically throughout the city to notify the population the event was indeed ready to take place. People crowded the streets for miles around to witness the first flight in America. Blanchard gave the signal to release the lines and set the balloon free. The crowd cheered and went wild with delight as he rose to an altitude of 1200 feet in the cold morning sky. He drifted across the city and the Delaware River into New Jersey, reaching an altitude of over five thousand feet before landing in a farm field near Woodbury, much to the dismay of the farmer. The resourceful Frenchman, who spoke no English, offered the farmer a swig from a bottle of the wine so generously given to him at departure. He then showed the farmer the document prepared by President Washington. The first farmer calmed down, but there were others racing across the field with raised pitchforks and guns in hand. Fortunately, the first farmer, with the assistance of both bottles of Blanchard's choice wine, got the other farmers to quiet down. Once the wine was consumed, they agreed to carry his balloon to the shore of the Delaware River. There, sailors retrieved him and his balloon, and all returned to Philadelphia to a hero's welcome.

Ballooning would be the sole method of flying for a number of years to come. Ironically, none would fly in America for another thirty years. James Mills of Baltimore would be the first American to fly when he flew from Lancaster to Cooch's Mills in Delaware on November 1, 1834. Therefore it should be noted that James Mills was probably the first person to see Delaware from aloft.

Tethered balloons were first used for military reconnaissance in the Battle of Maubeuge in 1794. Another military application was during the siege of Paris in 1870-71, when 168 citizens escaped over the heads of the Prussian armies in 66 separate balloon flights.

By design, balloon-flying is the most restrictive method of flight. The pilot, if you could call him one, has no control over the direction of flight. A balloon is an aerostat. Therefore there is no airspeed. The balloon becomes part of the air mass and will move in whatever direction the wind blows. The velocity of the wind equals the ground speed, which can cause bodily harm when attempting to land in a high wind. Landing a balloon in a wind above eight miles per hour can be more than a bit challenging. When landing, the basket will be dragged across the ground, with the occupants hanging on for dear life, until the balloon is deflated. Altitude is controlled by the pilot, in a fashion, by dropping ballast of sand/water or valving gas. Unfortunately, once the decision is made to drop ballast, there is no way of wishing it back.

In spite of the hazards, men like Samuel King of Philadelphia and John Wise of Lancaster County spent most of their lives mastering the art of flying balloons. Fourteen-year-old John Wise, who started ballooning in 1822, made flights up to twelve hundred miles in his lifetime. He initially started to experiment with hot air balloons after reading about the Montgolfier brothers' success. Like the Montgolfiers, he also started with small models and worked his way up in scale. Unfortunately, he did not have a wide selection of animals available for ballast. He decided to load the family cat into a cage under the balloon and set it on its merry way. According to Wise, the flight was a complete success. Unfortunately, the landing destroyed the balloon. The pussycat survived the ordeal but was not aeronautically inclined to repeat the event. The cat kept a healthy distance from John from that day forward. In another incident, John Wise and his balloon experiments came to a complete halt when he almost set his father's barn on fire. The particular balloon John built had the ability to carry its own fire in the gondola which theoretically would supply heat to sustain longer flight. Unfortunately, young John did not consider the wind direction that was blowing toward the barn filled with dry hay. His heart pounded in disbelief as he watched the balloon heading for the family barn. Only the shouting of an excited John alerted his father and other farmhands in time to extinguish the blaze before much damage was done. Needless to say, this event brought John's ballooning experiments to a complete halt.

Once Wise reached the age of twenty-one and was no longer under the control of his parents, he resumed flying balloons. During the 1840s John Wise probably made more balloon flights than any other person in America. He wrote two books on the subject and was honored with the title, "The Father of American Aeronautics." He made his last flight at the age of sixty-four from St. Louis, Missouri, when he accepted an invitation to make "one more balloon flight" from the place he had established a record 1200-mile flight thirteen years earlier. The balloon lifted off without incident and rose several thousand feet while drifting slowly toward Alton, Illinois. He was never seen again.

Even after powered flight was achieved, Samuel King flew for more than sixty years until he was eighty years old and proclaimed proudly that he was "the

Baron von Abercrom (in cap) supervises the recovery of his balloon from the S. H. Lowber cornfield two miles north of Little Creek, Delaware, October 23, 1907. Photo by Herman C. Taylor, from the George Jones collection.

World's Oldest Living Aviator." His statement was never challenged. He taught many students the art of flying a balloon expertly by removing ballast and valving gas. He taught them the art of feeling the wind, even though a balloon pilot in flight never experiences anything but dead calm. He made balloons by the hundreds and sold them with free flight instructions.

In an effort to obtain controlled flight, Dr. Fredrich Wolfert designed the cigar-shaped balloon and mounted a Daimler gasoline engine to the frame. He obtained limited success, and the design led to the rigid airship designed by Count Ferdinand Zeppelin of Germany. The Count built the monstrous 420-foot-long LZ1 which flew on July 2, 1900. The Zeppelin would control the skies until the destruction of the *Hindenburg* in 1937. The Germans were so taken aback by what they considered a minor achievement by Lindbergh, they cranked up one of their Zeppelins and promptly flew around the world. There were many in aviation who stood fast in their argument that the airplane would never attain any prominence in aviation. Airplanes were too fragile and could never offer the comfort and range of the Zeppelin.

Contests to see who could fly their balloon the longest distance were established around the world to determine the best pilots. Such a contest was organized in St. Louis, October 22, 1907. One of the contestants was the Baron Hugo von Abercrom, from Germany. After forty hours in the air and traveling over 800 miles he and his aide, Hans Hiedermann, landed in a cornfield near Dover. The account was reported by Herman C. Taylor, staff reporter for the Wilmington newspaper, *Every Evening,* October 23, 1907. In part, Mr. Taylor reported;

> The German balloon 'Düsseldorf', piloted by the Baron von Abercrom and Hans Hiedermann, which left St. Louis, Monday, October 21, at five o'clock landed five miles east of Dover this morning after flying for over forty hours. The balloon was first sighted flying over Dover shortly after 8 o'clock as it

passed northeastward. The occupants at that time were close enough to the people in their fields to talk to them.

Both pilots reported a safe voyage after they landed in the cornfield of S. H. Lowbar, two miles northeast of Little Creek. They said they had flown as high as 10,000 feet before landing. After resting and smoking a cigar, the Baron supervised the loading of the balloon onto a farm wagon by many sightseers. The airship and its attachments were then transported to Dover and loaded on the afternoon express for New York.

Three other balloons also departed St. Louis for the race, one of which flew over Wilmington heading northeast, according to city lineman Vaughn. The third balloon was reported over Goldsboro, Maryland, but was not reported being sighted again.

Sighting a huge balloon flying over Dover in 1907 must have created a year of conversation at the very least. At this point in time, sighting an automobile would have been a spectacular event in Dover. An honest to goodness airship over Dover! Who would believe it?

On May 27, 1931, Auguste Piccard and Albert Stevens set an altitude record of 51,775 feet, to penetrate the stratosphere in a helium-filled balloon. The record for a parachute jump was from a balloon in 1960. Air Force Captain Joseph W. Kittinger left the gondola of his balloon at 102,800 feet and fell for four and a half minutes before his chute opened.

There is something special about free balloon flight in that there is no noise from a power source. There is a feeling of being suspended in space. The sounds from earth are not muffled or distorted by trees, houses, and other obstructions. Once high enough that you are out of range of earth's manmade noise, there is no sound. The silence becomes absolute. Total silence is something humans do not adapt to easily.

It is not surprising that the sport of free ballooning is alive and well today. For the most part, free balloon flights are kept below two thousand feet. Balloon flying today is covered by the same Federal Air Regulations (FAR) as any other aircraft. A pilot's license is required. Balloon inspections are also required in the FAR. A thorough knowledge of micrometeorology is extremely important. Most balloon flights are made three hours before sunrise or three hours before sunset to avoid the unstable air of the day. Landing any aircraft on someone's property without permission is still considered trespassing. Damaging a farmer's crops, power lines, or other property becomes the pilot's liability. A thoroughly trained ground crew is equally important to help launch and retrieve the balloon. Hand-held, two-way, radio communications are also a must. There are flight schools available to teach one the art of flying balloons. The price for a private pilot's license runs about $1500. It would be advisable to join the Balloon Federation of America before attempting to venture into this field of aeronautics.

Flying balloons in Delaware is a rare sight today. Annually, the Delmarva Hot Air Balloon & Craft Festival holds a spring meet in Milton, Delaware, each May. Weather permitting, hot air balloons take off at dawn and late afternoon.

THE AIRPLANE IS BORN

Leonardo da Vinci, who dabbled in the theory of aeronautics came to the conclusion that the screw (now called the propeller) used to drive a boat could be modified to become an airscrew and propel an airship. His theory was indeed correct. Unfortunately, the only power available at the time was a steam engine which was far too heavy to use on an airship. Ultimately, the source of power would be discovered in Pennsylvania. That power was *petroleum,* discovered in 1859 by Edwin Drake of Bethlehem.

Oil had been in use for more than five thousand years by the Greeks prior to the discovery of petroleum. That source was whale oil which was used for lubricants and lamps. Although the discovery of petroleum was made in 1859, it took twenty years of research to separate the components to create *naphtha,* an early form of gasoline.

The first internal combustion engine was invented by Nikolas August Otto who received a patent in 1877. As with any new development, it was in its infancy and far too heavy and short on power to be used in an airship. As more research and knowledge were applied, these problems were ultimately solved. There is no doubt that petroleum and the internal combustion engine created the industrial revolution.

Langley's aerodrome floating in the Potomac River. Delaware Public Archives photo.

Without them, we would still be ballooning or jumping from high places with hand-held gliders.

Octave Chanute, Alexander Graham Bell, and Samuel Langley (all highly educated men of the time) spent much of their lives studying the science of aeronautics. Chanute was a civil engineer who supervised the laying of railroad track across the country. He is also credited with the design of the famous Chicago Stockyard. Langley was a man of many talents which included engineer, physicist, astronomer, and aerodynamicist. He also served as an astronomer at Harvard University in Cambridge (Massachusetts) before moving to a new position as director of astronomy for Allegheny Observatory near Pittsburgh in 1867. Once there, he accepted a position at Western University (now University of Pittsburgh) as professor of physics and astronomy.

Chanute's and Langley's interest in aeronautics was sparked by international recognition of Otto Lilienthal's successful experiments with man-carrying gliders in Germany. Langley designed a wind tunnel to observe wind effect on wings. From these observations he created numerical graphs and charts to set standards which would be used to determine centers of gravity and wing loading. His first attempts to master powered flight were with rubber band-powered models built of lightweight balsa wood and covered with tissue paper. Over the years, Langley would build over thirty test models with wingspans reaching over seven feet.

In 1887, in addition to all his other responsibilities, Langley accepted an appointment as Secretary of the Smithsonian Institution located in Washington, D.C. He held this position for over twenty years while still living in Pittsburgh and meeting his other commitments.

As his models began to improve with testing, he wrote in a scientific journal that heavier-than-air mechanical flight was possible. Langley built a twenty-eight foot wingspan, steam-powered model that flew an astonishing 4200 feet in forty-five seconds. Proof of the event was captured on film by Alexander Graham Bell, president of the Smithsonian. Bell, a renowned inventor in his own right, had been experimenting with tetrahedral and geometric kites since 1900. Some of his kites were so large and complex that it took three sturdy men to hold them in rein. Alexander Bell now joined William Thaw, University of Pittsburgh, in financial support of Langley's aeronautical experiments. However, it took the war between Cuba and America to generate the vast amounts of funds Langley and Bell needed.

Langley's aeroplane is perched above the floating aerodrome in the Potomac River. A catapult mechanism would launch the aeroplane to flying speed to keep it from stalling. Delaware Public Archives photo.

Workers in 1903 rescue of Charles Manley and the aeroplane from the cold waters of the Potomac River. Delaware Public Archives photo.

At Santiago, a U.S. Army tethered observation balloon was shot down, and the observer was killed. This action prompted the War Department to urge President McKinley and Congress to appropriate $50,000 to have the Smithsonian and Langley research and develop a suitable aircraft for military use. Langley and Bell had always struggled with funding, and their newfound wealth created great enthusiasm. The major obstacle to overcome was power. A steam engine capable of developing forty horsepower was far too heavy. The internal combustion engines used in automobiles of the day were balky, unreliable, under-powered and also too heavy. The only engine source that met the weight requirement, but fell short on power, was the lightweight motorcycle engine of the day.

A meeting with Stephen Blazer, a motorcycle manufacturer, finalized an agreement to build a lightweight internal combustion engine to Langley's specifications and to be delivered by February 1899. Unfortunately, when the engine was delivered it fell far short of the requirements. Langley, now sixty-four, was angry and assigned his assistant, Charles Manley, the task of improving the engine. Failure followed failure until success was finally reached in 1903, when Manley's 187 lb engine developed a remarkable fifty horsepower. By October 1903,

"First Flight" Kill Devil Hill, North Carolina, December 17, 1903

Manley was at the controls of an airplane set upon an aerodrome floating in the middle of the Potomac river. Highlighted by much press and fanfare, a large contingency of government officials and the public watched as the craft roared off the platform with whirling propellers and flopped unceremoniously into the Potomac.

Criticized by the press, Langley stood firmly behind his design and attributed the failure to the launching mechanism. The press hounded Langley unmercifully for more details. He refused any further comment which only made the press more inquisitive. By December 8, the airplane was repaired and returned to its perch atop the aerodrome in the middle of the Potomac for another attempt at heavier-than-air powered flight. Once again pilot Manley revved the engines to full power, released the launching mechanism and emulated the first attempt. This time the ice-cold waters of the Potomac almost cost him his life.

Needless to say, the press were not kind to the Smithsonian, Bell, or Langley. They demanded a full congressional investigation of the squandering of $50,000 of taxpayer money. Langley, now sixty-nine, despondent over the failures and publicity, withdrew from public life.

CHANUTE AND THE WRIGHT BROTHERS

Octave Chanute, world renowned glider pilot, had been corresponding with the Wright brothers, in Dayton, Ohio, for several years, and they had become very close friends. The Wrights also corresponded with the Smithsonian and Langley. Three years before Langley's failed attempt to fly off the aerodrome on the Potomac, the Wrights were flying tethered kites at Kill Devil Hill in North Carolina.

Orville and Wilbur Wright owned a successful bicycle shop in Dayton, Ohio, where they seriously studied the theory of flight. The Wright brothers required two ingredients when they searched for a test site; steady winds and solitude. Unlike Langley and others, who professed their ability to fly to the press or anyone else who would lend an ear, the Wright brothers conducted their tests quietly. They concluded that Lilienthal's and Langley's mathematical computations appeared correct on paper. However, they did not hold up in practice.

It was Chanute who suggested to the Wright brothers they accept, without compensation, the assistance of George Spatt, a medical student from Coatesville, Pennsylvania, who also delved seriously into aeronautics and airfoils. Reluctantly, the Wright brothers agreed to accept his presence on a trial basis to satisfy a long-time friend's request. In the end, George Spatt became a valuable asset to the Wright brothers because of his knowledge in the field of aeronautics. Thereafter, he was invited to return year after year. In fact, it was George Spatt who concluded that the approach to lift vs. drag computations, developed by Lilienthal and Langley, was inaccurate. George's approach to the problem was to balance the lift of a surface against the drag and then determine the relationship of the two factors directly to the aircraft. This theory is still valid today.

In the fall of 1903, the Wright team was ready to attempt the first powered flight. Unfortunately, when the engine was tested, the sprockets became loose and caused breaks in the propeller shafts. The only person whom the Wright brothers considered capable of welding the sprockets to the shafts was Charles Taylor. Taylor was the master machinist who built the engine and now ran the Wright Bicycle Shop while the Wrights were away. George Spatt volunteered to hand-carry the broken shafts back to Norfolk, then forward them via Railway Express to Dayton. He then boarded a train for a return trip to Coatesville for Christmas. His decision to return to Coatesville caused him to miss the historic "First Flight" by a controllable, powered aircraft on December 17, 1903.

The Wright brothers made four flights that cold and windy December day. The first flight, by Orville, was 120 feet long and lasted about 12 seconds. The longest flight was made by Wilbur, which covered 852 feet in 59 seconds.

One would think that this historic event would create unprecedented publicity. On the contrary, the complete opposite is true. The news media, apparently after witnessing the two failures of the renowned scientists Samuel Langley and Alexander Bell, decided any statements made on controllable, powered flight were mere hogwash. This was further fortified by the fact that Langley's attempt was supported by Federal funds, the Smithsonian and the University of Pennsylvania. How in the world could two unknown bicycle mechanics from Dayton, Ohio, achieve such a feat when others with more experience and money had failed?

In spite of the fact that the Wright brothers returned to Dayton and flew their airplanes for two years in front of thousands of spectators, the press held steadfast in their belief that it simply was not possible. The first documented published news that the Wrights actually flew was written two years after the first flight by the editor of *"Gleanings of Bee Culture,"* January, 1905. By this time, the Wrights were flying twenty-five miles, turning in circles and returning to the point of departure. Once the news finally hit the newspapers, along with undisputed photographs, the Wright brothers became international heroes.

CURTISS ENTERS THE FREY

The barrier was now broken. Alexander Graham Bell, founder of the Aerial Experiment Association (AEA) invited Glenn Curtiss, manufacturer of motorcycle engines, to join the fray. The AEA consisted of highly recognized men such as John McCurdy, Casey Baldwin, and Army Lt. Thomas Selfridge. Curtiss was developing an engine for Thomas Baldwin, who experimented with dirigibles. The AEA knew that a light-weight engine which could develop suitable power was the major obstacle to prolonged, heavier-than-air, controlled flight.

Glenn Curtiss went to Ohio to visit the Wright brothers, who freely discussed their technical revolution with anyone who would listen. Armed with mountains of aeronautical information, the AEA started to build an aeroplane. Their first successful flight was February 23, 1909, from Canada's frozen Baddeck Bay. John McCurdy flew directly off the ice with the tricycle landing geared airplane, *Silver Dart,* for over one-half mile. The event, witnessed by nearly 150 people, was the first powered flight in Canada. The AEA also built an aeroplane called the *June Bug* and entered it in competition for the Scientific American Trophy.

The Wright brothers were angered immensely when their airplane was disqualified from competition because it was launched from rails rather than a takeoff from the ground. The AEA won the coveted trophy with their Curtiss-built aeroplane.

Undaunted, Wilbur Wright crated the Wright *Flyer* and shipped it to France where he would demonstrate the aeroplane. European leaders, and the military, were more than anxious to witness first-hand what they had heard so much about. Simultaneously, brother Orville shipped the second Wright *Flyer* to Fort Meyer in Virginia to demonstrate the Wright *Flyer* to the War Department.

Within five years the public interest in aviation would grow to mountainous proportions. Rodman Wanamaker, a Philadelphian, purchased a French-built Bleriot XI in 1909 and flew it locally. Within two years he would help finance a Curtiss multi-engine flying boat to fly the Atlantic. This aeroplane would make history by being the first aeroplane to cross the Atlantic via Newfoundland. When plans for the French-designed *Demoiselle* were offered in *FLY* magazine, the entire supply was sold out in seven days. The sale of plans to at home builders created a new industry of aircraft suppliers which still flourishes today.

The business world took advantage of the publicity. Cash prizes were offered to accomplish flights which appeared impossible. Sponsors were eager to lend financial support if their products were displayed and supported. Armstrong Drexel from Philadelphia took an hour and ten minutes to capture the world altitude record of 9970 feet in his 50-hp Bleriot in 1910. Unfortunately, the Federation Aeronautique International (FAI) refused to honor the record because he did not break the old record by 100 meters. Cal Rogers won the prize for duration at the Chicago International meet in August of 1911. He flew his Wright *Pusher* for twenty-two hours and collected $10,000 for his efforts. He later flew his Wright *Flyer, Vin-Fizz,* the name of a popular soda, from New York City to Pasadena, California, in a gallant effort to win the $50,000 prize offered by Randolph Hearst. The 4,231 mile trip required a supply train to guide the way and carry spare parts. The railroads were referred to as "the iron compass" by pilots in the early days of navigation. Crowds gathered along the tracks at every small town to witness this unbelievable attempt to fly from coast to coast. The prize stipulated the flight be made in 30 days. Rogers survived 19 crashes and missed the prize money when he arrived in San Diego in 49 days. America's first woman pilot, Harriet Quimby, crossed the English Channel alone in a French-built Bleriot on April 16, 1912. Unfortunately, her extraordinary feat took a back seat to the tragic sinking of the *Titanic.*

Although the airplane was now accepted as a flying machine, few people felt it had any practical application. Most pilots were considered daredevils, or hell-bent on committing suicide. In spite of many failures, competition grew between individuals, groups, and countries, all of which accelerated development and publicity. Contests and prizes were offered to test the daring aeronauts and their flying machines. After each prize was won, a more difficult challenge was offered with a prize which often included money and a coveted trophy. It was cheap advertising for the sponsor who received national press coverage. The pilot of course would enjoy the financial reward, and an ego boost that would surely motivate him to accept the next challenge. Flying was here to stay.

THE DELAPLANE

A 1909 Dover Attempt

After the announcement that the Wright brothers flew, it appears the flying fever extended into Kent County as it did in New Castle. The brothers Enoch S. LaFevre and W. D. LaFevre of Dover, on August 30, 1909, were granted a patent on their flying airship. It is not known if the airship was built and tested. However, an elaborate analysis of its capability was printed and submitted to major newspapers for publicity.

> The LaFevre Airship has a vertical mast securely fastened to the bottom of the ship. Near the upper end of this mast are stationary aeroplanes, with controllable slats that can be thrown open when the airship is to ascend, and can be set at any angle or entirely closed when going forward or descending. Underneath these stationary aeroplanes are eight vertically lifting aeroplane propellers, four of them revolving like large kites around this central mast in one direction, each of the four discharging their air underneath each other, thus affording each aeroplane a stronger body of air to float upon, than the mere air it gathers in itself. And then there is another set of four counter-revolving, lifting aeroplane propellers which are revolving around the

LeFevre Brothers' Airship
The artwork and following text were copied from an original printed flyer dated 1903. From the George Jones collection.

mast in an opposite direction. These two sets of counter propellers keep the ship from twisting or whirling around and enable the ship to go steadily on its way. And these two sets of counter-revolving propellers are also constantly feeding each other with discharged air and thus adding greater lifting power to the ship. If each of these revolving-lifting planes were only eight feet by ten feet in size that would give it eighty square feet of lifting surface, and there being eight of them gives these lifting planes eight times eighty, or six hundred and forty square feet of lifting surface. These eight revolving planes alone have therefore considerably more lifting surface than any of the airships now in existence. And when we take into consideration that these are at once driven by the engine at greater speed than any of the other aeroplanes, it is evident that the ship must at once ascend into the air.

When as high as desirable to go, the slats on the stationary aeroplane can be closed, and then you have again added to the ship whatever amount of lifting surface is contained in the stationary plane, so that the LaFavre Airship will be capable of carrying large and heavy loads. And these it will carry with ease and safety. The stationary aeroplanes, revolving-lifting aeroplanes, and parachute are all made of light materials and are all overhead and above, and the carriage, engine, machinery, and passengers are all central underneath, so that their weight will always keep the ship balanced and in an upright position. It will require no attention whatever to keep the ship balanced. And if anything gets out of order or machinery ceases to work these aeroplanes and the parachute would enable the ship to come down slowly and in perfect safety.

An upright mast also works the side propellers, one set on either side of the ship. When these two sets are run at equal speed the ship will move straight forward. But by the mere movement of a lever the velocity of these side propellers can be changed, so as to revolve slowly on one side and rapidly on the other, and thus like oars on a boat can change the course of the ship in any direction desired, or turn it clear around.

Both the lifting propellers and these side propellers are so arranged that both can be worked simultaneously or either set alone and separately. The lifting set alone and go vertically up into the air, and can then turn off the lifting propellers, and turn on the horizontal and go forward, or can go upward and onward at once.

The LaFevre Airship is comparatively small and compact and can start off or come down almost anywhere. And the parachute will turn off the wind and rain and sunshine and afford a comfortable voyage.

Although a thorough search was made for photographs of the finished product, none were found. This does not mean that the LaFevre Airship was not built. Photographs of this era are virtually nonexistent. It also does not mean it never flew. It must be remembered that in the early days of aviation, achieving an altitude of one foot was considered a great flying accomplishment. Based on today's aeronautical knowledge, it is doubtful that this invention ever fulfilled the aspirations of the inventors. However, it does document the forward thinking of those committed to aeronautics. Many early pioneers, including Paul Goddard the father of rockets, were branded as lunatics.

To enter into a basic proof of concept, most airplane builders of this period first constructed model aeroplanes. Early model aeroplanes were used as miniature test vehicles for the full scale aeroplane which could carry the person aloft. It could be said with confidence that those involved with building full-scale aeroplanes, were also highly skilled model builders.

This was the case with Robie Seidelinger, who built and flew model aeroplanes of his own design. He accomplished this remarkable feat without having

Robie Seidelinger at the controls of Delaware's first aeroplane the Delaplane *at Horse Show Park, now Wawaset Park, Wilmington. Historical Society of Delaware photo.*

prior knowledge of aircraft design and construction. There is no supporting evidence that he corresponded with anyone seeking advice on aeronautics. His first model was a helicopter, which, after testing, gave way to fixed-wing models. He powered his models at first with rubber bands, then with reworked clock mechanisms.

Robie was working as the manager for the Monarch Typewriter Company in Maine in 1908 when he was transferred to Wilmington, Delaware. At night he would spend his idle hours building model aeroplanes as a hobby. Seidelinger's models caught the attention of George Crowe, a friend of Robie's and a salesman for the National Cash Register Company, at 825 Market Street, Wilmington. Crowe, inspired by the quality of Seidelinger's models, allowed him to display one of his flying models in the window of his store.

The three and one-half pound biplane model, with a wingspan of three feet, created considerable attention. The model had elevators just behind the wings, and the rudder was placed on the very rear of the fuselage. The model was of such quality many thought it was the prototype of a full-scale aeroplane under development. These comments caused David Snellenberg, manager of the Snellenberg Men's Clothing store located on seventh and Market street, to consider a full-scale aeroplane.

David Snellenberg was so impressed that he requested more information from Crowe as to its origin. He then invited Robie Seidelinger, along with Crowe, Grantly Postles (president of American Leather), William Dockstrader (owner of the Garrick Theater), John Gray, esquire, and John Montgomery (treasurer for Gilpin-Van Trump & Montgomery Insurance) to accept an invitation to create a flying club.

On February 10, 1910, the Wilmington Aero Club (WAC) was incorporated. The principals included David Snellenberg, Robie Seidelinger, George W.

Crowe and John Montgomery. The WAC address was listed as 11 West 10th Street, Wilmington; -the residence of John Montgomery. Collectively, the four raised $1,000 to start the business. Capital stock was $10 per share, with 10,000 shares being offered. A purchase of one or more shares of stock included membership in the WAC.

Their objective was to commit their combined resources and skills into a world-renowned aviation experimental group. A by-product of their efforts would be publicity for Wilmington which would draw more business to the area. The long-range plan included highly publicized air shows to be held in Wilmington to raise capital. Although a Wright *Flyer* could be purchased for $5,000, the WAC elected to have Robie Seidelinger design and build an airplane in Delaware at a cost estimated to be $6,000.

The WAC built a barn at the Gentlemen's Driving Park (also known as Horse Show Park and Wawaset Park) where Seidelinger would start construction of their first aeroplane, and name it the *Delaplane.* Using the existing technology of the day, the entire biplane was built from spruce with the exception of fittings and flying wires. The wings and control surfaces were covered with cotton and painted with dope for rigidity and tension. The initial timetable included finishing and test flying the aeroplane in time to enter an air meet in St. Louis in July. They learned in a hurry that building and test flying an aeroplane was not a simple task.

Robie Seidelinger was a perfectionist. He would weigh each piece of wood first. Then he would carve it and weigh it again. If a duplicate was needed, it must be an exact match. Although the WAC produced many press releases, Seidelinger was very secretive about many of his ideas (to which he gave consideration when applying for patents).

While the aeroplane was being built, the remaining members of WAC started planning a flying event which would create the publicity they so ardently sought. Other aviation clubs throughout the country were given considerable press coverage for their events which drew scores of other clubs and individuals to share their avocation. The event was to be held at the Gentlemen's Driving Park for three days starting October 7, 1910. The newspaper reports presented the idea that the city would soon see aeroplanes in the sky as often as automobiles on the street.

By August 31, 1910, the *Delaplane* was complete except for hooking up the controls, installing the engine, and mounting the propellers. The engine of choice was the forty-five horsepower, water-cooled Elbridge, which was awaiting arrival from the factory. The aeroplane was rolled out to take advantage of the publicity that came with the New Castle County Fair which was held at the park. Much to the surprise of many, the *Delaplane* did not look anything like the model displayed in the window of the Snellenberg Men's Store. The design utilized the pitch control in the front, the rudder in the back, and the ailerons under the lower wings. No one, however, questioned if it indeed would fly because it looked like an aeroplane. It was expected to do so.

The officers of the WAC were invited to Washington, D.C., to discuss an upcoming air meet where they happened to meet J. D. McCurdy, assistant to Glenn Curtiss. The Officers convinced McCurdy that he should look at the *Delaplane.* McCurdy gave the aeroplane a thorough inspection and commented with approval on many of the new features. His final comment was what everyone was longing to hear;

> Gentlemen, it will fly. This is the best constructed aeroplane I have ever seen and I have inspected all aeroplanes of reputation. The workmanship is excellent and I congratulate you. (Quote from Delaware History, *Delaware Birdmen,* Allen Hawk.)

Word spread rapidly about the *Delaplane.* A week later, Todd C. Schriver, who had gone into the airshow business after being a pilot for Glenn Curtiss, came to see the aeroplane for himself. His comments were also favorable and he gave his resounding assurance that it would fly.

Cloward, manager of the event, contacted other aeroplane owners, including Glenn Curtiss and McCurdy, to perform at the WAC flying event. The negotiations with Curtiss and McCurdy did not materialize, so they sought out others who flew in air shows. Cloward went to a Long Island air show where he met Todd Schriver. He signed him to a contract to fly his *Nighthawk* October 5 thru 8, 1910. He also signed up John Frisbee, from Rochester, New York, with his *Skylark,* as well as Thompson & Flaxon and their balloon. Man-carrying kites were also shipped from California to round out the gala four-day event.

WAC later announced the pilot of the *Delaplane* would be Todd C. Schriver, a well-publicized pilot of the day who had flown the Wright *Flyer* in Ohio and flew for Glenn Curtiss. Schriver had also made the news on September 19, when he broke the record for staying aloft at night. The announcements drew favorable reaction from the community and the press, which thirsted for more knowledge about aviation. WAC and the newspapers portrayed the event to be the greatest success the city had ever known.

The final work to be done on the *Delaplane* was the installation of the Requa-Gibson propellers. A consulting engineer from New York helped Seidelinger install them on the airplane. Schriver and Frisbee agreed to conduct taxi tests on the *Delaplane,* but wanted more room for a test flight.

The well-planned WAC event started off in a series of disasters which were beyond anyone's control. The day before the meet, Schriver made a test flight in his *Nighthawk.* When he decided to land, he could not shut off the engine for a glide to land. When he tried to land the plane with full power, it hit hard and flipped over on its back. Schriver ended up with a broken ankle and was now unable to fly. Why he did not fly until the fuel was exhausted and glide to a landing is not known. The Wilmington newspaper *Every Evening* commented that, "Wilmington not only recorded its first flight, but also its first aeroplane accident."

Wednesday, October 5, the opening day of the event brought winds too brisk to fly. To appease the crowds, John Frisbee flew his *Skylark* on a few low hops of a couple of feet. On his last hop the wind carried his airplane into a tree breaking the elevator. The next day the wind was accompanied by a driving rain. The third day it was too windy for anything to fly, including the kites and the balloon. The last day brought more bad weather and the event was canceled. The Aero Club announced they would schedule another event later in the month, but it was never held.

The taxi tests on the *Delaplane* revealed some problems which needed attention. The engine was removed from under the pilot's seat and reinstalled back of the main wings to improve balance. The twin propellers were removed and a single prop installed. The rear fuselage was also shortened. On October 21, 1910, the aeroplane was rolled out of its hangar once again. In spite of a dreary evening, Seidelinger and Eddie Bloomfield, the test pilot, decided to do some high speed taxi runs. On the last run Eddie pulled back on the elevator control and the *Delaplane* rose smartly from the ground a few feet, leveled off and flew about 300 yards before landing. Bloomfield was a Delawarean and could now claim to be the first Delaware aeronaut. The few spectators who witnessed the event jumped with joy at the sight. A few more flights were made on succeeding days, then the *Delaplane* was returned to its hangar for further modifications and winter storage.

In February 1911, the WAC Annual Report was filed with the Delaware Secretary of State listing David Snellenberg (Wilm.), president; George W. Crowe (Dover), vice president; John A. Montgomery (Wilm.), treasurer; and W. Manley (Dover), secretary. Directors were Snellenburg, Crowe, Montgomery, John Hartmann, and R.W. Crook. Total shares sold to date were $1290. The WAC at this point must have accrued considerable debt. Research to date failed to uncover why Robie Seidelinger was no longer listed on the board.

During the winter, the WAC planned another flying meet which was to be held June 6 through June 9, 1911. In spite of the huge failure of the first event, the Aero Club signed on the Wright brothers and their pilot, Walter Brookins. Schriver was engaged in another meet and could not attend. The meet would also include motorcycle races, track meets with local runners, and a drill demonstration by a newly formed organization called the Boy Scouts. The press was more than eager to place editorials in their papers since flying machines were considered a phenomenon most people still considered fictional. Philadelphia alone had a population of more than a million people, many of whom had never seen an aeroplane, much less seen one fly. The event would be a major draw from this populated area.

The weather on June 6 was the complete opposite of the fall event. Brookins, flying the Wright *Flyer* put on a spectacular show of steep turns, dives, figure 8s, and low passes across the field. The spectators were spellbound by the performance. The weather held out until the last day, when Brookins could only make two flights.

The air meet was such an overwhelming success the WAC began planning another event to coincide with the fall New Castle County Fair. While the meet was being planned, the *Delaplane* was moved to a shed on the New Castle rifle range. As luck would have it, the shed was struck by lightning and burned to the ground consuming the *Delaplane.*

The February 1912 WAC Annual Report filed with the Secretary of State listed Snellenberg president, R.W. Crowe vice president, and John Montgomery treasurer. No secretary was listed. Directors were Snellenberg, Crowe, Montgomery, Crook, E. A. Van Trump, and John Hartman.

Although the WAC planned to hold more flying events to raise capital, they never materialized. The last annual report filed by the WAC was January 1913. David Snellenberg was still listed as president; George Crowe, vice president; John Montgomery, treasurer; and William Mosley, secretary. Directors were Snellenberg, Crook, Crowe, Hartman, Montgomery, Van Trump, and James Cane. Total shares sold to date were $6290. This was far below the $100,000 the WAC hoped it could raise.

There were no other annual reports filed with the Secretary of State after January 1913. The WAC had disbanded but left a history in Delaware which could never be duplicated:

> The WAC funded the first aeroplane built in Delaware.
>
> The WAC flew the *Delaplane* seven years after the Wrights' flight.
>
> The WAC held the first two air events ever held in Delaware.

The *Delaplane* would never have become a reality if it were not for the vision of the members who put forth the money. Although the WAC needed Robie Seidelinger to build the aeroplane, one could not have accomplished the feat without the other. The plans for the *Delaplane* have never been located and are assumed to be lost. Seidelinger and the WAC faded from the limelight and were never in the news again. It was the end of a great effort.

CHAPTER 2

World War I (1914-1918)

Warfare has undergone three great revolutions: (1) the introduction of the Cavalry in the fourth century; (2) the introduction of field artillery in the fourteenth century; and (3) the use of mechanized equipment coordinated with aerial offensive and defensive aircraft in the nineteenth century.

The European population regarded early development of aircraft as a thrilling experience to witness. Large crowds would gather around the pilots and treat them as heroes for their defiance of the laws of gravity in their flimsy machines. European governments were also supportive of aviation development and encouraged those who delved in the sport.

A list of certified pilots at the beginning of 1911 showed: France led with 353; England was next with 57; Germany, 46; Italy, 32; Belgium, 27; United States, 26; Austria, 19; Holland, 6; Switzerland, 6; Denmark, 3; Spain, 3; and Sweden, 1.

A major increase in registered pilots occurred by 1913 when it was estimated there were over 2500 licensed pilots worldwide. This escalation was due primarily to the war which loomed on the horizon. In 1911 in an incredible display of airmanship, Eugene B. Eely flew a Curtiss pusher from the deck of a U.S. Navy battleship. This event gave the airplane a new dimension in mobility. It opened the eyes of the Navy to the unlimited possibilities the airplane could contribute to their forces.

Germany had warned the United States that shipment of munitions by sea to England and France would not be tolerated. Germany's warning included advertisements in major newspapers throughout the U.S. The British passenger ship *Lusitania* was specifically mentioned as a possible target. Germany claimed the passenger ship *Lusitania* carried munitions in her hold. War materials continued to flow from America to Britain in spite of continued threats from the Germans. The warning went unheeded by the allies and the *Lusitania* was sunk by a German submarine May 7, 1915. Over 1,100 lives were lost of which 126 were Americans. Much propaganda was distributed by the British to cover up the fact that munitions were carried in the hold of the ship. The loss of the *Lusitania* and the many civilian lives created considerable hostility between America and Germany. On April 6, 1917, the war in Europe expanded when the United States of America entered into the fray.

Glenn Curtiss hired B. Douglas Thomas, an English aeroplane designer, in 1913, to come to America to help in the design of his tractor-type airplanes. Thomas is credited with the design of the Curtiss J and JN series aircraft trainers for Curtiss. The ungainly Curtiss JN-4 *Jenny* was the only U.S.-built military airplane produced in large numbers during WWI. A massive bi-wing airplane with numerous struts and flying wires, the JN-4B *Jenny* was powered by a Curtiss-built, 90 hp V-8, water-cooled, OX-2 engine. Although 3495 *Jennies* were built through the war years as B, C, and D models, they would

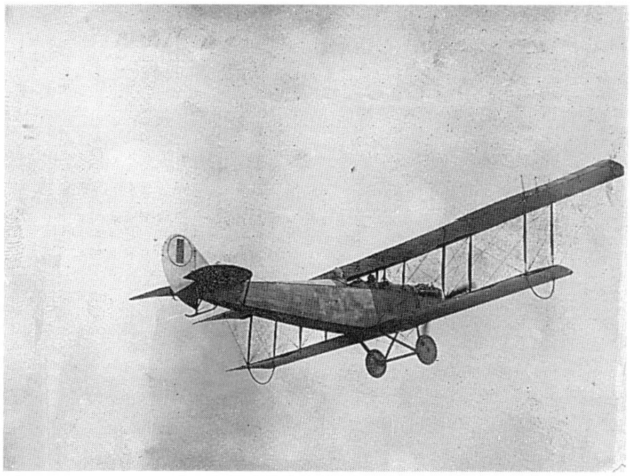

Smithsonian records reveal that ninety-five percent of all U.S. and Canadian pilots during WWI learned to fly in the Curtiss Jenny. *Photo courtesy of Ralph and Jo Cooper.*

only be used in the U.S. and Canada as trainers. They were far too slow (60 mph) and cumbersome to survive in a combat environment.

At this point in time, the U.S. entered the war with a total inventory of 110 military airplanes, none of which were war planes. Glenn Curtiss built the homely Curtiss JN-4B *Jenny* in Buffalo, New York, which was used for training pilots. The *Jenny* was also built in Toronto, Canada, under the name *Canuck* and used for pilot training. These airplanes were in such demand that six other U.S. manufacturers built them for Curtiss. To meet the demand, manufacturing of engines was subcontracted to Willis Motor Car Company. The airplane was an ungainly open-cockpit biplane which spent as many hours in maintenance as it did flying. A great deal of the needed maintenance could be attributed to student pilots who tended to abuse airplanes during flight training. In spite of its shortcomings, the *Jenny* would be used by the Army Air Corps until 1927. The decision to continue using the *Jenny* was strictly for economic reasons. After the war, military funding to purchase new airplanes was virtually nonexistent.

CURTISS JN-4B JENNY SPECIFICATIONS

Engine	Curtiss
Model	OX-2 (3, 4, 5)
Horsepower	90
Wingspan	43.8 ft
Wing area	352 sq/ft
Length	27.4 ft
Height	9.9 ft
Empty weight	1,390 lbs
Gross weight	1,920 lbs
Maximum speed	75 mph
Cruise speed	60 mph
Service ceiling	6,500 ft
Seats (in tamden)	2

The urgent demands of WWI pushed the airplane, which made its first flight a scant eleven years earlier, into rapid development in Europe. These wood, wire, and fabric airplanes were built entirely by hand. The first aerial war missions were reconnaissance flights. The airplanes were actually called "scouts" and then were designed to replace the function of the cavalry. The early flights were made with unarmed single-seat, open-cockpit aircraft. Later, two-seat reconnaissance aircraft replaced the scouts, wherein the observer took aerial photos of enemy installations. The pilots would fly over enemy territory and report back to the commanders any activities which might enhance ground movements. In some cases the observer attached notes to a carrier pigeon which was launched from the airplane while in flight. It must have been a bit of a shock for the pigeon to all of a sudden find itself airborne at several thousand feet at 100 mph. In any event, pigeons were used as messengers quite successfully.

Another method of surveillance during the war was the observation balloon. The balloon was tethered and allowed to rise to heights of 4,000 feet for the observer to take photos and report enemy activity. Thomas F. Naylor, Wilmington, Delaware, served in an Army Observation Balloon Squadron in France during the St. Mihiel, Argonne, and Meuse operations. Thomas Naylor was very fortunate to return home unharmed since balloons made excellent targets for enemy aircraft as well as ground fire. Naylor was a career military man and enlisted in the U.S. Navy in 1928. He then served aboard a submarine tender in the Caribbean, Iceland, Europe, and the Pacific during WWII. He passed away on June 15, 1945, from a heart ailment in a base hospital on a small island in the Pacific.

The very first aerial offensive and defensive weapon was the hand-held pistol or rifle. Pilots simply tried to fly alongside their opponents and take potshots with their pistols or rifles. By 1915, the observer was given a machine gun mounted on a swivel to return the salvos. He would stand erect in a cage which allowed him to make a 360-degree swing of his gun. Elevation stops were installed to eliminate any chance of the gunner getting carried away and shooting into his own airplane. The first single-seat

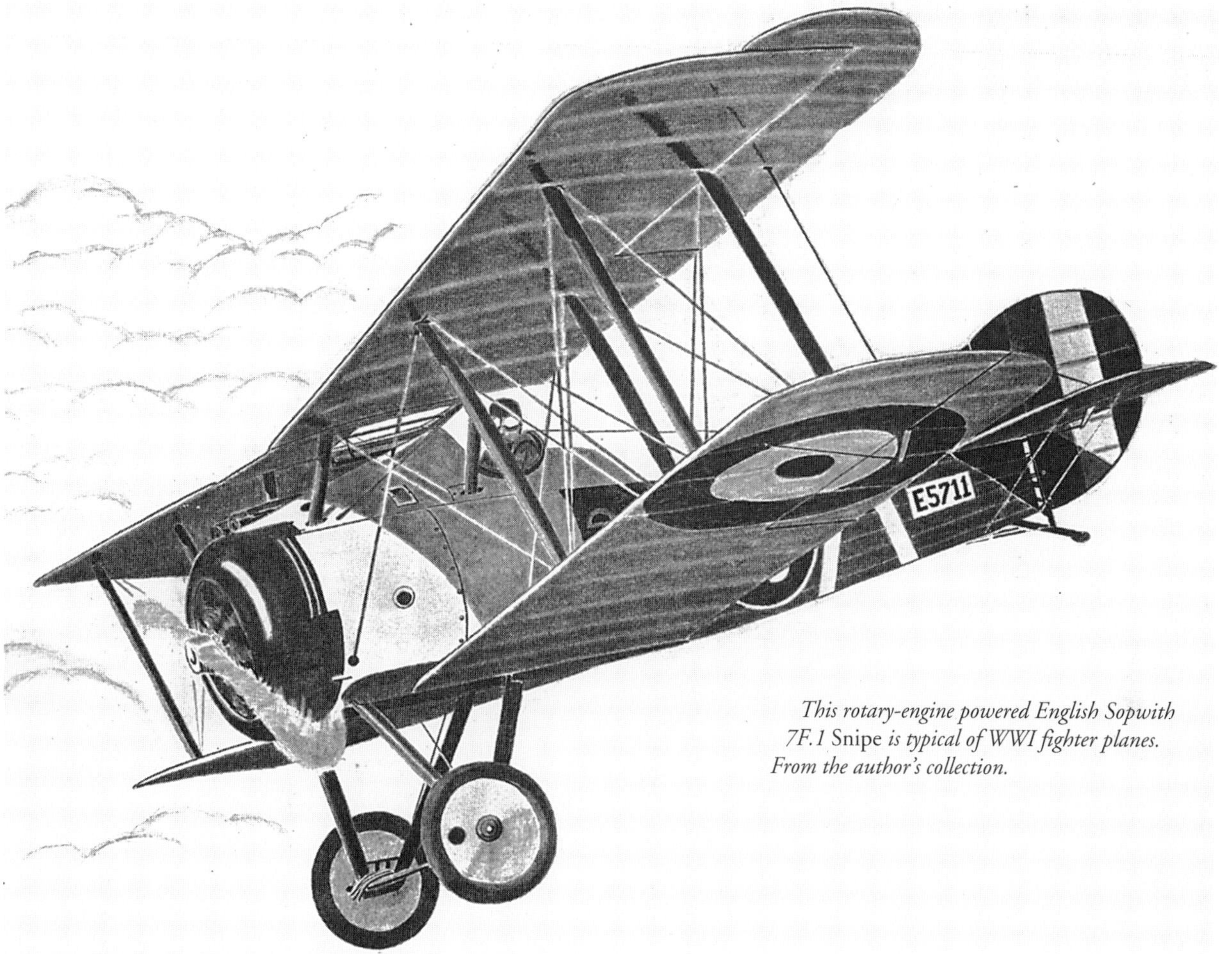

This rotary-engine powered English Sopwith 7F.1 Snipe *is typical of WWI fighter planes. From the author's collection.*

fighters mounted their machine guns above the top wing to allow the bullets to miss the propeller. A small cable, affixed to the trigger and looped down inside the cockpit, allowed the pilot remote control firing of the gun. Striking the enemy was more luck than skill. The accuracy would be similar to a rifleman holding his weapon over his head in attempting to hit his target.

In late 1915, Anthony Fokker developed an engine synchronizer which allowed the bullets to pass through the whirling propeller without hitting the blades. This allowed the machine guns to be mounted directly in front of the pilot. It increased accuracy to the degree that opponents were given the title "Fokker Fodder." The French and English saw the cowl-mounted guns and surmised the Germans placed metal plates behind the wooden propeller to keep from shooting it off. They promptly removed their propellers and mounted steel plates to the back side. The pilots ability to aim was greatly improved, but they complained bitterly about the shower of lead fragments which failed to make it through the gap. Fortunately for the allies, a German Fokker fighter mistakenly landed on a French aerodrome which revealed the secret. Naturally, it was promptly copied by the French and the British. In spite of the great achievements of Manfred von Richthofen and Eddie Rickenbacker, the average life of a WWI pilot was a mere three weeks. History fails to mention the pilots who paid the price of early death in their attempts to become aces.

The war would include aviators from Delaware. John J. Earner, 1321 North DuPont Street enlisted in the Army Quartermaster Corps and later transferred to the Air Service. Unfortunately, on June 18, 1918, during flight training at Ellington Airfield,

Lt. William Prickett
Photo from the William Prickett family.

Houston, Texas, the airplane crashed and he was killed. His flight instructor, also aboard the aircraft escaped with minor injuries.

Lieutenant William Prickett was a flight observer and gunner attached to a French squadron. He was responsible for directing ground fire from 155-mm guns via a wireless aboard the airplane. The twin-engine observation airplane, which carried a crew of three (rear gunner, pilot, and observer/gunner), would be covered by three fighter escorts. Prickett was later transferred to a fighter squadron in the same area. Reported in the June 29, 1918, Wilmington newspaper, Lt. Prickett detailed a recount of his dangerous missions.

> My ship was a new British-built Newport. The first few days of our stay here, we were taken up by members of other squadrons to learn the ground marks and how to stay in formation properly. We were also taught the proper method of 'essing' to obtain maximum visibility and avoid anti-aircraft fire.
>
> On my first mission, we crossed the lines at 4500 meters and the photo fellow got to work taking pictures of enemy positions. Now I saw some real archy. Fritz now had a group to shoot at and a slow moving one at that. Fritz especially gets riled when we are doing photo work. It must be verboten in Germany. Anyway Fritz started to fill the air with high explosives. It came so close I could hear them over the roar of the engines. I was glad I was not in the photo plane as it had to fly straight, which made for an easier target. We essed our fighters violently to avoid any strikes. After about a quarter hour, I heartily wished air photography had never been invented. The photo plane then broke a pump and headed back to our aerodrome. Unfortunately, he crash landed in a field about one mile short. No one was hurt and the airplane was repairable. A funny thing about it all was, I completely forgot to look for enemy Albatrosses.
>
> In another mission we had a real fight. We were to meet three French Salmson bombers about 14 kilometers into enemy lines at 5000 meters. I led three other Newports. When we arrived at altitude and looked up the valley, I saw a lot of anti-aircraft fire. Climbing higher into the sun, I could now see the anti-aircraft fire had ceased because enemy fighters were now after the Salmsons. I signaled my flight to dive and attack. There were only two Salmsons left and heading rapidly for our lines and safety. I counted at least six German Albatrosses after the bombers. It reminded me of sharks in some frenzy after their prey. I dove on the leader and the rest followed. I shot at him with both guns as he rolled away in a dive with me still on top on my back. I followed him down to 3500 meters still shooting with both guns when he went into a spin. As I climbed back to altitude, another Hun tried to get over me. Two more Albatrosses joined in to get me. Fortunately, my Newport could out-climb the Albatross or Plafz. As I outdistanced them the other two left the chase. Since I was slightly higher than the remaining Albatross, I decided to turn and dive at him head-on. He quickly fell in behind my circle where neither one of us could get a shot. The other two saw the melee and turned back again. I broke from my turn and dove east for our lines. The three

> Germans were in hot pursuit. Once I crossed the lines, they broke off and went home.
>
> We saved the two Salmsons but lost one of our pilots. We later learned we downed two of the German fighters.

The letter was quite lengthy and covered, in detail, much more of the war and the carnage it left on both sides. Lieutenant Prickett wrote the account when he was recuperating from wounds in a French hospital after his airplane was shot down in combat. His wounds included a punctured lung, broken ribs, and broken wrist. When he returned to Delaware he resumed the practice of law. Lt. William Prickett was the son of Judge William S. Prickett of Wilmington.

Another Wilmington aviator was Lieutenant John MacArthur, who on July 1, 1918, was credited with downing his fourth enemy aircraft flying his British-built *Newport* fighter. He was attached to the 27th Squadron Airplane Service. His commanding officer made the following press release.

> The German *Rumpler* bi-plane was pursued by Lt. MacArthur below the cloud level in his *Newport* fighter at about 4500 meters. The two machines exchanged fire when the *Rumpler* went over in a loop followed by MacArthur, firing as he flew within thirty yards of the enemy. He saw tracer bullets enter the enemy plane and it started to catch fire. He had fired 220 rounds to bring the Hun down.

Lieutenant MacArthur was the grandson of Mrs. Martha Ward of the Donaldson Apartments, 1004 Jefferson Street and nephew of Christopher L. Ward, a well-known Wilmington attorney. His father, Charles, was the chief engineer of the Pennsylvania Railroad. MacArthur, a graduate of Yale, was employed by the DuPont Powder Company as an electrical engineer prior to entering into the service. He received his flight training at Fort Worth, Texas, before being sent overseas.

Reconnaissance aircraft led to bombers of a sort. The bombardier sat in an open cockpit surrounded with bombs. The pilot flew over enemy territory while the bombardier simply dropped each bomb over the side as he saw fit. Not very accurate, but quite unnerving to anyone on the receiving end. In defense, anti-aircraft guns, machine guns and rifles sent salvos into the air as a method of retaliation.

As the war progressed, airplanes grew larger, heavier, faster and more destructive. They also carried more lethal offensive and defensive weapons. The airplane was now considered an important piece of war machinery capable of major destruction.

War needs machinery and personnel which are necessary to make the parts needed to build engines, guns, tanks, trucks or airplanes. On the other hand, war also breeds prosperity. Unemployment was not something one faced during wartime. The young went to war and the others stayed home to produce the weapons of destruction in order to win. All airplanes during this period were manufactured, for the most part, from wood and fabric. The larger the airplane, the heavier the weight. The additional weight required engines to develop more horsepower. And so it went on.

Most European single- and two-seat airplanes were equipped with rotary engines. This unique design allowed the *entire engine* to rotate around the stationary crankshaft. An unorthodox idea, but it worked quite well. Manufacturers on both sides built rotary engines, including Bentley, Oberursel, Le Rhone, Gnome and Clerget. When an airplane crashed, German or French mechanics often removed the rotary engine and re-installed it in one of their own airplanes. The highly prized Swiss-built copies of the rotary engine were considered a real find. It would not have been uncommon for a mechanic to replace an engine before it reached fifty hours of operation. Larger aircraft generally used water-cooled engines which were, for the most part, adaptations of surface engines.

In Europe, over 170,000 airplanes were built during the war of 1914-1918. This is an astonishing number when one considers that the Wright brothers first flew a scant fifteen years earlier.

In late 1915, Thomas resigned from Curtiss to join the Thomas Brothers Aeroplane Company (no relation) in Bath, New York. It was here that Douglas Thomas designed the Thomas T-2 biplane trainer. The Thomas T-2 was purchased by the U.S. Navy, Canada and England for training pilots. In January 1917, the Thomas Brothers merged their company with the Morse Chain Company. Thus was born the Thomas-Morse *Scout.* This conven-

The Atwood aircraft engine. The 522-cubic inch geared engine, with electric starter and dual carburetors developed 120 hp. Photo from the Hagley Museum & Library.

tional biplane used the 100-hp French Gnome rotary engine which was built under license in the USA by the General Vehicle Company. The United States Army purchased 100 of these aircraft and the U.S. Navy ordered 10.

Another Thomas-Morse MB-3 was also ordered by the Army. The MB-3 used the American-built version of the 300-hp French Hispano-Suiza water-cooled engine. The V-8 engine vibrated so badly that structural problems soon surfaced. The Navy ordered 11 MB-3s for the U.S. Marines. However, the Marines were so displeased with the vibrating engines they gave them to the Army. MB-3s were equipped with two, 30-caliber machine guns which fired through the propeller gap. In an effort to get rid of the beasts, some were sacrificed in the 1927 movie *Wings.*

In the spring of 1917, Pierre S. du Pont, Irenee du Pont and John J. Raskob sought to make a contribution to the war effort by creating a flying school to train pilots. All three were executives of the DuPont Powder Company and incorporated the school under the name "Delaware Aeronautical Company." The objective of the school would be to train young flyers for government air service. The three agreed to contribute $20,000 each to finance the project.

The trio then hired Harry Atwood for $650 per month on April 14, 1917, to establish the curriculum. In August 1911, Harry Atwood piloted a Wright Model B, 1300 miles, from St. Louis to New York in 12 days. Atwood also designed an overhead valve, V-12, water-cooled aircraft engine in Williamsport, Pennsylvania. The engine was installed in an experimental flying boat and tested on the Susquehanna River in 1916. Apparently he was unable to sell the design and joined the Delaware Aeronautical Company.

Atwood was the type of person who paid great attention to detail. He accounted for every minute of flying time each student flew. He also filed a highly detailed weekly report to the directors outlining accomplishments and projections. Frank L. Garey, secretary-treasurer, documented every penny spent which could pass the scrutiny of the best IRS agent today. The Delaware Aeronautical Company was run as professionally as could be done.

For a location, John J. Raskob allocated a portion of his Claymont estate in Claymont, Delaware, to be cleared of trees for a landing field complete with three hangars and a weather station. The airfield was typical of airfields of this era in that it did not have designated runways. The airfield was circular in design so pilots could take advantage of landing or taking off directly into the wind. The airfield was dressed, seeded and rolled to allow for smooth operations. The three hangars were built by the J. A. Bader Company of Wilmington for $2600 each. The location had an additional advantage as it also bordered on the Delaware Bay to allow seaplane operations.

The first order to be placed was for a Curtiss JN-4B *Jenny* for $9,800. The *Jenny* was shipped in a crate, and thus required assembly upon arrival. On the same order dated May 14, 1917, was a long list of spares that included OX-2 engines, radiators, propellers, etc. The total bill came to $30,066.00. On a separate order were two Thomas flying boats which were scheduled for delivery in July.

On May 31, 1917, two of the flying instructors, Thomas Birt and B.R.J. Hessell made the first test flight in the *Jenny* from the airfield. Birt was a former member of the Royal Canadian Flying Corps and Hessell a former instructor for the Curtiss Aeroplane Company. The instructors were paid $50.00 every week, plus $10.00 per hour for flying time. Mechanics were paid $25, and a guard was paid $14 weekly. Sunday, June 10, Birt and Hessell made a flight in the *Jenny* to review the surrounding flying area for possible landing sites which could be used in an emergency. The flight attracted considerable attention from spectators when they passed over the city in the late afternoon at about 1000 feet.

By June 12, twenty-one students were enrolled in the first class to study the theory of flight, aeromechanics, and meteorology. Harry Atwood sent the first list of students to the Directors as follows:

Bacon, Milton E.	Gouverneur, F. H.
Bard, William H.	Jones, Harold A.
Brewer, Russell P.	Murphy, John J.
Colleton, Thomas E.	McTernan, Edward
Carpenter, Allen W.	Nelms Jr., Frank
Denham, W. B	Reed, Harvey T.
Darling, Phillip G.	Randall, Charles H.
Denning, Matthew J.	Sands, Austin L.
Eaton, Thomas W.	Smith, Allen O.
Gadsden Jr., P. H.	Talbot, Ralph
Witt, August F.	

There were so many applicants, a complete new group of students would be added during overlapping classes. All of the students were college material and were required to have made previous application for entry in the United States Aviation Corps. Most signed up as Naval aviators, although several chose the Army Air Service.

It was a spartan and rigorous lifestyle at the Delaware Aeronautical School, and the men had many restrictions which included no drinking of alcohol or smoking. Civilians were not allowed on the premises to visit students or request flying rides. The day started at 5 a.m. and lasted until sundown.

It is interesting to note that there was no charge for taking the flying course. The du Ponts and Raskob felt it their patriotic duty to "do their bits" for their country. In spite of their executive responsibilities to the DuPont Powder Company, they dedicated additional time and energy to create a school worthy of their efforts. The school appeared to flourish when over 300 applications were on file after the first class graduated. Unfortunately, of the many who applied, few were chosen because of space and training personnel limitations.

Harry Atwood announced to the press that the school would fly a mock air attack on Wilmington by dropping leaflets over the city. On June 26, 1917, the Wilmington paper reported the event.

> Thousands of Wilmingtonians witnessed a real demonstration of "bombing" a city as practiced on the war-stricken towns of Europe when Harry Atwood, today, in charge of the aviation camp at Claimant and Thomas A. Birt, a former member of the Royal British Flying Corps soared over this city and dropped "Bombs" in the form of handbills. The demonstration by these daring airmen was made in the interest of the drive for recruits by the Delaware National Guard and the "bombs", which were eagerly sought by everyone, read as follows:
>
> "This handbill was dropped from an aeroplane over your head. It is not a German bomb, but it might have been. This is an appeal to the men of Wilmington between the ages of 18 and 45 to enlist in the Delaware Regiment before a bomb strikes us. Apply at the State Armory, Twelfth and Orange Streets, or

at recruiting stations, old City Hall, Court Hall law, tenth and Market streets, or No. 207 Maryland Avenue."

> Announcement of the sham air raid had caused everyone to be on the lookout for the aeroplane and shortly before 9 o'clock. A cry of "Hear she comes" was heard on all sides. Far to the northeast of the city, high in the heavens, could be seen a small speck and as it drew nearer and nearer, the planes of the air monster become visible. Soon the aviators were directly over the DuPont Building and the chug, chug of the motors could be heard distinctly.
>
> All businesses in the office buildings in the heart of the city ceased for a time. Trolley cars came to a halt and housewives flocked to the streets to see the machine and secure handbills which were dropped over the city by the aviators.
>
> The machine made two complete circles over the city, turning each time over Front Street and as the last turn was made and the airmen started to ascend on their return journey, one of the aviators waived his hand over the side of the machine at the populace below.
>
> Today's flight was arranged by the officers of the National Guard, in conjunction with Mr Atwood to demonstrate to the citizens how easy it is for a city to be attacked from the air and to show that by rapid recruiting, this country could expect a victorious ending of the war with Germany.

Another event which made the news was one of deep sorrow for the Delaware flying school. The school announced to the press that on July 21, 1917, Allen Ashley Smith, a 25-year-old student from Brooklyn, New York, on an unauthorized flight, lost his life when he crashed a Thomas flying boat into the Delaware River. The Wilmington newspaper reported;

> Smith, who also was head gardener in the employ of John J. Raskob, treasurer of the DuPont Powder Company, went up on his flight without a companion. At a height of 1000 feet, his machine was observed to be unsteady and wavering, and suddenly it started to fall and descended rapidly to the river.

On May 21, 1918, the School participated in a patriotic Red Cross parade by allowing one of their airplanes to be trailered through the procession. This would be the first opportunity for the local population to see one of the airplanes closely. The parade started at Fifth and French Streets and included twenty-four separate organizations displaying their service to the country.

The summer flying season ended on November 30, when the students of the school held a dinner in honor of the founders. The dinner was held at the Hotel DuPont, where Allen Carpenter would act as toastmaster. Harry Atwood, manager of the school, would address the guests of honor.

On July 18, 1918, the directors announced the school would not open for the coming flying season. The government flying schools by this time had caught up with demand. Additionally, the war was now coming to an end. All of the equipment would be donated to the School for Airplane Mechanics, operated by the West Side YMCA in New York. The list of equipment was quite impressive. A Thomas-Morse D-5 biplane with only five flying hours was the newest of the airplanes donated. Other airplanes included a Curtiss JN-4B *Jenny,* and one Thomas B-4 flying boat. Additional equipment included spare engines, airplane parts and all the tools required for maintenance.

A six-page letter sent to Pierre S. du Pont on December 28, 1918, by Frank Helms, Jr, outlined his life after leaving the Delaware Aeronautical School. A portion of his letter follows:

> I went to Boston Tech and after graduation I was assigned to Key West Naval Air Station for further training. From the Navy, I transferred to the 1st Marine Aviation Force where I received instruction in stunt flying, gunnery, bombing and formation flying before obtaining my commission June 5, 1918 as 2nd Lieutenant. On July 17, I sailed for Brest, France, where our squadron was assigned to the Northern bombing Group at Oye. We were then assigned to the English Royal Air Force, Squadron 218 where we now went to work. Our first work was day bombing, and the first place was Bruges. There were twelve planes carrying four 50 pound bombs each, convoyed by 18 scout planes. Our force was met by seven Hun machines. Our scouts shot down one Hun in flames, but the others managed to break through our formation. We quickly dropped our bombs and turned to fight the Huns. We drove three other Huns down. The fight was a most furious affair, all the machines on our side being riddled with bullets. We didn't lose any machines on our

side and the Huns lost four. The entire flight took three hours.

After many similar missions I was recalled to the Marine Squadron and made Ferry pilot between England and France for awhile. Next, I was made a War Pilot and was assigned a new observer, F. J. Gibbs, who was also a graduate from the du Pont School. On October 27th, there were six Liberty deHavillands used to bomb the railway at Lokeren. Each machine carried four 112 lb. bombs, two fixed Marlin machine guns and one Lewis swivel machine gun for the observer. The clouds were quite low that day and we had to fly under them. Our six machines reached the target and obtained 100% hits. We started to return and my engine developed ignition trouble at the same time four Huns saw I was in trouble and proceeded to attack us. I could not make our airfield from 1000 ft. altitude, so I turned north to land beyond the German lines in Dutch territory. The four Huns were right on my tail and the machine was riddled with bullet holes when we landed. The Dutch authorities kept us for three weeks before allowing us to walk back to rejoin our squadron. It was a long walk.

Every machine I took up had bullet holes in it when I came back. The nearest I came to being injured was when a bullet passed between my thumb and first finger and scorched it.

Before I go back to the service I would like to stop in and shake your hand, and thank you for what you have done. I know you are a busy man, and I would appreciate it if you would let me know when it would suit your convenience. It won't take two minutes of your time.

Yours truly,
Frank Nelms, Jr.
2nd Lt. U.S.M.C.

Frank Nelms was from 552 South 48th St., Philadelphia. Another student from the Delaware school who would serve in the same RAF squadron with Nelms and Gibbs was Lt. Ralph Talbot. Another student who graduated from the school was Walter Josephs. Walter's mother was Marion Willoughby Josephs from Seaford and later Smyrna. Walter was a graduate from the University of Delaware and enrolled in the Delaware Aeronautical school. After graduation, he joined the U.S. Navy, but did not like flying boats and transferred to the Marines. He was later sent to France where he piloted a French Caproni bomber. After the war he returned to the U.S.A. for a short period before returning to France once again to study art. Walter "Buzz" Speakman, Smyrna, who is related to Josephs, affectionately calls him "Uncle Walter." Buzz has retained much of Walter's works of art which include a WWI painting.

Walter Willoughby Josephs poses next to his Curtiss JN-4B Jenny *while a student at the Delaware Aeronautical school in 1917. Photo from Walter "Buzz" Speakman.*

The final list of thirty-four graduates of the Delaware Aeronautical School were located in the Hagley Museum & Library.

Abeling, H. H.	Knauer, H. P.
Bacon, Milton	Lonaberger, R.
Bard, W. H.	McTernan, E. K.
Brewer, R. P.	Mills, R. H.
Carpenter, A. W.	Nelms, F.
Colleton, T. E.	Olmstead, F. M.
Cushing, W. H.	Randall, C. H.
Darling, Phillip	Reed, Roland
Dean, T. F.	Reed, H. S.
Denham, W. F.	Reinhardt, C. G.
Denning, M. J.	Remey, J. T.
Eaton, Thomas	Smith, Allen O.
Gibbs, J. F.	Sternbergh, L.
Gouverneur, F. H.	Talbot, R.
Huhn, Jno.	Tallman, F. G.
Jones, H. A.	Witt, A. F.
Josephs, Walter	

Although the Delaware Aeronautical Company closed in 1918, it did achieve history in its brief existence. The Delaware Aeronautical Company created the first flight school in Delaware. Although the flying field was not considered a public-use airfield, it was the first *operational* airfield in Delaware.

By April 1917, the war in Europe was considered a stalemate. Both sides hammered each other unmercifully with no end in sight. When America sent in fresh troops to support the French and British, it created an imbalance of power and the war came to a bloody end in nineteen months. The loss of human life was staggering. Many soldiers lost their lives when poison gas was used by the Germans as a weapon. Americans paid a great price for entering into Europe's conflict, and it would not be soon forgotten. No American-built airplanes entered WWI combat. The U.S.A. never saw the huge Gotha, Vickers, Handley-Page, Fokker, or Sikorsky multi-engine bombers used during the war in Europe. Unfortunately, this left America far behind the Europeans in aircraft design, development, and production. This gap in technology would not be closed until mid-WWII due to our government's position that the airplane had no real application beyond a war machine. The great fighter planes of WWI like the Albatross, Fokker, Pfalz, Spad, Sopwith, Nieuport, Avro, and Bristol were all European-built fighter planes. All of these aircraft outclassed the Curtiss-built *Jenny* and the Thomas-Morse *Scout* by a large margin.

After the war, President Woodrow Wilson toured the entire country campaigning for the United States to join the League of Nations. He campaigned so vigorously he exhausted himself and had a stroke which left one side of his body paralyzed. In spite of his handicap, he pressed on until the vote came before Congress. Much to President Wilson's dismay, the bill was soundly defeated. Wilson failed to understand that we paid a huge price in lives lost and wounded by entering into the war, and it was now Europe's responsibility to keep the peace. The American people were more interested now in taking care of our own concerns.

POSTWAR AVIATION

The end of the war in 1918 brought a worldwide economic collapse. The voracious appetite for war machinery and supplies came to an abrupt end. This shut down factories and put millions of people out of work. The airplane, in spite of its rapid escalation in technology during the war, was perceived by many to have little value in peacetime applications.

Available as surplus after the war, the Curtiss *Jenny,* which sold new for $10,000, was available for as little as $300 in flying condition. Dedicated pilots tried to start flying schools using the *Jenny.* Others used it for building flying time to gain experience. Pilots who would literally fly for food, turned to "barnstorming." Barnstormers were flying gypsies. They used their airplanes to land near any small town farm field they could find and sell rides and perform aerobatics for the public. They virtually lived with their flying machines much like gypsies would live in their wagons. Charles Lindbergh mastered the art of flying by barnstorming around the country before flying the U.S. mail. Clarence Chamberlin was another nationally-known aviator who carved out a living barnstorming. A good day of barnstorming would sometimes end with the farmer inviting the pilot to stay for dinner. An overnight invitation was an added luxury which was more than welcome when most nights were spent in a sleeping bag under the wing.

Prior to WWI, attempts had been made to establish a U.S. Air Mail Service. Although tests were successful, Congress did not approve funding. It was not until after the war when flying was at a virtual standstill that Congress appropriated $100,000 for testing a single mail run between Washington, D.C. and New York. This route would have mail planes flying directly over Wilmington, Delaware. However, since there were no airfields in Delaware at this time, no stops were planned.

Although the U.S. Army Air Corps initiated airmail flights in May 1918, the Post Office Department assumed complete control by midsummer. This event marked "the first scheduled air transportation" in the United States. The complete success of the Washington-New York run broadened by 1920 to include transcontinental airmail service to Cleveland, Chicago, Omaha, and San Francisco. This network of long-distance flying in all sorts of weather made it necessary to expand the ground support system. Major General Charles Menoher, Chief of the Army Air Service, publicly announced;

> Landing fields are as essential to flying as yards and stations are to railroads, as docks and harbors are to steamships—-The transverse of the United States by regular lines running airplanes and airships carrying passengers, express, mail, and light freight is dependent upon the speed by which our country provides these facilities.

Menoher's vision of aviation was right on the mark. Achieving his goal in depressed times was a challenge that many thought impossible. Although the Post Office had installed some lights for landing and navigational aids, their funding was limited. In many cases, bonfires were lit by citizens to guide pilots on course. More auxiliary airfields were needed for emergencies, and more beacons were needed to light runways and airways. The Post Office Department asked cities, counties and states to provide more landing facilities. A national airport and airway chart was planned by the federal government, and most states and cities desired to be included. Any community that did not have an airfield would lose out on any economic development aviation brought.

New Castle County lay beneath the crucial Washington-New York airway. The Department of Commerce approved three unmanned auxiliary landing fields in Middletown, Newark, and Wilmington in the early 1920s. At least one airway beacon was known to have been erected on Iron Hill, four miles south of Newark near Cooch's Bridge. This beacon served pilots until 1942 when it was replaced by a new beacon erected on the new Army Air Corps base in New Castle.

With the more than ample supply of surplus airplanes available, aircraft manufacturers entered into a deep recession. Many simply shut down and filed bankruptcy. Others merged to consolidate their meager sales and workforces. Great names like Lockheed, Martin, Boeing, Ford, Curtiss, Beechcraft, Stinson, Waco, Aeronca, Fleet, Ryan, and Bellanca all struggled for sales and popularity.

General Billy Mitchell, a very vocal voice in promoting military aviation, continued to pressure the U.S. Government to continue development of military aircraft after the war. When President Calvin Coolidge, with little funds available, was asked if he would support funding of aircraft for the military, he made the infamous remark, "Why don't we just buy one airplane and let the pilots take turns flying it?"

Glenn Curtiss was one of the fortunate ones to secure some military contracts and start production of pursuits for the Army and Navy. By July 1921, Mitchell astonished the military by sinking a captured German battleship by aerial bombardment with Curtiss-built aircraft. This action caused the Navy to reconsider its options in terms of defensive warfare, which ultimately led to Congressional funding to build the first United States aircraft carrier, the U.S.S. *Langley,* the following year.

Another milestone for aviation was established in July 1924. Army Lt. John McCready and Lt. Oakley

G. Kelly completed the first nonstop flight across the United States in a single engine Fokker T-2. They departed Mitchell Airfield, Long Island, and landed at Rockwell Airfield, San Diego, in just under 27 hours. Records of this nature made headlines in the national press, which was a much-needed stimulant for aviation progress.

In these early days of aviation, Delaware consisted mostly of farmland. Although the Capital was in Dover, the state's major population was located in Wilmington. Although aviation made newspaper headlines, it failed to make an impression on the Delaware legislators. Therefore Delaware fell far behind the rest of the nation in terms of aviation development after WWI.

THE GOLDEN AGE OF AVIATION

To encourage aviation development and competition after WWI, the private sector held air shows which included contests and prize money to encourage participation. The most prestigious annual flying event was The National Air Races (NAR). Started in 1920, it drew large crowds and media publicity. It was every manufacturer's dream to win an NAR trophy and the publicity that followed.

H. B. du Pont was an avid supporter of the NAR and attended all their annual meets. In 1925 he witnessed the Bellanca-built WB-I compete at the Philadelphia National Air Races. The six-seat WB-I, powered by the new 200-hp Wright J-4 *Whirlwind* engine, outperformed every airplane "in its class." Little did H.B. realize that he would be the driving force to bring Bellanca to Delaware. In 1930, he was given a commemorative bronze medal by the NAR for the DuPont Company's support of the annual NAR events. In the September 1927, Wilmington Chamber of Commerce official publication, H.B. du Pont submitted a feature article on aviation in which he asserted:

> It is only natural that the public should have an exaggerated idea of the dangers of flying because of the type of publicity given to aviation accidents. In the case of automobiles, if a serious accident occurs due to reckless driving nobody condemns motoring as dangerous, but when an airplane lands in a farm field it makes front page news and the public is given the impression flying is very hazardous. The development of the modern high-speed automobile motor has been greatly accelerated by racing car competition, but nobody worries about the danger of driving on the highways because of the great number of speedway injuries and fatalities. The Wright Aeronautical Corporation covered the performance of all their 200-hp Whirlwind engines used commercially in 1926. During 175,000 miles flown with these engines, only three were forced to land with mechanical trouble. In all three incidents, there was no damage to the airplane or injuries to the occupants. As a means of rapid long distance transportation, either for business or pleasure, the airplane has no rival.

The article left little doubt of H.B. du Pont's commitment to aviation. However, the economic times only allowed the financially well-off to indulge in flying activities.

The NAR ushered in what is recalled today as "The Golden Age of Aviation." The races were held each Labor Day weekend with events that included; Pylon Races, the Bendix Trophy Race, the Thompson Trophy Race, Shell Speed Dash, Mixed Free-For-All, General Tire Trophy, the Goodyear Trophy Race, and the Ford Reliability Run. Giuseppe Bellanca would be a fierce competitor with his aircraft and would win many of the events. The trophy, prize money, and publicity could spell immediate success to the winner of any one of these major events. In many cases, winning determined the future existence of a company.

The final and most exciting event of the day was the "Mixed Free For All." The class consisted of "unlimited" category aircraft. It was a pylon race whose winner was decided by a checkered flag on the final lap. The most outstanding race of the time was the 1929 race held in Cleveland, Ohio, which included Army and Navy entries. The Army's Pratt & Whitney *Wasp*-powered Curtiss XP-3A was piloted by Lt. Breene, who was carefully selected for his skill. The Navy countered with their Curtiss-built F6C-6 *Hawk,* piloted by Lt. Commander Clark. Both airplanes were highly modified versions and were considered the best fighter planes of the day. Everyone suspected a military rivalry battle to the finish.

Barely noticed was a small red and black low-wing airplane being pushed to the starting line. The airplane was listed as the *Travel Air Mystery Ship*. Surely a small unknown airplane built in Wichita would not have the slightest chance against the might of the Army and Navy. The starting flag dropped and the sleek little *Mystery Ship* shot into the lead immediately. It was thought that pilot Doug Davis was out of the race when he cut the first pylon. He turned sharply and rounded the pylon for the second time and promptly passed the leader before the end of the first lap. Adding insult to injury, he later cut a second pylon and turned back once again and passed the field. He flashed by the checkered flag with an average speed of 194.9 mph. His fastest lap was recorded at a blistering 208.69 mph, which meant he was flying 235 mph on the straightaway. Needless to say, everyone was shocked, especially the military, who were literally shot down by an unknown civilian pilot and airplane.

The Delaware-built Bellanca airplanes were a major force to be reckoned with for the competition in many NAR events. Giuseppe Bellanca over the years established records that were the envy of the competition. A list of the Bellanca records can be found in the Bellanca section of this book. The Bellanca victories created worldwide publicity for the company and Delaware which would otherwise not have been obtainable.

When America entered into WWII, all air races and most civilian flying came to an end for the duration. After the war was over, the National Air Races and the Bendix Trophy Race were revived. Once again, thousands of surplus military aircraft became available at bargain prices. Stripped of their armor, the fighter aircraft became the choice for racing pilots who had many hours of combat training. The races were now dominated by high-powered, low-time military fighters that were available for a small investment. A Bell *Kingcobra,* stripped to the bone of wartime necessities, won the first NAR *unlimited* category after WWII. The National Air Races, unlimited category, now took on a new perspective that ended an era of racing that was previously driven by creative individuals and struggling companies with limited funds.

The last time a civilian-built airplane would win the unlimited class in the NAR was the little *Travel Air Mystery Ship* in 1929. The golden age of aviation had ended when military aircraft and service rivalry entered the competition. The last Bendix Trophy Race was flown by Captain Bob Sowers in an Air Force Convair B-58 *Hustler,* flying from Los Angeles to New York in 2:01.

This photo of the Travel Air Mystery Ship *was taken at Bellanca Airfield on a visit to Delaware by Frances Rees Hayford in 1930. Photo from the Frances Rees Hayford collection.*

CHAPTER 3

Atlantic Aviation Corp

Henry B. du Pont (1898-1970) poses next to his Buhl Airster. *This 1927 photo was taken on the DuPont Airfield. Photo from the Hagley Museum & Library.*

Without question, the name du Pont is legendary in Delaware. The legend was created by the close-knit family that believed in education, invention, innovation and progress. The family and the company which bore their name were aggressively involved in virtually every facet of the American industrial revolution. The company history includes the production of gunpowder, paint, chemicals, plastics, automobiles, motorcycles, boats, and, yes, even airplanes. They were visionaries and opportunists. Once a commitment was made, they fought brilliantly to have it turn into reality.

The du Pont-sponsored Delaware Aeronautical Company (1917-1918) provided the young 19-year-old Henry Belin du Pont, nephew of Pierre, the opportunity to take his first airplane ride. The Delaware Aeronautical Company, although short-lived, was a nonprofit organization dedicated to train pilots for WWI. A graduate of Yale and MIT, Henry Belin du Pont would grow to be one of the most influential figures in the future development of Delaware aviation. In the summer of 1923, Henry was sent to Reinhart-Whelan flying school in Dayton, Ohio, where his instructor, Ben Whelan, taught him to fly a Dayton-Wright *Chummy.* His pilot's license would be signed by the then famous Orville Wright.

A pilot license, at this point in time, was not a government-issued certificate. Flying schools, that were members of the National Aeronautic Association, issued licenses to students who successfully completed their course.

From 1923 through 1925, Henry was employed by General Motors in their research department in Detroit. While in Detroit, he became acquainted with Alfred (Fred) Verville, a talented aeronautical engineer, who built his first successful airplane and seaplane in 1915. By 1920, Fred also designed and built the Verville-Sperry Racer which won the Pulitzer Speed Trophy for flying 191 mph. The first American airplane with a welded steel fuselage was also built by Fred Verville in 1921. In 1924 a new cantilever, low-wing, retractable landing gear, Verville Racer won the Pulitzer Speed Trophy for the second time with a blazing 232 mph top speed.

Henry du Pont had given up his job at General Motors in 1925 and became vice president of E.I. DuPont deNemours when he returned to Delaware. Although H.B. du Pont served on several corporate boards, his general position was to avoid personal involvement. His responsibilities grew to the point where he needed a business manager to keep things in order.

Stewart E. Poole was H.B. du Pont's right arm throughout his entire life. Photo from the Poole family.

Stewart E. Poole was 23 when he responded, with several other applicants, to an advertisement for a personal business manager to H. B. du Pont in 1927. Stewart was the best of the best, and a choice Henry would never regret. Stewart E. Poole was a graduate from Beacom College and later Temple University with a degree in business management. Henry du Pont would place Stewart Poole in the position as president, or secretary/treasurer, of most of his corporations in which he was involved. To confirm the faith and judgement Henry had in Stewart overseeing his business matters, a signed power of attorney was located in the files of the Hagley Museum and Library. Simply stated, the one-page legal document declared, "Stewart E. Poole has the legal right to sign any document in my behalf." No higher praise can be given this remarkable man who would spend his entire life managing H. B. du Pont's enormous wealth, business holdings, and day-to-day affairs.

Stewart E. Poole passed away in 1995 at the age of 91.

Fred Verville was in the same financial straits as most of the other airplane builders during the early days of aviation. Although he designed and built several open-cockpit biplanes, they never achieved fame. The VCP-1 was built in 1918. The Verville-Sperry *Messenger* was built in 1920, which was followed by the Verville DW-1 in 1921. His most beautiful biplane was the Verville *Sportsman* built around 1929. Recognizing Fred Verville as a brilliant aeronautical engineer, Henry B. du Pont assisted him financially by holding promissory notes to produce these airplanes. Although the Army Air Corps purchased four *Sportsman* biplanes in 1929 as trainers, the domestic market failed to produce sufficient sales to attain profitability.

Fred's final design was the four-seat, V-151A *Air Coach* monoplane for the business and private market in 1928. Henry du Pont managed to put a deal together with Fairchild Aircraft Company to purchase all the airplanes built, along with exclusive

The Verville Sportsman *(ATC-323). Photo from the Hagley Museum & Library.*

worldwide distributorship rights. All that was needed was a Department of Commerce approved type certificate (ATC). Once again, Henry accepted promissory notes to build the prototype.

By the 6th of August 1929, the airplane was completed and was test-flown by test pilot Myron Zeller. He reported that the airplane handled beautifully through all aerobatic maneuvers. Zeller reported rolls and loops were a sheer joy to perform. Controls were light and the airplane responded to the slightest touch. It trimmed out stable in level flight without pilot intervention. For the spin test at 3200 pounds gross weight (the empty weight was 2050) lead buckshot was secured to the rear floor in place of the seats. At 4000 feet, Zeller stalled the airplane and entered into a half-turn spin to the left and recovered easily. Then repeated one to the right with the same result. Then a full-turn spin left and right with clean recoveries. The final test was a two-turn spin from 4000 feet. On the second turn the fabric from the right wing tore loose. Zeller fought to save the airplane down to 1500 feet before he bailed out.

The Department of Commerce investigation of the accident revealed that a mechanic had cut the top side wing fabric to gain access to service a control cable pulley. He had stitched the fabric back in position and glued additional fabric over the seam. This was the approved procedure for this type of repair. Unfortunately, it did not hold up in the high stress-loading spin test.

Fred Verville was devastated. He was not in a financial position to build another plane and fund government certification costs. Henry du Pont came to the rescue once again. On August 29, 1929, Henry offered to trade his promissory notes for stock in Verville Aircraft Company. In addition, he would purchase additional stock to finance construction of another *Air Coach*. Along with some cash to keep the company solvent, Henry now owned the controlling interest in Verville Aircraft Company. (It is perceived that this action resulted in H.B. du Pont's decision to submit his resignation from the board of Bellanca Aircraft.) Verville Aircraft Corporation now became a Delaware corporation with Fred Verville, president, R.H. Curtiss, vice president and Stewart E. Poole, secretary-treasurer. (R.H. Curtiss was a major supplier of aircraft materials to the aviation industry. He also traded some Verville debt for stock in the company.) It can be assumed that du Pont and Curtiss opted for this solution to keep the company alive in hopes of some day redeeming their investment in promissory notes.

The Verville V-151A Air Coach. *Photo from the Hagley Museum & Library.*

The second *Air Coach* received its ATC and production began immediately. To say that sales of the *Air Coach* flourished would be an exaggeration. The depression was in full force, and aircraft sales were anemic at best. The problem lay in Bellanca, Fairchild, Waco, Stinson, Beechcraft, Cessna and others who were also competing for market share. The market was simply too small to support all the manufacturers.

By September 5, 1931, with the company on the verge of going into receivership, Fred Verville submitted his resignation as president. The board, recognizing his engineering talents, requested he accept the position of chief engineer. Curtiss was now named president and Stewart Poole retained the position of secretary-treasurer. By now Henry du Pont had invested $240,000 in Verville, and there appeared no end in sight.

Stewart Poole recommended to Henry du Pont the Verville Company be liquidated. On December 5, 1931, Henry du Pont sold all the rights, drawings, and spares to the *Sportsman* for $38,000 to the Army Air Corps. The Army had previously purchased four they were using as military trainers and were satisfied with their performance. Purchasing the rights would allow the Army to have additional aircraft built as well as to produce spares for the existing airplanes. The remaining Verville assets were sold to John L. Alexander, December 29, for $3,000. The Delaware corporation ceased to exist after December 31, 1931.

Fred Verville went into deep depression over Henry shutting down the business. He became very angry at Henry for not wanting to continue financing the operation. Henry pleaded, in vain, for Fred to accept the unavoidable and make a new life. The anger boiled over to the point Fred threatened a law suit against Henry. The law suit, which went no further than one deposition, severed their long, personal friendship permanently. Verville's anger and depression also resulted in his not being able to secure employment. Henry, out of the kindness of his heart, paid for both of Fred's daughters college educations and sent funds to his wife to keep the family together. Fred Verville finally accepted a job with the Department of Commerce working on ATC approvals, but he never fully recovered from his depression.

THE DuPONT AIRFIELD

In 1924, William and Anne du Pont allowed Henry to use part of their farm for a private airfield. Located alongside Route 141 and Route 48, the airfield, during its existence, would play a major role in Delaware aviation history. The airfield consisted of a single 1800-foot-long grass runway and a small, single airplane hangar with an attached workshop. At this point in time, Henry used the single grass airstrip for his personal flying pleasure. Over the years, Richard,

This is the original hangar built by H.B. du Pont, circa 1924. The wing and tail of du Pont's airplane can be seen to the left of the photo. The tall poles on the roof held the windsock and the wind velocity indicator. Photo from Hagley Museum & Library.

Emile, A. Felix Jr., Lammont III, Paul Jr., Frank, Alice and others would learn to fly from this airfield.

On April 24, 1926, the airfield property, consisting of approximately 45 acres, was transferred by William and Anne du Pont to the Delaware Land Development Company. The Delaware Land Development Company was a holding company established by the du Pont family. This change was probably made to eliminate legal repercussions, since Henry's flying friends would often fly in to visit.

April 6, 1927, H.B. du Pont purchased a new Buhl *Airster,* a three-seat, open-cockpit biplane, which was designed by Fred Verville. The open-cockpit Buhl *Airster* offered Henry purest flying pleasure. However, what it did not offer was creature comfort in inclement weather.

As 1927 began to unfold, Henry B. du Pont would become involved in aviation beyond his wildest dreams. The *Spirit of St Louis,* and its pilot, Charles Lindbergh, would be guests on his airfield. Henry would also be the driving force to bring Bellanca to Delaware. During this time he would also own controlling interest in Verville Aircraft. The original hangar was completely rebuilt and painted to spruce up for Lindbergh's arrival. Many scrub trees were also removed around the airfield to enhance the appearance. After all, the most famous aviator in history was about to make a visit, and the place should not look shabby.

LINDBERGH COMES TO DELAWARE

The Daniel F. Guggenheim Fund for the promotion of aeronautics sponsored a Lindbergh-*Spirit of St. Louis* tour of the United States. The tour accepted offers for Lindbergh to stop in every state starting Wednesday, July 20, 1927, at Mitchell Field, Long Island, New York, and ending there on Sunday, October 23, 1927. Upon completion of the tour, the Foundation estimated that 30 million people saw Lindbergh and the *Spirit of St Louis.*

The Guggenheim Foundation reserved Friday, October 21, 1927, as the date for Lindbergh's arrival at the DuPont Airfield from Atlantic City, New Jersey. Lindbergh would spend the night at the DuPont Hotel and would depart the DuPont Airfield the fol-

Henry B. du Pont and Charles Lindbergh pose for photographers when Lindbergh visited Delaware October 21-22, 1927. Photo from the Frances Rees Hayford collection.

lowing morning for Philadelphia. Mayor Forrest established a committee to oversee the elaborate arrangements for Lindbergh's reception.

The committee felt Lindbergh's safety was of paramount importance. The last thing they wanted was a repeat of the mob scene Lindbergh was confronted with when he landed in Paris. Therefore, no civil aircraft were allowed to rendezvous with Lindbergh during his arrival and departure. To control the anticipated crowds, the Delaware National Guard, the State Police, and the City of Wilmington Police were called to duty.

Lindbergh left Atlantic City and passed over Salem, New Jersey, to arrive over H.B. du Pont's airfield precisely on time. He circled the field several times to be sure he could land the *Spirit* on the short, single runway. Once on the ground, everything went as planned.

After the reception committee formally met Lindbergh, they proceeded by motorcade to the Municipal Building on King Street to pick up Governor Robinson and Mayor Forrest. Throngs of people lined the route which included Kennett Pike, Delaware Avenue, 10th Street, and Market and King Streets. The motorcycle-escorted entourage then proceeded to the Bayard Stadium where Lindbergh addressed the waiting crowd of supportive Delawareans.

After the public ceremonies, Lindbergh and the motorcade of dignitaries proceeded to the DuPont-Biltmore Hotel where a banquet in his honor was held at 7 p.m. The Chamber of Commerce sold the entire 275 tickets for the occasion in a single day. The next morning Lindbergh was picked up at the hotel by motorcade for a planned departure from the DuPont Airfield to his next stop, Philadelphia.

After Lindbergh and the *Spirit of St. Louis* completed their North American tour, Paul Garber, curator for the Smithsonian Institute, suggested that the airplane be donated to the Smithsonian. In spite of its monetary value, Lindbergh and the investors agreed, and the final flight was made to Washington, D.C. Total flying time on the airplane (since the first test flight) was 489 hours.

The Spirit of St. Louis *parked outside the hangar at Henry du Pont's private airfield. The* Spirit *could not fit inside due to the long wingspan. The airplane in the hangar, with folded wings, is H.B. du Pont's Buhl* Airster. *Note the security guards and the chain-link fencing. Photo from the Hagley Museum & Library collection.*

In 1928 Henry purchased a Bellanca CH-200 (#105) to fly to the Los Angeles air races in the fall of that same year. In the $13,000.00-agreement with Bellanca, a mechanic and Pilot would also accompany Henry on the trip. Once in California, the pilot would enter the Bellanca in two efficiency contests. The Bellanca won first place in both events.

Henry B. du Pont did not encourage transit aircraft to land on his airfield. Several letters were located in the Hagley Museum and Library files requesting permission to hold air-shows on the airfield. A reporter from a Washington D.C. newspaper also requested permission to do a feature article on the facility in their Sunday edition. All were politely refused. The du Ponts were not interested in suggesting the airfield was a public facility.

In the summer of 1928, Henry built a tall steel tower to mount a beacon to enhance locating the airfield on hazy days and at night. The beacon flashed a high intensity green light as it slowly rotated 360 degrees. On March 6, 1930, the Delaware Land Development Company transferred the airfield property to the Trust Department of Wilmington Trust Company. The DuPont Airfield was still listed as "private" on aeronautical charts, although it would eventually grow over the years into one of the most modern, privately owned, public-use air facilities in the country.

On February 9, 1932, fire destroyed the new hangar on the north side of the DuPont Airfield, and three airplanes were destroyed. *The Wilmington Morning News* reported that a mechanic welding on a fuselage to make a repair started the fire. Lost in the blaze were a Keystone *Commuter,* Stinson, Pitcairn, Verville *Sportster* and a Loening amphibian. The first two airplanes belonged to du Ponts. The Verville *Airster* belonged to Angus Echols, treasurer of the DuPont Company. The paper went on to say that the hangar and airplanes were insured, and H.B. would immediately start rebuilding the loss.

When the debris from the burned hangar was removed from the airfield, the small original hangar was also torn down. A single, large, multi-level hangar was built with new office space and workshop. The airfield beacon and windsock were mounted on the roof of the hangar.

August 12, 1933, Henry purchased a new classic Waco UIC, (serial #3785) enclosed-cabin biplane (NC13438). The Waco was equipped with every available option; including, Lear radio, electric starter, generator, landing lights, Hamilton metal propeller, brakes, and full instrumentation. The base price of the airplane was listed as $5985. The original bill of sale, located in the H.B. du Pont files at Hagley Museum & Library, disclosed the final price rose to $7540. Today, Waco airplanes are considered some of the most highly prized, collectable aircraft built during this era. Another small note located in the H.B. du Pont files suggests Henry's Bellanca CH-200 (#105) was sold to J. Allison Buck.

On December 27, 1935, New Castle County land records show Henry B. du Pont became sole owner of the DuPont Airfield. The date suggests that H.B. du Pont, who had shown a great love for flying, received a nice Christmas present from the family.

The name Atlantic Aviation Corporation first found its way into Delaware history when a group of businessmen from Jamaica, New York, incorporated in Delaware on January 7, 1930. The Delaware Secretary of State, Department of Incorporations files list: Francis Rogers, president; Milford Simie, vice presi-

This aerial view of Henry B. du Pont's airfield was taken May 11, 1929. The round circle on the field was a federal requirement to enhance recognition of the airfield location from the air. Photo from the Hagley Museum & Library.

Henry Belin du Pont flying his beautiful Waco UIC cabin biplane over Delaware October 14, 1933. Photo from the Hagley Museum & Library.

dent; Michael Halloran, secretary; Frederick Krant, treasurer; and John Carl, assistant treasurer. Twelve others from New York were included on the documents as directors. The Corporation Trust Company, in Wilmington, acted as their resident agent.

In spite of the impressive size of the board of directors, apparently Atlantic Aviation Corporation failed to survive the depression years. The Delaware Department of Incorporations records further indicate Atlantic Aviation Corporation was proclaimed void January 1934. The Department's action acknowledged that Atlantic Aviation no longer existed as far as they were concerned. H.B. du Pont had no connection with this company whatsoever.

By 1938, the private haven of Henry B. du Pont's airfield began to change significantly. The decision to expand the airfield was directly related to Richard du Pont buying controlling interest in All American Aviation from Lyle Adams. Richard needed manufacturing and engineering facilities to redesign and manufacture the airmail pickup system. Richard also needed hangar facilities to maintain the fleet of Stinsons. Henry du Pont's airfield became the place of choice. Henry du Pont, on October 25, 1938, incorporated Atlantic Aviation Sales, Inc. in Delaware. Atlantic Aviation Sales, Inc. listed: Stewart E. Poole, president; Oliver L. Davis, vice president; Chester Charles, secretary; Edward Beacon, treasurer. An interesting observation is that the name Henry B. du Pont is not listed on the board. This confirms Henry's confidence in Stewart E. Poole. Since Henry owned the controlling interest, Stewart reported to him on any matters of importance.

Atlantic Aviation Sales secured the Fairchild and Beechcraft aircraft franchises after their incorporation. A satellite Beechcraft dealership was established on Roosevelt Airfield, New York, and managed by Ollie Davis. A new multi-level operations building

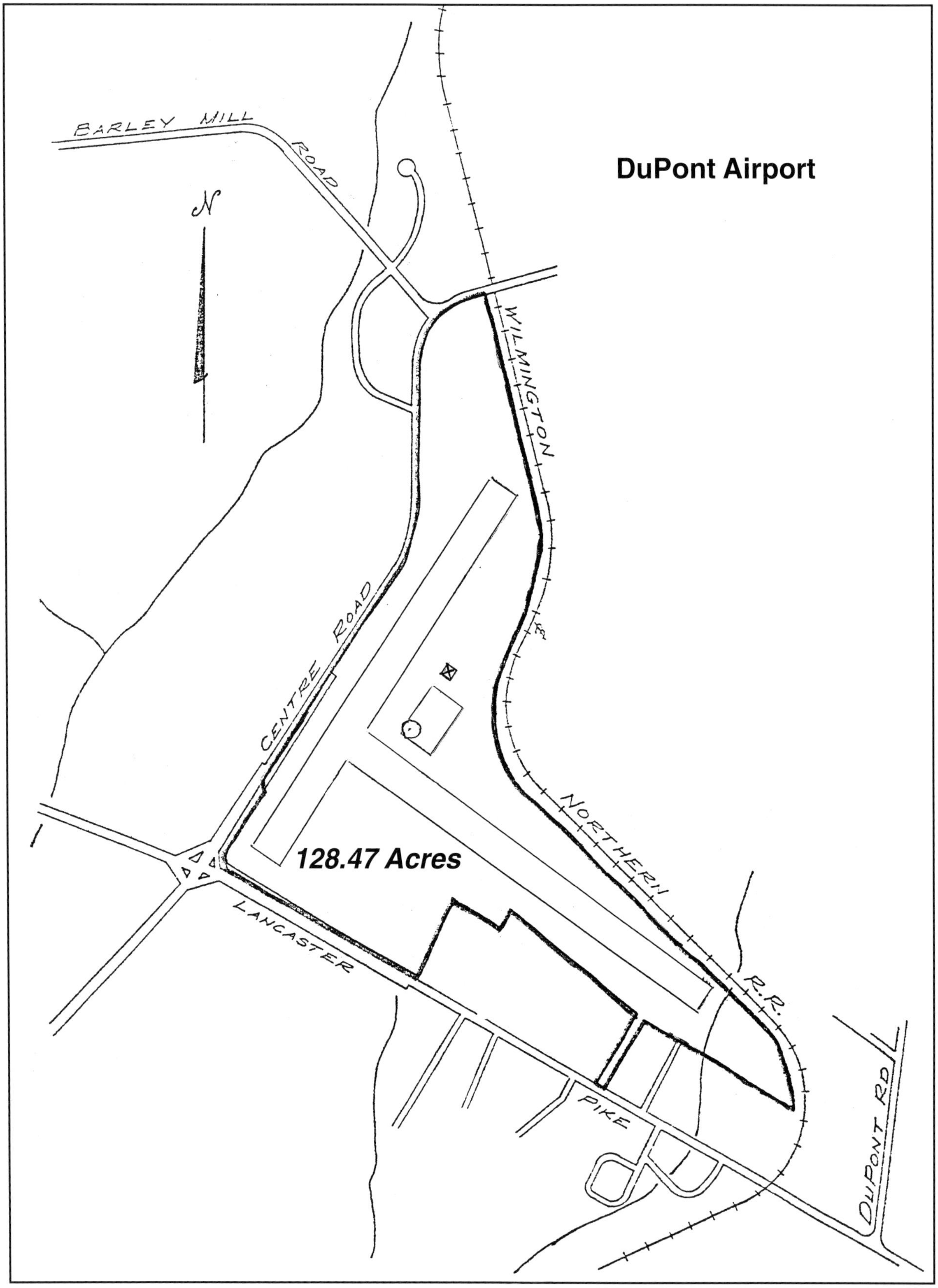
DuPont Airport
128.47 Acres
BARLEY MILL ROAD
N
WILMINGTON NORTHERN R.R.
CENTRE ROAD
LANCASTER PIKE
DUPONT RD

was constructed on both sides of the existing hangar. This would accommodate H.B. du Pont's Atlantic Aviation Sales, Inc., his newly acquired Beech 17 *Staggerwing* and Richard du Pont's All American Engineering Division. The land to the north, which can be seen in the early photo with the baseball diamond (page 38), was also purchased to lengthen the runway. Later, additional land was purchased on the east side to add an additional runway. With these additions, the airfield now consisted of 128.471 acres.

The DuPont Airfield by 1938 was listed in the Bureau of Air Commerce Directory of Airports as follows:

> Three miles NW of the city, Altitude 185 ft., T-shaped runways; 2600 ft. NE/SW and 2700 ft., NW/SE, sod, level, natural drainage, DUPONT and WILMINGTON on hangar roof. Beacon, 24" rotating green. Flood, boundary and approach lighting. Aircraft service day and night. Waco, Fairchild sales/service. RCA radio dealer. Airport transmitter/receiver, low-powered radio station WDUP, operating frequency 278 kc, with radio range operated on request "DP". HF transmitter 25 watt on 3072.5, 4937.5 and 4967.5 kc. Fuel 73 and 80 octane 24 hrs.

All previous listings on aeronautical documents listed DuPont Airfield as "Private." This listing leaves little doubt that the DuPont Airfield was now a commercial operation. In 1939, Henry du Pont incorporated *Atlantic Aviation Service, Inc.* in Delaware as a separate corporation to service aircraft. This was followed by a third supporting corporation named *Atlantic Aviation Supply, Inc.* Atlantic Aviation Supply, located three minutes from Teterboro Airport, New Jersey, would maintain an inventory capable of shipping parts the same day orders were received. Once again, Stewart E. Poole was named president of the corporations. Atlantic Aviation *Sales,* Inc. sold aircraft and RCA radios. Atlantic Aviation *Service,* Inc. would concentrate on maintenance of aircraft while Atlantic Aviation *Supply,* Inc. offered aircraft parts and materials for sale.

Two of All American Aviation's three divisions, All American Manufacturing and All American Engineering, were also established on the DuPont Airfield in the fall of 1938. The third division, All American Airways, operated from the all-weather airport in Martinsburg, West Virginia.

When war was declared in December 1941, the Department of War restricted all airfields located within 150 miles of the Atlantic coast from civil flying. The enforcement of the rule was given to the Civil Aeronautics Board (CAB). Although the DuPont Airfield was banned from any civil flying activities, military-approved flights prevailed. Henry and Richard du Pont, avid supporters of the CAP,

Mike Guididas, chief pilot for Atlantic, accepts delivery of Atlantic Aviation's first new Beechcraft Bonanza *from Walter H. Beech on February 15, 1947. Photo courtesy of Mike Guididas.*

The automobiles and airplanes in this DuPont Airfield photo date it circa 1951. To the far right is All American Engineering. Photo from the Hagley Museum & Library.

This close-up photo of the control tower on DuPont Airfield was also taken circa 1951. Photo from the Hagley Museum & Library

offered the airfield to the newly formed headquarters of the Delaware CAP Wing. All American Aviation, Inc., headed by Richard C. du Pont, was also able to secure several U.S. Navy contracts to teach flying from the DuPont Airfield. Aircraft listed under the government contract included 12 Pipers, 4 Beechcraft, 3 Fairchilds and 1 Waco. An allotment of 18,500 gallons of aviation fuel each month was made available to Atlantic Aviation Service, by the Office of Price Administration (OPA), to fulfill the obligations of the contract. While all other civil airfields in Delaware were shut down during the war years, the DuPont Airfield prospered.

Also located on the DuPont Airfield during the war was Richard du Pont's All American Manufacturing, and All American Engineering divisions. They played an important role in the development and testing of many military experimental projects. All American's Engineering Division grew to the point they occupied a major portion of Georgetown Airport as a test facility after the war.

After the War, the DuPont Airfield resumed full civil operations. Atlantic Aviation Sales, Inc. reinstated the Beechcraft dealership as well as RCA radios. At auction, All American Aviation liquidated All American Manufacturing to raise needed capital and was no longer on the DuPont Airfield. Although the Engineering Division remained, most of the testing was carried out on Sussex County Airport, Georgetown, Delaware.

According to Michael Guididas, who was employed by H.B. du Pont since the mid-1930s, the last airplane to fly from the DuPont Airfield was at 9 p.m., May 10, 1958. The DuPont Airfield site was later transferred to the DuPont Company, who completely demolished the buildings and reconstructed a

This photo shows the new hangar occupied by Atlantic Aviation on NCCA in 1948. What appears to be a runway to the left of the building is actually the taxiway. Photo Courtesy of Reynolds Jones.

All American Engineering has now occupied the hangar and operations building. Note the fence and cars parked on the ramp. The Airfield is no longer operational. Photo from the Hagley Museum & Library.

research and development center named "Barley Mill Plaza." More than likely the hundreds of people who are working in the research center have no knowledge of the historical background of the land they occupy. The tenants may change, but the history will remain intact.

ATLANTIC AVIATION CORP.

It is interesting to note that the last photo taken of the airfield reveals some interesting insights. There are no airplanes on the facility. The tall steel tower that housed the airfield beacon is gone. Also, a fence has been built separating the ramp from the airstrip. Grooves can be seen where cars have been driving across the runway. It is obvious the facility is no longer an airfield. Note the postwar housing development behind the hangars. The main hangar door is permanently sealed and has windows installed. The upper part of the former hangar is now painted "All American Engineering Company," whose testing operations were moved to Sussex County Airport. All these indications give the impression the airfield is no longer operational.

Henry du Pont had never planned for the DuPont Airfield to become a public-use facility. The airfield simply evolved with the times. Henry's attempts at entering into the manufacturing of aircraft failed. The war had created a quantum leap in aviation technology and it now surpassed the ability for the airfield to expand. Realistically, aviation had outgrown the airfield, and it was time to move on.

After the New Castle County Airport was returned to New Castle County on August 3, 1946, Henry started making plans to close the DuPont Airfield. Henry knew the aviation technology created by the war would filter down to the private sector, and the DuPont Airfield was too small to expand. Now that New Castle County Airport was open to the public and provided all-weather flying, H.B. moved his three Atlantic Aviation corporations to NCCA. It was a window of opportunity that Henry would not miss. At this time Atlantic Aviation Sales listed $1.6 million in annual sales, $1.5 million in assets, and $500,000 in liabilities.

Henry du Pont started a new company named "The New Castle Hangar Company" in 1947. Stewart Poole was named president and secretary/treasurer.

Land was rented from the New Castle County Airport Authority for $82.95 per quarter for the purpose of building facilities to accommodate Atlantic Aviation's move from the DuPont Airfield. The hangar, 148'x162', was completed by April 1948 and leased to Atlantic Aviation Sales, Inc.

The first order of business was the acquisition of a military surplus Douglas C-47 to convert into an executive version of the DC-3. The airplane was stripped down to bare bones and everything brought up to new specifications. When completed, the airplane was used as the DuPont Company's corporate aircraft. It was also used in promotions to show what the new Atlantic Aviation could offer in the way of custom executive aircraft interiors and exteriors.

It would take a corporate lawyer and timeless research to document H.B. du Pont's game of corporate hopscotch over the years. To what end is a matter of personal opinion. To illustrate: The board of directors of Atlantic Aviation were changed on numerous occasions over the years. On January 25, 1943, changes included Stewart M. Ayton, treasurer and H.B. du Pont as a director. On November 28, 1943, Barbara Tuffs was added as secretary. January 25, 1946, Stewart E. Poole resigned as president and Sydney Nesbitt was named president. Stewart E. Poole was moved to treasurer. Henry merged Atlantic Aviation Sales, Atlantic Aviation Service and Atlantic Aviation Supply into a single corporation *Atlantic Aviation, Corp.* in December 1956. Atlantic Aviation was now headed by Stewart Pool, president; Stewart Ayton, vice president; John Erskine, secretary; and Watson Richards, treasurer. Although Henry B. du Pont still owned controlling interest, he was listed as a director only. H.B. du Pont never sought notoriety, he sought people who had the corporate skills to operate his business interests profitably.

In 1958, the Grumman *Gulfstream* line of aircraft was added to Atlantic Aviation's corporate list of aircraft franchises. Although the Beechcraft franchise was doing well, Atlantic Aviation added the higher profile Grumman line of executive aircraft. Grumman offered the twin-engine, turboprop G-1, which was appealing to corporations. These aircraft were flown, from the factory in Long Island to Wilmington, with bare interiors and flight decks. Atlantic installed custom-fitted interiors and state-of-the-art flight systems for the discriminating buyer. By 1963, Atlantic Aviation became the corporate distributor in the U.S.A. for the British-built Hawker-Siddley DH125 corporate jet. To accommodate the new growth, $60,000 of inventory was stocked by the service department alone.

The postwar period experienced phenomenal growth when additional locations were established at Teterboro, Philadelphia, Worchester, and Lynchburg. Atlantic Aviation sold more Beechcraft, Grumman and Hawker-Siddley aircraft than any other franchisee. The corporate list of customers included, Bell Telephone, Remington Arms, Phillip Morris, DuPont, AT&T, Ford, and Aluminum, Ltd.

In 1967 Atlantic Aviation purchased Timmins Aviation, with the largest general aviation facilities in Canada. Bases included; Montreal, Toronto, Clargy, Edmonton, and Vancouver. Following this acquisition, a major marketing study was initiated regarding the use of business aircraft in North America. The result of this study targeted air-minded Houston, Texas, as the next Atlantic facility. This multi-million-dollar facility would become the largest general aviation facility in the world. At the same time, the Wilmington facility added a new 40,000-square-foot facility to expand office and service areas. With the general aviation fleet now exceeding scheduled passenger airliners by more than fifty times as many aircraft, sales soared to over two billion dollars annually.

It is interesting to note that H.B. du Pont never allowed Atlantic Aviation Corporation stock to go public. He also never issued dividends to stockholders. Therefore the stock was only worth what someone was willing to pay for it. Finding a buyer for nonpublic traded stock can be a frustrating experience. Most stockholders simply held on to their stock in hopes that some day Atlantic Aviation would be purchased by some conglomerate, and the new issue would become public. That day would not come until June 1997, when Atlantic Aviation was sold to Legg Mason Merchant Bank, in Baltimore. At that time Atlantic Aviation boasted 500 employees and $73 million in annual sales.

The Ford Motor Company aviation director came to Atlantic Aviation to visit their service department. For a contrast in transportation technology, John Hite offered his 1925 Ford touring car to take the entourage to lunch. From left, Don Redpath, V.P, Atlantic Aviation; Ken Spinney, sales manager, Atlantic; Watson Richards, president, Atlantic; Bob Hixson, Ford Motor Co.; and John Hite, Atlantic customer service supervisor.

Henry Belin du Pont died in 1970, at the age of 72. During his lifetime he served as a research engineer for the DuPont Company, directed the Technical Division of Engineering, served on the board of directors, served on the executive committee, and was vice president. He also held the office of president of Christiania Securities. In 1933, he was elected as a director of North American Aviation. In 1935, he was elected as Chairman of the Board for TWA. Two career-pilots of Atlantic Aviation, Mike Guididas and Kenneth Spinney, confirmed that H.B. would call in the late evening and simply say, "What's new?" From that point the conversation could go on for hours. In spite of his enormous wealth, "H.B." was considered a regular guy who held his employees in high esteem. Without question, he was an icon in Delaware Aviation History.

CHAPTER 4

Government and Aviation

In today's world it would be inconceivable to think that an entire industry would actually seek government regulation. As we look back on history, we will see why and how this actually happened in aviation. The complexity of accomplishing the goal would be expressed as a multitude of bills sent to Congress. It would take six agonizing years before all the lobbying turned into law.

Although the Post Office Department had been carrying mail since 1920, they did so on a very limited budget. In an effort to support the ailing aviation industry and support the Post Office, the U.S. Congress passed the 1925 Kelly Air Mail Act which created airmail routes throughout the country. If the Post Office subcontracted airmail routes, companies would be paid by the pound. In many cases the mail sack weighed more than the contents. Most of these small companies operated on a shoestring budget. Before the Kelly Air Mail Act there were no government funds to install en route navigational beacons or a beacon on airfields. Many beacons were no more than small bonfires set by civilians when they heard the sound of an airplane in the distance. Most of the funding was used to purchase and modify the Post Office war surplus de Havilland DH-4B aircraft. These modifications included closing off the forward cockpit and converting it to carry mail sacks. The few beacons which existed were installed by the U.S. Post Office, whose funding was sparse, to say the least. Although in the beginning the Air Mail Act provided jobs for mechanics and pilots, it did little to expand aviation manufacturing due to the flood of surplus airplanes left over from the war. Biggs Airfield was the only airfield in Delaware at this time and was used only in an emergency.

The Post Office Department chose the war surplus de Havilland DH-4B biplane with two open cockpits. However, it was modified for postal use by covering the fuselage with plywood and leaving the pilot in the rear cockpit. Additional modifications included installing an extra fuel tank in the place of the forward cockpit and constructing a mail storage compartment behind the engine. Giuseppe Bellanca modified the wings and the struts of several DH-4s for the Post Office Department to improve performance. In spite of improved performance in takeoff, speed, load, range, and landing, the Post Office would not increase orders due to financial limitations. The DH-4B was a cumbersome airplane, powered by a 12-cylinder, 400-hp *Liberty,* water-cooled engine that would require considerable maintenance to keep it airworthy. In spite of its 400-hp engine, the de Havilland would be hard pressed to reach 100 mph. It was totally void of any creature comforts, such as a cabin heater. It took dedicated airmen to tolerate the conditions of flying the mail in all sorts of weather. Initially, all flights were restricted to daylight operations. Night flights began on February 22, 1921, when a transcontinental pony express flight from New York to San Francisco was made. This flight, although considered successful by many, cost the life of one pilot and two airplanes.

THE NINETEEN TWENTIES

In 1920, *Aero Digest,* the most popular aviation publication of the time, reported that, "Aviation would never reach its full potential until the public was convinced flying was safe. Aviation had only itself to blame for its public perception of the fear of flying." Another publication, *Aviation Yearbook,* published an article defending the need for government controls. Among other points that were made, one concerned insurance. The insurance companies shied away from writing any policies that were aviation-

The Post Office choice of airplanes was this 350-hp Liberty-powered de Havilland DH-4B. U.S. Postal Service photo.

It was antics like this that thrilled spectators at air shows. Unfortunately, many airmen died in their attempts to perform on the very edge. Photo from the Hagley Museum & Library.

related. Their primary concern was writing policies where there were no rules or laws to follow. The dismal track record of pilot performance and airplane crashes would prompt the insurance companies to pass-on the costs to the policyholder. Thus, in most cases, aviation insurance was not affordable.

The 1920's air shows and barnstormers made the headlines almost daily. Unfortunately, those headlines often were accompanied by photos of fatal crashes. Daredevil pilots thought nothing of flying rooftop-high down Main Street to announce their arrival. Others would think nothing of doing loops around bridges, and other low-level flying stunts. One barnstormer was asked, "What is your biggest fear of flying?" He was quick to reply, "Starving to death!" Rules for flying were nonexistent. Even worse was the lack of maintenance of many of the airplanes. The owner of the airplane decided when, if, and how the airplane was maintained. Many would simply come apart in the air under the stress of aerobatics. To make matters worse, manufacturers and individuals could build and sell airplanes that were far from airworthy.

Even Hollywood directors, who seemed to have an insatiable quest for thrilling their audiences on the silver screen, contributed to the problem. Ormer Locklear, the first of the fearless wing-walkers, plunged to his death while filming a movie in the 1920s. Films like, *Speed Girl, Broken Wing, Sky Highwayman* and *Wings* were made to thrill the audience and included death defying stunts. The movie *Wings*, with its aerial antics and destruction of men and their machines, would be the first to win the Motion Picture Academy's prestigious Oscar with its aerial antics and destruction of men and their machines.

Although there are many documented cases of early airmen trying to prove they were made of the right stuff, two stand out. At the dedication of the Lincoln Memorial on Memorial Day, 1922, an Army pilot, who had been relieved of his duties for reckless flying, buzzed the crowd. With President Harding in attendance, the crowd was huge. The pilot flew the airplane so low over the audience, the President's speech was drowned out. Fortunately, the pilot flew off without killing anyone. In another incident, a pilot buzzed an open automobile driving down a country lane. The driver was terrified, lost control of the car, went over an embankment and was killed. In both of these events, there was no law that the pilot violated because there were no laws.

In the mid-1920s, Connecticut, Massachusetts, California, Kansas, Maine, Minnesota, New Jersey, Oregon, and Wisconsin enacted aviation laws in a vain attempt to address the problem. Fortunately, the America Bar Association (ABA) supported the position that the air space was within the jurisdiction of the Federal Government, not the state. It proclaimed, rightfully, that if every state enacted its own laws, it would become a nightmare for aviation and the legal profession as well. This position placed the problem and the solution squarely in the lap of the U.S. Congress. In spite of this support, there were opposing factions. The military wanted to be excluded from any civil laws. But how could the government create laws controlling the air space and allow the military loose to do as it pleased? There were other dissenters who felt government regulation was unnecessary and would only choke the industry to death with regulations.

In spite of this sorry state of affairs, there was one group which set an example above all the others. The U.S. Air Mail Service, although unregulated, flew the mail the entire year of 1922 without a fatality. Their record was not a lucky streak of good fortune. It was earned. Their remarkable achievement was accomplished through an extraordinary, self-imposed set of rules for maintenance and pilot proficiency requirements. Airmail pilot applicants were required to have a minimum of 500 hours flying time and to pass a qualifying examination. Pilots were also required to have periodical medical examinations. Aircraft were given a thorough inspection of 180 items after every flight. The airplane was considered grounded until the inspection sheet was signed off. The engine was removed from the airplane and given a complete overhaul every 100 hours of operation. (Today's modern aircraft engine exceeds this by more than 20 times.) The airframe was completely overhauled every 750 hours. This rigid program required 353 mechanics to keep 96 airplanes flying. Of the 745 people employed by the Post Office, 700 were ground personnel. All this effort paid off handsomely in terms of safety when compared to the private sector. The private sector posted one flying fatality for every 13,500 miles against the Post Office record of one fatality for every 789,000 miles flown. The 1922 achievement also earned the Post Office Department the Collier Trophy for outstanding achievement. It was this shining example that proved to the public that flying could be safe. The Collier Trophy was awarded annually for the largest contribution to aviation progress.

Although the private sector could not afford the luxury of 3.7 mechanics for every airplane in service, it was obvious that rules governing airmanship and maintenance of airplanes would be necessary. The National Aeronautic Association (NAA) was formed in 1920 by a group of concerned pilots who recognized the problem of safety as early as 1919 when military surplus *Jennies* were selling for $350. You paid your money and flew it away. No one bothered

FÉDERATION AÉRONAUTIQUE INTERNATIONALE

NATIONAL AERONAUTIC ASSOCIATION OF U. S. A. INC.

Certificate No 7314

The above named Association, recognized by the Fédération Aéronautique Internationale, as the governing authority for the United States of America, certifies that

Forrest P Wenyon

born .... day of 1896

having fulfilled all the conditions required by the Fédération Aéronautique Internationale, for an Aviator Pilot is hereby brevetted as such.

Dated October 15, 1929

CONTEST COMMITTEE

Orville Wright, Chairman

Executive Vice-Chairman

[SEAL]

Signature of pilot: F. P. Wenyon

National Aeronautic Association, a private organization, pilot license issued to Forrest Wenyon. Note the Orville Wright signature.

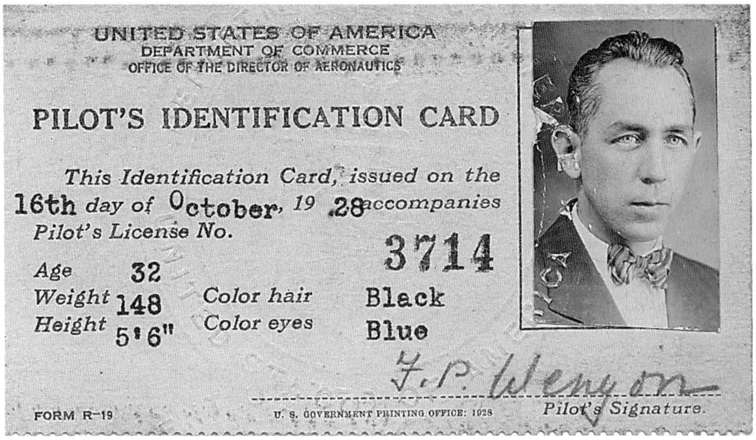
UNITED STATES OF AMERICA
DEPARTMENT OF COMMERCE
OFFICE OF THE DIRECTOR OF AERONAUTICS

PILOT'S IDENTIFICATION CARD

This Identification Card, issued on the 16th day of October, 1928 accompanies Pilot's License No. 3714

Age 32
Weight 148
Height 5'6"
Color hair Black
Color eyes Blue

F. P. Wenyon
Pilot's Signature.

FORM R-19
U. S. GOVERNMENT PRINTING OFFICE: 1928

U.S. Department of Commerce, Aeronautics Branch, issued this license to Forrest Wenyon. Photos courtesy of Jack Beach.

to ask if you knew how to fly. The self-imposed rules set down by the NAA included licensing of pilots, which was a step in the right direction. Unfortunately, the NAA failed to have the power to police the industry.

The 1920s saw America's entire surface transportation system supported by government. The railroads laid down over 250,000 miles of track, enough to span the globe ten times. In 1925 the railroads carried 902 million passengers and 417.4 billion ton-miles of freight. This transportation network would not have been possible without the assistance of government. The automobile industry in 1925 also became the largest industry in America. If the government had not built the network of highways for automobiles, the industry would not have achieved this goal. The waterways were also supported by government. Channel markers and beacons were installed to keep ships from running aground. Nautical maps were printed by the Coast and Geodetic Survey for captains to navigate the waterways safely. Lighthouses were also built and manned to guide ships in inclement weather. The government also made laws to govern the use of the transportation systems it supported for the public good. It was reasonable to accept the fact that if the government would support the aviation industry in the same manner it supported surface transportation, aviation would be safer. If aviation were to survive and grow in an orderly manner, a bond between government and the private sector would have to be forged. The basic infrastructure of airports, airways, navigational aids, controlled air space, and uncontrolled air space needed to be addressed in much the same manner as surface transportation. One can only imagine the chaos and confusion on our highways if there were no traffic signals, rules, or regulations.

It was these issues that brought Congress, with the support of the aviation industry, to the point of passing the Air Commerce Act of 1926. The Act would become the cornerstone of the Federal Government's regulation of civil aviation in the United States.

The legislation did not pass without its share of dissenters. Many rugged individualists in aviation felt the Federal Government would regulate aviation completely out of the sky. The sky was the pilots' symbol of freedom. If birds could fly wherever they wanted, why restrict airplanes? Although the dissenters lobbied hard, they failed in their attempt to kill the final bill.

THE BEGINNING OF GOVERNMENT CONTROLS

The landmark legislation charged the Secretary of Commerce with fostering air commerce, issuing and enforcing air traffic rules, licensing pilots, certifying aircraft, establishing airways, and improving aids to aerial navigation. These duties would be bestowed upon a newly created department under the Department of Commerce called the "Aeronautics Branch" May 20, 1926.

On August 11, 1926, William P. MacCraken, Jr. took office as the head of the Aeronautics Branch. MacCraken set in motion a program to establish intermediate landing fields for pilots in distress. The airfields were required to meet the following conditions to warrant placement on the forthcoming government-published aviation charts.

> At sea-level the airfield should be no less than 400 feet wide and 1,800 feet long consisting of two runways, giving the pilot four directions in which landings can be effected. The surface should be fairly level, with clear approaches, with good drainage. The airfields must be located near a highway to provide transportation and commercial power and telephone extensions. Each intermediate airfield shall also be equipped with a standard beacon light 50 feet high and a network of lights around the field to mark boundaries, hazards, and approaches for night flying. A 50 foot diameter white circle must be placed to mark it for daytime identification.
>
> Airways are to be marked with beacons placed 10 miles apart between intermediate airfields. The base of each beacon tower would be of concrete with an arrow pointing in the direction of the airway.

Aircraft also needed ground support facilities which included runway lighting, parking, fuel, and hangars. The lack of proper facilities created many dangers for mail pilots like Charles Lindbergh, who bailed out on several occasions when his fuel became

H. B. du Pont purchased the first Aeronautics Branch-approved airplane, the Buhl Airster. *Photo and art from the Hagley Museum & Library.*

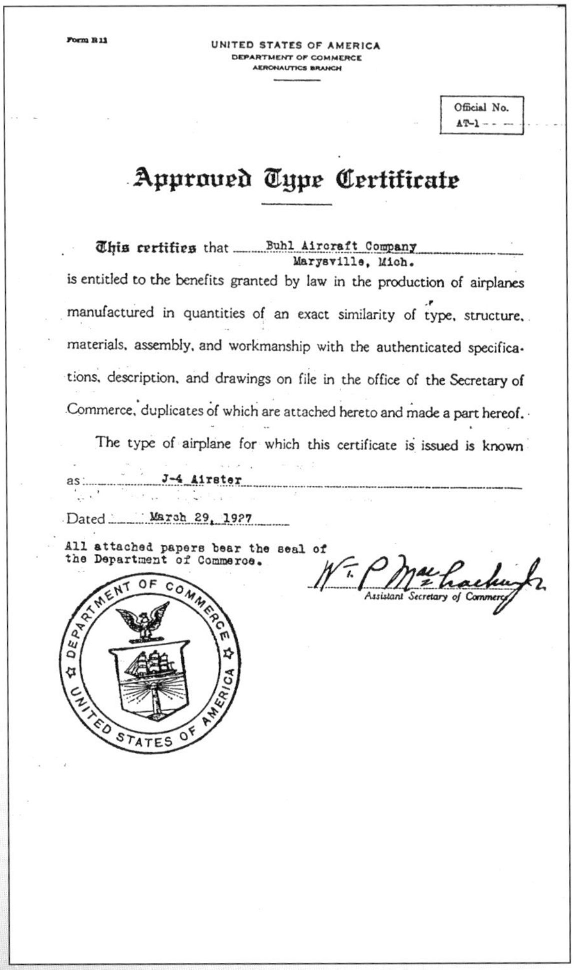

Form B11

UNITED STATES OF AMERICA
DEPARTMENT OF COMMERCE
AERONAUTICS BRANCH

Official No.
AT-1

Approved Type Certificate

This certifies that Buhl Aircraft Company
Marysville, Mich.
is entitled to the benefits granted by law in the production of airplanes manufactured in quantities of an exact similarity of type, structure, materials, assembly, and workmanship with the authenticated specifications, description, and drawings on file in the office of the Secretary of Commerce, duplicates of which are attached hereto and made a part hereof.

The type of airplane for which this certificate is issued is known as: J-4 Airster

Dated March 29, 1927

All attached papers bear the seal of the Department of Commerce.

W. P. MacCracken Jr.
Assistant Secretary of Commerce

DEPARTMENT OF COMMERCE UNITED STATES OF AMERICA

exhausted and ground visibility was impaired. Others were not so fortunate when they gave their lives in desperate attempts to arrive at their destinations.

Following the example of other states, Delaware enacted its first aviation legislation in 1926. The legislation simply required all Delaware pilots to adhere to all rules and regulations issued by the Federal Department of Commerce, Aeronautics Branch. Since Delaware only had two airfields in 1926, Biggs Airfield and DuPont Airfield, legislators felt the State did not require a separate aeronautics administrator or policing agency. The total number of airplanes in the state could be counted on one hand. Aviation in Delaware would be overseen by the existing Delaware State Police.

Because of the many aircraft companies rising like the Phoenix from the ashes, the Aeronautics Branch issued regulations in early 1927 governing the sale of new aircraft. A new aircraft would require testing to determine if it were airworthy before being offered for sale to the public. This became known in the aircraft industry as the "Approved Type Certificate" or "ATC."

The first Manufacturer to be issued an Approved Type Certificate (ATC) was Buhl Aircraft Company, Marysville, Michigan, March 29, 1927, for the Buhl *Airster.* The *Airster* was an open-cockpit, three-seat biplane of conventional design for the era. The airplane included a unique feature of folding wings for more compact storage. The first *Airster* was sold to Henry B. du Pont, on April 6, 1927.

After the issuance of the ATC, the manufacturer was allowed to produce duplicate airplanes to sell to the public. If any change or improvement was requested pertaining to the original ATC, a Supplemental Type Certificate (STC) would be required. The STC had to meet the same standards of approval by the Aeronautics Branch before it could become an addition to the airplane. These rules are still applicable today.

1929-1950's Airplane Markings — what they meant

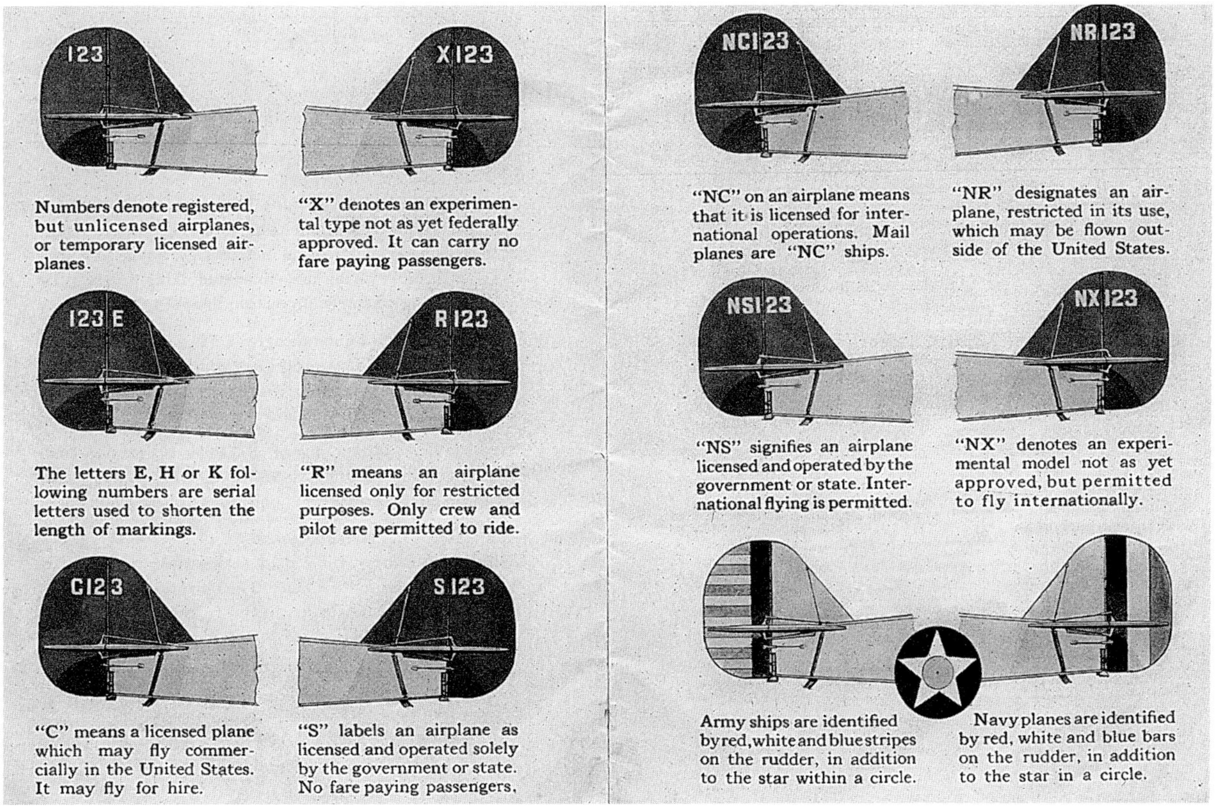

J. Allison Buck, Delaware's first federally licensed pilot. Photo from the Hagley Museum & Library.

The Aeronautics Branch also issued identification letters and numbers that were assigned to each airplane before it was allowed to fly. Initially, the identification numbers were painted on the top of the right wing, the bottom of the left, and both sides of the vertical tail. Today they are displayed on each side of the aircraft only.

More important than support facilities and sale of aircraft, etc., the Aeronautics Branch concentrated foremost on safety. The initial rulings would be to license airplanes and pilots nationally and would require pilots to obey the rules set down by the Aeronautics Branch. Any pilot who was caught disobeying the law, or endangering the life of others, would have his/her license revoked. All aircraft were also subjected to an annual inspection to determine if they were airworthy. Annual inspections would be carried out by certified aviation mechanics. Aircraft that were not able to pass inspection were grounded until approved repairs were made. Any work performed on the airplane was required to be entered into an aircraft maintenance log and signed.

By the end of 1927, the Aeronautics Branch announced that the combined municipal, commercial, Army, Navy, and private airports in the United States which met their requirements had reached 1,047 in number. Delaware was on the map with two airfields, Biggs and DuPont. Both of these airfields were privately owned. Bellanca airfield would not open until the spring of 1928. In the 1927 Aeronautics Directory of licensed pilots, Delaware listed three;

J. Allison Buck, #T720, 701 Van Buren St. Wilmington.

Henry B. du Pont, #T1644, DuPont Bldg., Wilmington.

Howard P. Young, #T3799, 1600 Washington St., Wilmington.

The Stockley-VanCamp Rehoboth Canning Factory complied with Lindbergh's request in 1929 and painted their roof so pilots could get their bearings if they became lost. Photo from Jack Beach collection.

The Aeronautics Branch also offered support to aviation in the way of financial help with the installation of airfield boundary lights and beacons on privately owned, public-use facilities. The Branch also required all airfields to display a fifty-foot white circle on the landing site so pilots could readily identify it as an approved landing facility from the air. These white circles can be seen in aerial views of airfields printed in this book. They also set minimum requirements for airfields and published an airman information manual which was available to pilots. All this effort to improve aviation was rewarded in 1928 to the Department of Commerce, Aeronautics Branch, by being awarded the prestigious Robert J. Collier Trophy.

An innovative idea Charles Lindbergh presented to the Daniel Guggenheim Foundation, in 1929, was the marking of rooftops and water towers throughout the United States to assist pilots in navigation. Lindbergh's idea was sponsored by the Foundation, and eight thousand postmasters throughout the country circulated leaflets to towns with populations between 1,000 and 10,000. The leaflets suggested the towns paint their name in large letters on a prominent rooftop. The circulars also requested the inclusion of a large arrow pointing north and a smaller arrow pointing to the nearest airfield. Communities that responded to the request were given a certificate signed by Charles Lindbergh. When the campaign ended in December 1929, more than 8,000 towns had completed the markings. Remarkably, not one single copy of the certificate could be located for reproduction.

Charles Lindbergh's visit to Delaware in 1927, and Bellanca's completion of an aircraft-manufacturing plant in New Castle in 1928, caused a belated reaction from Governor C. Douglass Buck. On December 10, 1929, Governor Buck notified Henry Belin du Pont that he would serve on a State of Delaware Aviation Committee along with James Morford and Angus Echols. Governor Buck's apprehension, outlined in his letter to du Pont, aired his concern about other neighboring states creating aviation-related business which could come to Delaware. Governor Buck also made it clear in his letter that he would support a state-owned airport. He stressed the importance of the committee to have a report ready for the next General Assembly. This letter is the first recorded indication that the state of Delaware had any interest in aviation.

The committee completed the following report by December of 1930 and submitted it to the Legislature January 6, 1931, as follows.

> Up to three years ago, Delaware lagged behind most states in aeronautical progress. During the last three years, however, there has been an increase in flying in Delaware that it now has 43 aircraft registered and now ranks sixth in respect to the number of aircraft per 100,000 population. Delaware, with 24 licensed pilots ranks twenty-first per 100,000 population. Practically all the aviation activity is at the Bellanca Airfield where Bellanca Aircraft Company carries on its manufacturing activities and where there is a commercial flying service. In addition, there is the DuPont Airfield, a non-commercial flying field where a few private owners operate. These are at present the only two established airfields in the State. *(Editor's note: It is unknown why Biggs Airfield was not mentioned in this report. Biggs Airfield was operational from 1922 to 1954.)* The New York-Washington Airway, on which more passengers are carried than any other airway in the United States, passes over the northern tip of Delaware. We believe that it will be only a short time before airmail and passenger service will make scheduled stops in Wilmington. We believe also that there will be a considerable increase in the amount of flying, both in commercial and private, in the State during the next few years owing to the high per capita wealth of Delaware, its excellent terrain for flying, and because such a heavily traveled airway crosses the northern part of the State.
>
> In respect to aeronautical legislation, Delaware has kept pace with other states. On March 23, 1926, there was approved a Uniform State Law for Aeronautics. This law requires all aircraft and pilots to conform to Federal regulations concerning ownership and flight of aircraft. These laws to date are enforced by the Delaware State Police. We believe these laws can be enforced by the State Police without the appointment of special enforcement officers. At the present time cities and towns in Delaware are legally unable to acquire land outside of their corporate limits for the purpose of establishing airports. In order to facilitate the development of municipal airports in Delaware, we earnestly recommend that the Enabling Act which follows be submitted to the Legislature.

> In regard to the State's embarking on a program of constructing airports and emergency landing airfields, we believe that the present volume of air traffic does not justify the expenditure of State funds for such a purpose.

STATE ENABLING ACT OF 1931

> SECTION 1: The Legislative bodies of cities and incorporated towns in this State are hereby authorized and empowered to acquire, establish, construct, own, control, lease, equip, improve, maintain, operate and regulate airports or landing fields for the use of airplanes and other aircraft either within or without the limits of such cities and towns, and may use for such purpose or purposes any property suitable therefore that is now or may at any time hereafter be owned or controlled by such city or town.

The legislation passed, but did not spur the interest the Delaware Aeronautic Committee felt the cities and towns needed to stimulate the creation of air facilities in Delaware. This reluctance to get involved in airport construction was for several reasons. The legislation failed to include any financial support. With only twenty-five licensed pilots and 43 aircraft in the state, politicians would be hard-pressed to find support from their constituents to fund aviation-related programs. It was not until 1940, when the City of Dover purchased three farms to build a municipal airport (now Dover Air Force Base), that the legislation was ever used.

It should also be remembered that in these early days of aviation most people felt airplanes were expensive toys for the wealthy. (To a degree, they were right.) Pilots were still perceived as daredevils who seemed possessed with a death wish. Every aviation accident, or incident, was publicized by the press with the utmost sensationalism. The uninformed and uninvolved, who constituted the vast majority of the Delaware population, also controlled the vote. If the legislature were faced with funding one mile of roadway, or one mile of runway, the roadway would win every time. The vast majority of the people in Delaware, then and now, continues to accept aviation only as a necessary evil.

In May of 1932, the United States Coast and Geodetic Survey printed the first aeronautical maps for the Aviation Branch, called "Sectional" charts. The term "Sectional" came from the decision to divide various areas of the country into large sections for printing the maps. Cities, towns, roads, railroads, rivers, bridges, and navigational beacons were depicted in various colors to give pilots a bird's-eye view of the terrain below. Printed in an easy-to-read scale of 1.25" = 10 statute miles, pilots had little difficulty reading them in flight. Available to anyone for a mere forty cents, pilots cherished them for their clarity and accuracy. These maps confirm the location and installation of aerial navigation aids in Delaware. They also show the growth of privately owned, public-use air facilities throughout the state over the years. The first issue map (1932) indicates only one airway beacon installed at Iron Hill, just south of Newark, Delaware. This installation aided pilots in navigating across Delaware. It also confirms that airfield beacons were installed at the DuPont and Bellanca Airfields. These were the only navigational aids available in Delaware in 1932. The map also confirms that there were no airfields in Kent or Sussex Counties. Although there was one at Bethany Beach, it was marked "Military Auxiliary unmanned."

AERIAL RADIO NAVIGATION – 1926-1998

In the fall of 1926, the Ford Motor Company was engaged in ferrying air freight between Chicago and Dearborn airports. The first of the famous Ford trimotor airplanes was placed into service on this run in early 1927. It was in this year that a young Ford radio engineer named Eugene S. Donovan patented the first four-course, loop-type, low frequency (LF) radio range. The Ford Motor Company installed two ranges, one at Chicago Lansing Airport and the other at Detroit Ford Airport.

The Ford LF ranges proved so successful, the Federal Government requested permission to conduct further testing of the system. Ford, in the interest of the common good, allowed the Government to install their patented system nationwide without royalties. By 1950 the government installed over 400 stations operating in the 200-400kc range throughout the

Federal Airway System. These ranges became the backbone of aerial navigation throughout the country for a quarter of a century.

Delaware saw its first government-installed radio navigational aide installed by the War Department in 1942. The low frequency (LF) radio range station was installed about three miles south of the New Castle Army Air Base. Issued frequency 350kc, and Morse code identifier DNC, it allowed pilots to make non-precision instrument approaches for landing in New Castle. The LF radio ranges transmitted two signals in Morse code in the form of a cone. The letter A and the letter N. If he was dead center he would hear a steady "hum." When the aircraft passed directly over the station, sound ceased. This was referred to as the "cone of silence." The system was considered an *aid* to navigation due to its many shortcomings. If the navigator became disoriented, there was no way to determine if the flight path was "TO" or "FROM" the station. Storms made the system virtually useless due to static interference. Volume was the only way of judging how far one was from the station. A strong signal would indicate the airplane was close to the station, a weak one distant. Effective range was about 50 miles under the best conditions. If the compass was thrown out of whack due to electrical disturbance, there was no way of confirming which of the four quadrants one was flying. The pilot could identify when the station was passed when the A and N signal would shift. Another problem with the system was when weather broadcasts were transmitted 15 minutes past the hour and 15 minutes before the hour. These transmissions would completely block out the signal. This would be especially troublesome if the pilot was on the final leg of an instrument approach. By today's standards, it was a primitive system. Fortunately, technology replaced the low-frequency navigational system with VOR (Very high frequency Omni Range) and later with GPS (Global Positioning System).

The early 1950s saw the CAA (now FAA) supporting aviation with continued assistance to upgrade aerial navigation. The CAA funded the installation of VOR transmitters throughout the United States. Delaware's first VOR (DOV 111.4) replaced the LF

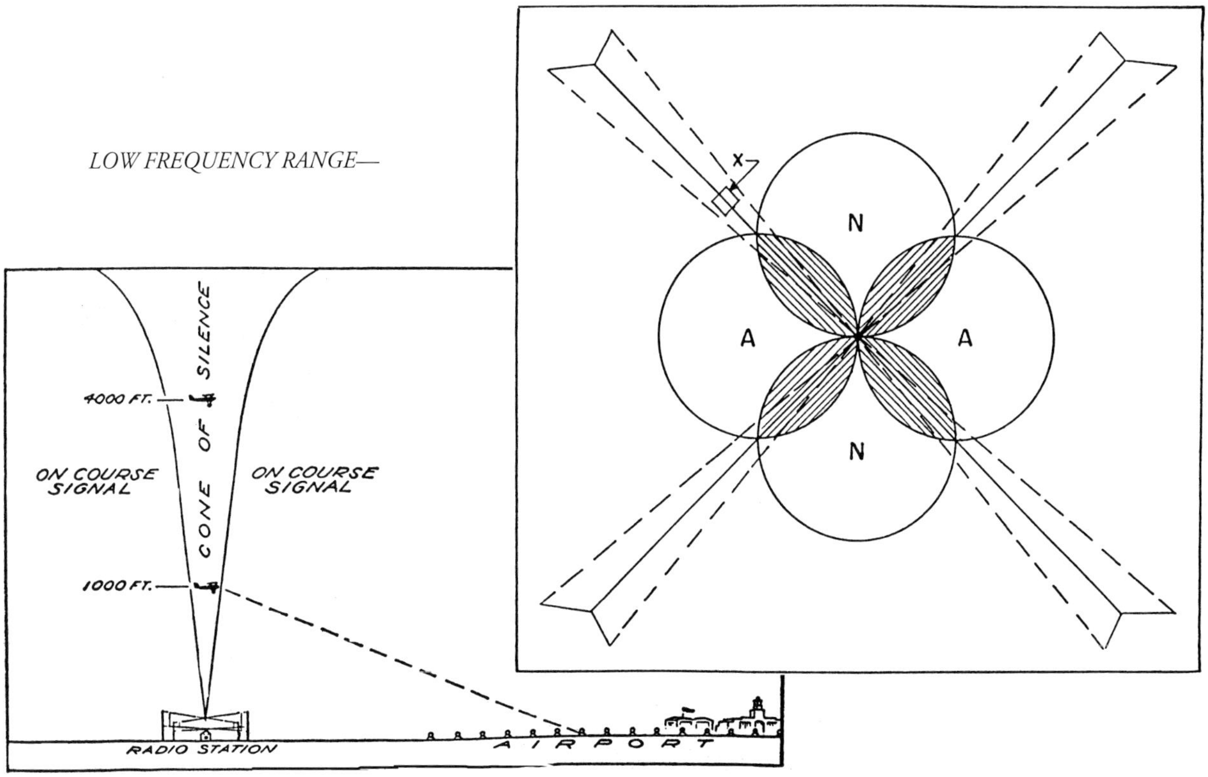

LOW FREQUENCY RANGE—

This VOR transmitter (ENO 111.4) is located off Route 9 south of Leipsic. Installed in 1952, it was the first of three to be installed in Delaware, one for each county. The distinctive white-coned building is also a visual landmark for pilots. Photo by the author.

range site south of Leipsic in 1952. The letters DOV would be transmitted continuously on the frequency in Morse code as an identifier. The name of the DOVER VOR would change over the years to KENTON (ENO 111.4) and later again to SMYRNA (ENO 111.4). The FAA does not give a reason for changing the names of station identifiers. (There is insufficient space to document pilots' responses for their action.) Following this installation, two more VORs would be installed in Delaware. The next would be on New Castle County Airport named DuPont (DQO, 114.0). As soon as this became operational, the CAA decommissioned the LF radio range three miles south of the New Castle Airport. The third VOR was located near Broadkill Beach, in Sussex County, and named WATERLOO (ATR 112.6).

By 1953, the FAA funded the installation of 400 VOR transmitters across the country and another 100 were planned. These stations form an interlocking network along the airways. A VOR transmitter sends a continuing signal 360 degrees on its assigned frequency which airmen home onto with their airborne receivers. Therefore it was possible to fly inbound from any direction, as well as outbound, with a high degree of accuracy. The aircraft VOR receiver had a centering needle which gave the pilot a visual reference to the course flown. The receiver also displayed a flag to alert the pilot as to whether the aircraft was flying "TO" or "FROM" the station. A shortcoming of the VOR was that the signal was effective only in *line of sight.* Pilots would simply "hopscotch" from one to another. In the event of surface electrical power failure of a VOR, a diesel-powered generator would self-start and run until surface power was restored. A major improvement over the LF range, it allowed an accuracy in aerial navigation never before known. The system was sufficiently accurate and reliable to the extent the FAA began issuing special VOR instrument-approach approvals to airports throughout the country. Delaware Airpark in Dover, Sussex County Airpark in Georgetown, and Summit Airport in Middletown all have FAA-approved VOR instrument approaches.

Although the VOR system of aerial navigation at this writing is still alive and well, its life is limited. The technology has been superseded by the new Global Positioning System (GPS) which operates from signals emitted from orbiting satellites. The GPS system is computer technology at its best. GPS provides current information (updated every second) to the pilot, such as: ground speed, time and distance to destination, heading, bearing, closest airport, and actual position of the airplane on a map, whether on the ground or in the air.

The world of civil aviation has taken a quantum leap in the last twenty years. Corporations have given preference to the jet engine over the whirling propeller. With the introduction of the computer in everything from flight training, moving map navigation, weather radar, and air traffic control, the tools available to the modern pilot are nothing less than awesome. A new chapter in aviation history is in the making.

GOVERNMENT AIRPORT DEVELOPMENT

The Aeronautics Branch and the Civil Works Administration combined their resources to sponsor an airport development program throughout the United States in 1934. On February 1, 385 new airports were approved for construction, and 688 were listed for improvements. The new sites were selected to provide air service to cities along the newly established airway routes. For reasons unknown, Delaware was the only state not to take advantage of the program. Therefore, Delaware was the only state not listed as a stop on the U.S. Airway charts. By 1934 the State of Delaware had six privately owned, public-use airfields (none hard-surfaced): DuPont, Bellanca, Biggs, Buck, Skycroft, and Dover. Apparently, Delaware legislators must have felt that these privately owned airfields were adequate for the state for they did not apply for funding for their improvements.

The Department of Commerce, Aeronautics Branch, in 1936, took over the important function of Air Traffic Control (ATC) in the United States. Pioneer controllers used radio contact, maps, blackboards, and mental calculations to determine the safe separation of aircraft flying the airways under instrument flight rules (IFR). ATC at this time was not involved in airport operations, such as takeoffs and landings. Low frequency radio communications with pilots offered PIREPS (pilot reports), weather, and verbal check points.

The evolution of the Department of Commerce, Aeronautics Branch, of 1926 included several name changes over the years. In 1938, Congress approved the Civil Aeronautics Act that transferred the central government's civil aviation role from the Department of Commerce to a new independent agency, the Civil Aeronautics Administration (CAA). That legislation also expanded Federal civil aviation responsibilities to include the issuance of air-carrier route certificates and the regulation of airline fares.

In 1939, there were only five privately owned, public-use, airfields in Delaware; Biggs, Bellanca, DuPont, Dover, and Rehoboth. Bethany still had an auxiliary field which was used on occasion by the National Guard, but was unattended. Not one of these airfields offered a hard-surfaced runway. The spring thaws would turn the runways to mud. Snow removal was non-existent since removal would scalp the much-needed grass. None could accommodate, on a year-round basis, the world renowned Douglas DC-3. Forty-seven states offered scheduled airline service and all-weather runways.

In contrast to Delaware, Rhode Island had a 411-acre, state-owned and operated airport in Providence, the state capital. The airport had three concrete runways, 3000' long and 150' wide, laid out in a triangle. The airport, which included a control tower and modern passenger terminal, was open twenty-four hours daily. American Airlines and Wiggins Airways offered scheduled air service to the public. Full aircraft service and maintenance was also available from two fixed base operators in modern adjacent hangars. Instrument departures and landings were enhanced by lighted runways, taxiways, airport beacon, low-frequency A-N range, and a localizer. Rhode Island

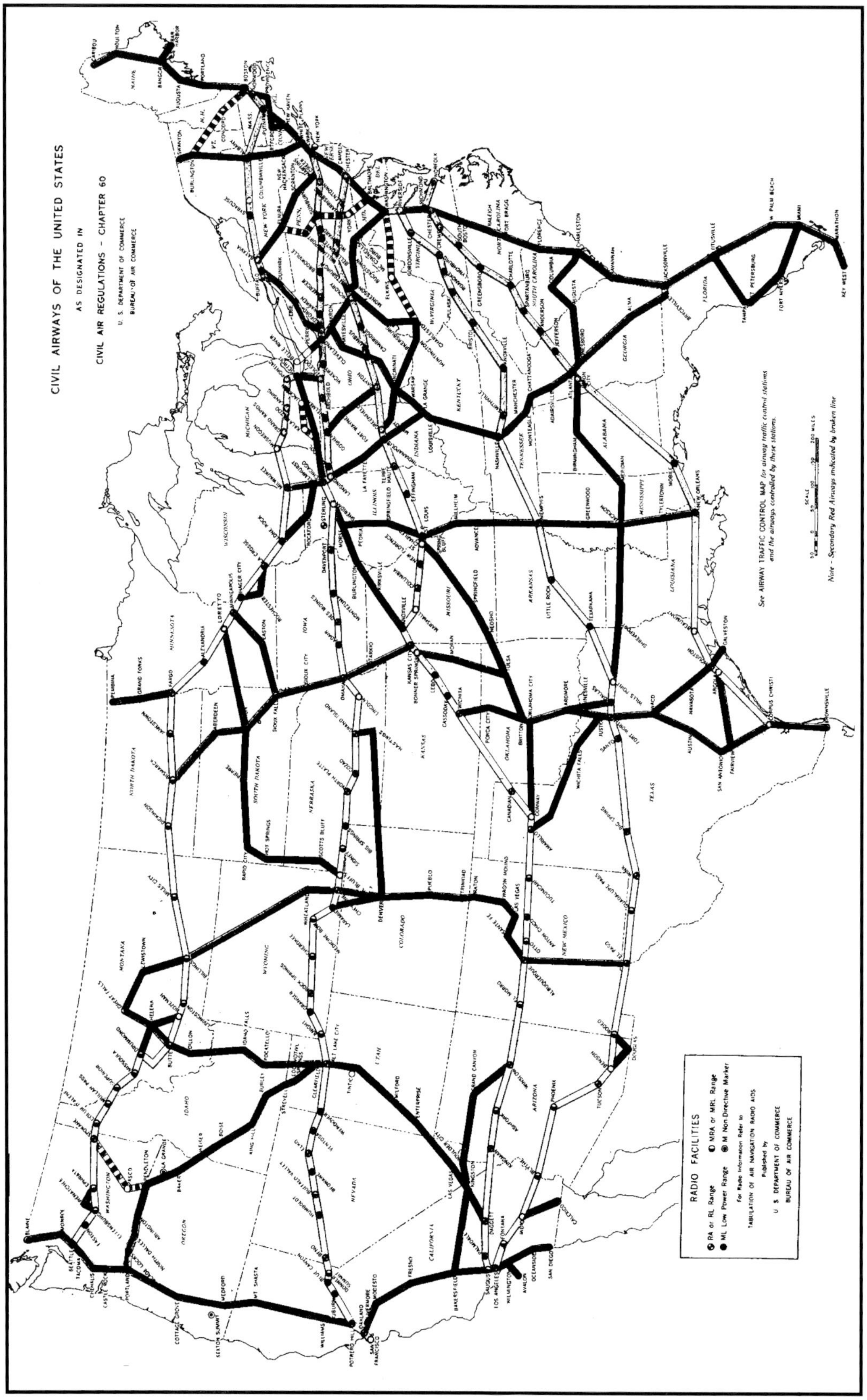

This 1934 airway chart connects every state in the union except Delaware. Map from the du Pont collection.

also had four other privately owned, public-use airfields, one of which included seaplane operations. Delaware was the only state in the union that could not offer airline service due to lack of proper facilities. In spite of this glaring shortcoming, no applications for Federal aid were requested.

The Post Office Department still controlled all airmail routes. The Congress expanded the Post Office responsibility by passing legislation, with funding for one year, to include experimental airmail service to small towns starting in December 1938. This legislation was brought about from lobbying efforts of Richard du Pont's All American Airways.

The next government restructuring was partially due to a law suit filed by Tri-State Aviation (Lytle Adams) against All American Airways, the Post Office, and the Department of Commerce in 1940. Adams had a falling-out with Richard du Pont over the way du Pont was leaving him out of All American management decisions. Du Pont was making requests to the Post Office for expanded use which would directly affect Adams-owned Tri-State Aviation.

Adams claimed Congress approved, and supplied, funding for experimental postal routes for one year. The year was up, and the Post Office no longer had the authority to approve the routes and the hauling of freight and passengers du Pont had requested. These requests were now under the jurisdiction of the CAA. Adams and his supporters won the suit, which forced Congress to create new laws governing aviation.

On May 14, 1940, Congress and President Roosevelt divided up the various functions of the Aviation Branch and the Post Office airmail system into the Civil Aeronautics Authority (CAA). The CAA, and its administrator, fell under the jurisdiction of the Commerce Department. However, a new Civil Aeronautics Board (CAB) (an independent body), became responsible for safety and economic regulation. The CAB would designate airway routes, timetables, tariffs, mergers, and many other aspects of air carrier operations. This action left the industry with two separate Federal agencies to deal with concerning aviation regulations. Needless to say, there were many occasions when the two agencies collided in their separate quests for more power and control.

On June 30, 1940, Franklin D. Roosevelt reorganized the CAA again. The five-member board was transferred to the Department of Commerce and renamed the Civil Aeronautics Board (CAB). The CAB was to report to Congress and the President through the Secretary of Commerce. The Secretary of Commerce exercised the functions of safety rulemaking, adjudication, investigation, and airline economic regulation. The Administrator, with a new title of Administrator of Civil Aeronautics, was transferred to the Department of Commerce and was placed under the supervision of the Secretary. Although the Administrator's functions included safety regulating, he did not have the power to suspend or revoke a pilot's license. Two months later, the Department of Commerce designated the functions of the Administrator as "Civil Aeronautics Administration" (CAA). In typical Washington style, the original Civil Aeronautics Authority continued to exist on paper for eight years as an entity embracing the CAB and the CAA, without any real teeth.

WORLD CONFLICTS and AVIATION

To provide some historical background for these, and other events, it is necessary to look at the world scene. The world events of this period had a direct influence on everyone involved in U.S. aviation.

The embryo of WWII began when Japan invaded Manchuria by force in 1931, in spite of the objections of the League of Nations. China had been in civil war for years trying to decide between a Communist or Nationalist government. Apparently, Japan felt it was an opportune time to takeover China while it was disorganized. On July 7, 1937, Japan invaded China with the ultimate goal of gaining complete control over all countries and islands lying between Manchuko and Australia, and from the Fiji Islands to the Bay of Bengal. The war would escalate to the point that it would be called the C.B.I. (China, Burma, India) theater of war.

The American public still felt that the League of Nations was responsible for keeping the peace. The lessons of WWI had been learned, and Americans felt the solution of international difficulties lay with arbitration. Unfortunately, the aggressors felt they were above international law, ignored the League of Nations, and moved at will to conquer the weak.

Franklin Delano Roosevelt became President of the United States of America in 1932. Roosevelt inherited the economic problems left by the disarmament of WWI. The entire country was mired in a deep depression which included massive hunger and an idle work force. He remembered well the lesson of his predecessor, President Wilson, that the people did not want America to solve European problems at our expense. President Franklin Delano Roosevelt, although concerned with the situation, was more hard-pressed to deal with the depression which had spread across the country. The President, from time to time, would continue to extend the hand of peace in an effort to reduce the flow of destructive armaments and support law and order. His gestures were wholeheartedly approved by the isolationists. The memories of our commitment in WWI were still too vivid for many people to accept involvement in another military conflict.

Nazi Führer Adolf Hitler, who had come to power in 1933, became a force to be reckoned with. One of his first actions was the withdrawal from the Disarmament Conference. Germany had successfully introduced a government-supported civil pilot training program to foster development and growth of civil aviation as early as 1934. It is now perceived that this was part of a long-range plan by Hitler to create military pilots. Since civilian pilot training was not in violation of the Treaty of Versailles, it progressed without objection. This civil pilot training program was the embryo of what would become the mighty Luftwaffe. Hitler then started building a massive war machine which allowed him to invade the Rhineland in 1936, occupy Austria in 1938, and Czechoslovakia and later Poland in 1939. Hitler's vicious attack on Poland on September 1, 1939, brought Britain and France into a European war that would last for six horrible years.

Italy's first aggressive action under Fascist Dictator Benito Mussolini was the invasion of Ethiopia in 1935, in direct contravention of the League of Nations rules. In 1939 Italy became more aggressive and invaded Albania, and by the summer of 1940 joined Adolf Hitler's forces in the occupation of France, Greece, and Egypt. In 1940 Italy entered the European conflict by announcing it would side with Adolph Hitler and Germany. The result of this combined war machine of Italy and Germany was the complete takeover of France in June, Egypt in September, and Greece in October. Germany, under Hitler's leadership, invaded Poland and Yugoslavia. To further solidify their efforts, the trio of aggressor nations (Germany, Italy and Japan), signed the Rome-Berlin pact September 27, 1940, which set the precedent for a ten-year military alliance. This alliance played a major role in the development of weapons-technology which was shared among the trio.

China was not an industrialized nation and had to look to other nations for military support to counter the Japanese movement. One aircraft manufacturer to respond to China's plea, in 1936, was Bellanca Aircraft in New Castle, Delaware, with the model 28-90 *Flash*. Under a covert operation of high security, intrigue, and misleading information, Bellanca built aircraft with 30-caliber machine guns and shipped them to China. Edward "Smitty" Smith, an aircraft mechanic for Bellanca, was sent to China to oversee their assembly and maintenance program. The administration in Washington, for the time being, simply ignored the sales which would continue over the years to also include Fleet, Douglas, and Curtiss. In reality, it was not assistance. Airplane manufacturers needed sales to stay alive during these depressed times. They would build airplanes for anyone.

This early support of China from Americans was not looked upon favorably by the Japanese military, and combat was assured if spotted. Six Bellanca 28-90 aircraft were destroyed in an air raid, at the site where they were being assembled, when the Japanese uncovered the operation. In any event, Delaware-built Bellancas were engaged in war with Japanese aircraft as early as 1936. Unfortunately, the Chinese did not seem to be inspired to keep aviation records and the combat history of the Bellancas is lost in time. The Bellanca 28-90 *Flash* that was shipped to China was also earmarked for shipment to France under the auspices of being a high-speed mail plane. In reality, it was destined to support the Republicans in the war with Spain. The U.S. Customs caught wind of the ploy and refused to allow the shipment.

It is believed the U.S.-built Bellanca 28-90 Flash *was the first U.S. armed warplane sold to China to combat the Japanese. Note the single wing-mounted machine gun and rear gun. Photo from the Daniel Koble collection.*

General Claire Chennault formed an American Volunteer Group (AVG) of pilots and ground service personnel to counter the Japanese advancement into China in 1939. The airplane used for this effort was the America-built Curtiss P-40 *Warhawk* fighter. The airplanes and the uniforms of the pilots would display the sunburst symbol of the Chinese Nationalists. Some of the volunteers would come from Delaware. Another AVG group who flew Douglas-built military C-47s supplied materials to China by flying over the treacherous Himalaya mountains, called the "Hump."

In 1938 the United States Navy utilized the aircraft carriers *Lexington* and *Saratoga* to plan a surprise mock aerial attack on Pearl Harbor. The mock attack was a complete success. The Navy gloated how they could destroy an enemy harbor with their carrier-based airplanes without warning. Unfortunately for us, the Japanese remembered the episode all too well. The Japanese attack on Pearl Harbor December 7, 1941, was a mirror tactic of the U.S. Navy's, with one exception. The Japanese raid was for real.

The war in Europe and China continued to escalate. The mood of the American people continued to lean toward staying neutral. President Roosevelt and Congress extended an olive branch of peace and sought freedom from European entanglements by introducing the "Neutrality Act" of 1937. It was a vain attempt to stay out of a war that we were not economically or militarily prepared to enter. This act was born out of pressure from isolationist sentiments lead by many Americans including Charles Lindbergh. Lindbergh spoke openly about his feelings that we would be entering a war which we could not win. The Neutrality Act forbade any American firm from *delivering* aircraft, or other war materials, to any belligerent nation. The Act hit the aviation industry hard since American-built airplanes were also being sold to Canada, who had aligned with Britain in the war against the Axis powers.

In an effort to circumnavigate the law, the aircraft industry chose a sight near Pembina, North Dakota, where the flat land bordered Canada. Aircraft were flown to the border airstrip where a line was drawn in the dirt to indicate the separation between the

U.S. and Canada. The aircraft were flown to the airfield and parked near the line where a rope was tossed across by a farmer who had the end tied to a team of horses. The pilot secured the line to the aircraft and walked away. The horses pulled the aircraft into Canada where awaiting Canadian pilots flew them to their destinations. The airplane manufacturers did not *deliver* the airplanes, they simply *abandoned* them at the border, thereby not breaking the law. This site became the first of many which would become transfer points for aircraft across the border. When the situation escalated in Europe and China, to the point where isolation from the conflict became a poor choice, pilots totally ignored the law and delivered their aircraft to Canadian airfields on a regular basis.

The European and Far East situation grew worse with time. The Germans moved into North Africa and Russia. London was being bombed day and night. The Japanese became more aggressive when they turned their eyes toward the unfortified islands in the Pacific. In spite of the escalation, there were many Americans who felt we should not get involved. President Roosevelt convinced the American people to support his lend-lease plan when he said, "If your neighbor's house is on fire, you would lend him your garden hose to put out the fire, wouldn't you?" His words were well chosen, and the people united behind the president. After all, there was no commitment to send our men to war. We would only supply materials to help our neighbors. What the administration did not tell the people was, our neighbors, to offset the balance of payments, would charge the U.S. for any docking fees, labor, food, shelter, use of facilities, landing fees, etc.

As the war continued to escalate, Washington began to change their attitude toward a military force to protect the Western Hemisphere. The industrialized might of the United States would slowly create an infrastructure of factories and a united work force beyond anyone's wildest dreams.

Early in 1939, the CAA offered 100-percent financial aid to build municipal airports throughout the country. The first to be constructed were airports near the Atlantic Ocean which could be used for military defense if the situation were warranted. The

The first student to graduate from the University of Delaware CPTP program December 21, 1939, was Charles W. Wagner. The airplane is a Taylorcraft. Photo from the Delaware Public Archives.

three airports in Delaware were not created out of interest from the state, but from national necessity. The war had escalated to the point where U.S. shores were vulnerable to attack.

In addition to the federal aid for municipal airports in 1939, the CAA offered financial aid for a Civilian Pilot Training Program (CPTP). This program would utilize existing civil aircraft and instructors to teach civilian personnel flying, meteorology, and navigation. It was felt this program would, in the event of increased conflict, create a pool of existing pilots who, at the very least, would know the basics of aeronautics. These pilots would be the first to be inducted if, or when, they were needed. The program was quite successful throughout Delaware because it gave the existing privately owned Fixed Base Operators (FBO) a financial boost that was sorely needed in the depressed economy. CPTP classes were held in local schools, and flight training was provided by DuPont, Air Service, Inc., Cherry Island, Dover Airfield, and Rehoboth. Most FBOs used the infamous Piper *Cub* for training because of its low investment and operating costs. Many Delaware Army Air Force personnel who went to war were graduates of the CPTP program.

In Kent County, under the direction of Harold Turpin, flight instructor, and Karl Dorzback, ground

JAMES MONTGOMERY FLAGG
I WANT YOU
FOR U.S. ARMY
NEAREST RECRUITING STATION

instructor, an official CPTP training program was started in Dover in 1940. Karl Dorzback taught classes in navigation, meteorology, and airplane maintenance at Dover High School at night. Harold Turpin trained pilots in a Porterfield airplane on weekdays, as well as weekends, at the Dover Airfield.

New Castle County instituted a similar program through the University of Delaware. Irene Gregory, wildly enthused over aviation all her life, joined the CPTP program in her sophomore year. She received her private pilot's license in 1940. By 1942 she acquired a commercial and instructor's rating. She joined the WASPS and was stationed at New Castle Army Air Base before being transferred to Brownsville, Texas, for pursuit school, where she learned to fly fighter aircraft.

WORLD WAR II IS DECLARED

The Selective Service Act was created by Congress and signed into law by President Roosevelt, September 16, 1940. Families were now broken up to meet the challenge of possible war. Many who served in our armed forces would not return. The ones who did, their lives were forever changed. The ones who stayed behind concentrated on supporting the war effort. Many of the material things which had been abundant, suddenly became scarce or unavailable. Scrap drives were initiated to recycle food cans, tin foil, kitchen grease, and aluminum. The Department of War created the Office of Price Administration (OPA) which controlled pricing of every commodity. Gasoline was rationed, and new cars became non-existent. Tires for existing cars required OPA approval before they could be purchased. If proof could not be documented that the tires were essential to the war effort, your request was denied. Few even took the time and effort to apply. Life made a drastic change for those who left home as well as those who were left behind.

The Axis powers of Germany, Italy, and Japan scoffed at all international law and rejected any mandates from the League of Nations or anyone else who opposed their aggression. By Roosevelt's second term in office, it was becoming more apparent that the Axis powers sought world conquest, and U.S. isolation became less an option.

Canada joined the British in their effort to bring the European conflict to an end. American industry shipped supplies, including war planes, across the border into Canada. The North American AT-6 *Texan,* (Named the *Harvard* by Canada) played a major role in training Canadian Pilots for combat. The sales of military supplies to Canada was the subject of much controversy since our government declared its position of neutrality.

The Japanese-built Mitsubishi *Zero* was introduced into combat in China during 1939. The *Zero* was an aeronautical engineering marvel of the time. None of the allied nations had an airplane which could match its performance. The Curtiss P-40 *Warhawk* was the best that could be offered for China's defense against the advancing Japanese. AVG pilots who flew the P-40s, with the noses painted like the jaws of a vicious tiger-shark, earned the title of the *Flying Tigers.* They learned very early that they could not engage in a dogfight with the *Zero* and win. Low speed maneuverability was the *Zero's* strongest offense. The Japanese pilots were also highly trained in aerobatics to further enhance the ability of the *Zero* to survive in a dogfight. However, it was learned the *Zeros* shortcoming was its inability to maneuver quickly at high speed. The higher the speed, the less maneuverable the airplane became. The *Warhawk,* which was faster, would only engage a *Zero* from higher altitude with a diving run. This aerial tactic was used throughout the war against the *Zero* with considerable success.

In May 1940, President Roosevelt asked Congress to appropriate one billion dollars for a military defense fund. He also stressed the need of an Army Air Force which would have 50,000 airplanes available. Secretary of State Roos met with King MacKenzie of Canada, in August of 1940, to ask Canada to join the United States in a joint defense force to protect North America.

Hitler announced publicly, in January 1941, that he would torpedo any United States ships carrying aid to Britain. This announcement was not good news for our unarmed merchant mariners. Hitler made good on his threat and sent his German U-boat captains into the North Atlantic in groups called, "Wolf packs."

By the beginning of 1941, countless merchant marine ships were being sunk by German submarines. We were supplying England with war materials under a lend-lease program for which Congress and the president earmarked a staggering $43 billion dollars. May 27, 1941, Roosevelt made a speech in which he proclaimed, "We will not allow the Axis Powers to shape the things to come." When Japanese warships appeared off Camrahn Bay in November 1941, President Roosevelt announced the freezing of all Japanese credits and assets in the U.S.

The Japanese found themselves hopelessly bogged down in their war with China. The Chinese-American supply lines would have to be severed and the Indo-China ports and the Burma-Chungking Road would have to be closed. To back down would be tantamount to losing face. The Japanese Embassy in Washington tried in vain to reach an agreement with the U.S. without losing dignity. That was totally unacceptable to the U.S.A., since Japan was the aggressor.

Major George Welch, from the original oil painting. Photo courtesy of Delaware State Museums, Dover.

Many historical journals place the December 7, 1941, Japanese *surprise* attack on Pearl Harbor directly on Washington. Pearl Harbor was not notified of the diplomatic breakdown of communications between the two countries. If the military had been on "full alert" for a possible attack, the outcome would have had a dramatically different ending. Personal interviews with Pearl Harbor survivors alone is convincing evidence that they were unaware of any possible aerial strike. U.S. military rivalry was so paramount, it caused a detrimental lack of communications between forces. While the attack was underway, many thought it was a military maneuver. When the realization struck, the military was in such disarray that they shot down some of our own aircraft and killed our pilots with ground fire when they bailed out.

One alert Delawarean from Wilmington, Lieutenant George Welch, and his friend Lt. Kenneth M. Taylor of Oklahoma, saw the red insignia on a low-flying bomber and immediately knew the raid was real. They jumped into Taylor's car and raced for Haleiwa Airfield, which was an auxiliary grass airfield about ten miles away. They knew there were P-40 *Warhawks* there ready to fly if they could get there. The Japanese pilots were so sure nothing got off the ground they were not looking for any aerial combat. Welch and Taylor spotted a string of single-engine dive bombers starting to make a run on Ewa Airfield and dove after them. As they approached from the rear, the enemy's focus was on the ground. It only took a five-second burst from their machine guns to send two down in flames. Welch would take down two more before the enemy was gone. The Pearl Harbor attack only lasted about one and one-half hours. The Japanese attack was made for one sole purpose. Cripple the U.S. Pacific Fleet while their main forces captured and took control of all the remaining islands in the Pacific. Although Welch downed four airplanes, it was a small revenge for the over three thousand people killed in the raid. Welch went on to fight in the Pacific with the famous *39th*

Fighter Squadron headed by Colonel Thomas Lynch and Captain Richard Bong. Welch would become an ace, flying a twin-engine P-38 *Lightning* in the New Guinea Theater.

Lieutenant Welch achieved national recognition, and stories of his performances were in the news media nationwide. He was brought home with several other war heroes to be greeted by President Roosevelt at the White House, was hailed in New York City in a ticker-tape parade and took part in War Bond Rallies. In Wilmington, he was greeted by Governor Bacon and 8,000 Delawareans at Rodney Square.

After the stateside tour, he returned to Hawaii and was transferred to New Guinea. On the anniversary of his first encounter at Pearl Harbor, Welch shot down three more Japanese airplanes in a ten-minute duel with his P-39 *Aircobra.* First Lieutenant Welch was next assigned the new Lockheed P-38 *Lightning* in the famous Major Bong Group. On his first mission on June 21, 1943, he downed three *Zeros,* and on August 21 added two more to his score. Following this melee, Welch, with only ten hours flight time in the P-38, led his squadron into battle, over Wewak Island, encountering thirty Japanese aircraft. In the ensuing battle his squadron downed fourteen aircraft of which three were credited to Welch. Following this encounter he was promoted to Captain and bestowed the title of "Ace." During his tour of duty in New Guinea, Welch flew a total of 218 missions, over enemy territory, where he escalated his total to sixteen enemy aircraft by the end of the war. Captain Welch earned the Distinguished Service Cross, the Silver Star Medal, the Air Medal, two oak leaf clusters, four battle stars, and a Special Service Cross from the state of Delaware. After the war, he was promoted to major and assigned to Edwards Air Force Base where he flight-tested jet aircraft. Noted painter Peter Hurd painted his portrait in 1944, which was hung in Legislative Hall among other Delaware heroes. The Dover Air Force Base also named their school "The George Welch Elementary School" in honor of his extraordinary service history.

George Welch was born in Wilmington on May 10, 1918, and grew up at Wawaset Park, where the first airplane was flown in Delaware. He attended Friends School and Perdue University before entering the Army Air Corps in 1939. George earned his commission in 1940, after his flight training at Randolph, Brooks, and Kelly airfields in Texas. He arrived at Wheeler Airfield, Oahu, Hawaii, his first duty station, in February 1941. He died test-flying a Lockheed F-100 Super Sabre jet at Edwards AFB in 1954. He is interned at the Arlington National Cemetery.

Another Delawarean who also kept his cool under fire at Pearl Harbor was Captain Stephen G. Saltzman, a member of the Coast Artillery, assigned to Schofield Barracks. He rushed to his post, crossing an open field. A Japanese plane started his strafing run and Saltzman opened fire with his hand-held, automatic rifle striking the pilot and killing him instantly. Captain Saltzman, from Wilmington, was given the Silver Star for his actions. He later transferred to the Army Air Corps and was sent to England where he flew the Lockheed P-38 *Lightning.* In a fight over Belgium, his right finger and wrist were fractured by enemy machine-gun fire. He was also wounded by antiaircraft fire which struck his other arm and leg. In spite of the injuries to himself, and his airplane, he limped back to a safe landing in England.

These two events are only examples of the heroic efforts Delaware servicemen displayed during the war. For a further in-depth account of all Delaware servicemen in WWII, refer to *Delaware's Role in World War II* volumes one and two. Only 500 copies of this title were published in 1955, by the Delaware State Archives, Hall of Records, Leon deValinger, Jr.

All three of Delaware's municipal airports were well under construction when the war broke out December 7, 1941. The War Department promptly leased all three airports for the duration and took over their control. The Army Corps of Engineers provided the manpower and the funds to complete all three sites and use them for military purposes for the duration.

Back home, Bellanca Aircraft ceased all civil aircraft manufacturing when war broke out and converted to manufacturing items for war. Employment at the factory would approach 3000 personnel working in

Bellanca-built AT-21s under construction at Bellanca Airfield. The AT-21 was built by Bellanca and used as a trainer for gunnery and bombing. Photo from the Daniel Koble collection.

The Civil Defense insignia can be seen clearly on this photo of the DuPont Airfield taken in 1942. Photo from Eleanor Wallin.

three shifts. Along with many other military contracts, a contract was signed with Fairchild to build the AT-21 (Advanced Trainer) for the Army Air Corps.

Eight days after the U.S. declared war on Japan, Hitler declared war on the U.S. The Department of War immediately grounded all civilian flying 150 miles inland from the East Coast. This action closed all airfields along the entire Coast to civilian operations. This included major airports such as Philadelphia, Trenton, and Roosevelt Airfield in New York. Flying in this air space was restricted to approved military operations. As the threat of war decreased, the 150 miles would change to 25 miles. The DuPont Airfield would assume the role as headquarters for the Civil Air Patrol (CAP). The Rehoboth Airfield would fly the first operational CAP flight to patrol the Atlantic for German submarines. The three Delaware municipal airports would now become military for the duration.

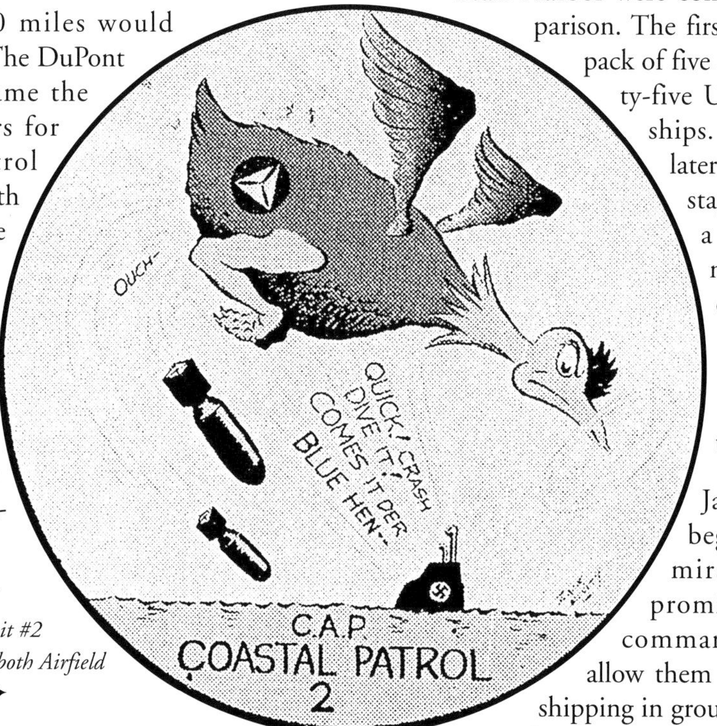

This is the insignia of the Delaware CAP unit #2 which flew from Rehoboth Airfield during WWII. ➨

CIVIL AIR PATROL, December 1, 1941

Submariners who went to sea during WWI had a code of honor which required the captain to surface and allow all hands to abandon ship prior to its being sunk. The theory was that the crew would no longer be a threat without a ship. The ships crew would man the life boats and watch in horror as their ship was sent to the bottom. This policy was abandoned during WWII when submarines sent anything to the bottom without warning. The British captured German Captain Karl Doenitz and his entire submarine crew in WWI when it was forced to surface due to mechanical problems. He spent the remainder of the war as a British POW, which left him with a lot of hostility.

One of the first U.S. ships to be sunk by the Germans on October 30, 1941, was the destroyer U.S.S. *Reuben James.* The *Reuben James* was named after a Delaware seaman who saved the life of Commodore Stephen Decatur in the war with Tripoli. When the German submarines patrolled our coasts looking for prey, they were so effective that our losses suffered at Pearl Harbor were considered small in comparison. The first month alone, a wolf pack of five submarines sank twenty-five U.S. Merchant Marine ships. The Germans would later escalate this total to a staggering ninety-one in a single month. A German Commander was quoted as saying after returning unhindered from a mission, "Don't these people realize they are at war?"

January 1942 saw the beginning of (now Admiral) Karl Doenitz's promise to his submarine commanders that he would allow them to stalk United States shipping in groups called "wolf packs." Our shipping lanes started in the vast Gulf of Mexico, wrapped around the tip of Florida, and followed the Atlantic Ocean north to Nova Scotia. Refueling and replenishment of stores to the diesel-powered German submarines was accomplished at sea at rendezvous points with sub tenders outside the range of American air and sea defenders.

The German commander was right – Americans along the coast had not yet taken the blackout seriously. The lights from shore created a silhouette of the ships against a night sky that outlined targets almost impossible to miss. People on shore witnessed explosions and fireballs in the night skies

This is the result of German U-boat's torpedoes on a U.S. Merchant ship. Many more would go to the bottom before aircraft began their counter attacks. Photo from Tom Murray, Jr.

while debris and floating bodies washed up on the shores. The beaches became black with oil and birds died by the thousands because of the pollution. The black specks of oil in the sands of Rehoboth today are a grim reminder of those ships that still lay on the bottom of the Atlantic. For the most part, the ships sunk were unarmed Merchant Marine types that supplied vital cargo and oil to the war effort. Our Coast Guard was totally unprepared with countermeasures. A handful of airplanes, a few Coast Guard cutters, four blimps and a squadron of unarmed patrol planes could do very little to cover such a vast area. German submarines lurked beneath the surface with little fear of being detected.

Although Delaware was ranked as the second smallest state in the Union, its strategic location bordered the Atlantic Ocean and the mouth of the Delaware River. The river was a major shipping lane from the industrial north of Philadelphia and Trenton which allowed Merchant Marine ships easy access to the open Atlantic Ocean. Needless to say, the cargo these ships carried was essential to the war effort. Defense of the river rested with Fort DuPont, Fort Delaware

Pilots from the Delaware CAP Coastal Patrol No. 2 pose for an official photo, circa 1943. Seated; Sykes Ewing, Hugh Sharp, Richard Roberts, Ed "Smitty" Smith. Standing L/R; Tom O'Day, Robert Ready, Bill James, Buzz Thompson, Walter Fullerton, Pappy Garrett, Jim Tegg, Tom Worth, Shelly Edmonson, Alex Szyky, Unknown, Harry Proctor, Howard Carter, Smedley Butler, Jr., Bill Nicholson, Ralph Fidance, "Cousin" Caulfield, Tom Sanschagrin, Bernie Mullin, Earl Wilson. Photo from the Hagley Museum & Library.

at Delaware City, and Fort Saulsbury near Milford. The Chesapeake and Delaware Canal was also a vital link between another industrial city, Baltimore.

General "Hap" Arnold suggested that a civil air patrol manned by civilian pilots flying their own airplanes could be used as submarine spotters. They could radio their position if they sighted a sub, and the Coast Guard or military could respond to the call. Vice-Admiral Adolphus Andrews, Commander of the North Atlantic Coastal Frontier, and *New York Times* editor, Gill Robb Wilson, supported General Arnold's plan. Wilson had been a pilot in WWI, and had toured Germany as late as 1936 where he saw first hand the German military build-up. Wilson, Fiorello LaGuardia, director of Civil Defense, and "Hap" Arnold became the trio who pressed for the CAP's establishment.

Wilson wrote an editorial in the *New York Times* imploring the War Department to utilize the CAP. He felt the CAP "spotter" airplanes would force the submarines farther out to sea and away from our sea lanes. After the second month and twenty-two ships sunk, additional support for the CAP came from the Maritime Union, which demanded that the War Department allow the CAP to fly.

The War Department was still anything but pleased with the idea of sending overage, unqualified, "Sunday pilots," with no military training, against a highly seasoned enemy. They tried one last desperate action to kill the entire idea by saying that there were no funds available. This argument was immediately dispelled when J. Howard Pew, president of Sun Oil, advanced $10,000 toward costs. Other oil companies followed suit until the fund swelled to the point the War Department had no other alternative but to give in. Their goal was reached and the Civil Air Patrol (CAP) was signed into existence December 1, 1941.

With the greatest amount of reluctance, the War Department agreed to a ninety-day CAP trial period. The U.S. Army Air Corps would have the responsibility for organizing and training the civilians in the military way of doing things. CAP volunteers were issued military clothing with their own distinctive identification patch. Although still civilians, CAP members took an oath to accept all orders issued by the military.

Richard C. du Pont, President of All American, immediately started to hold meetings with Major General John C. Curry, CAP National Commander, and Gill Robb Wilson of the National staff. These meet-

This old 1931 Chevrolet truck was converted into a gas truck to service the airplanes at the Rehoboth CAP. Photo courtesy of Jack Beach.

Jack Beach's mother, far left, worked for the CAP during WWII in the Rehoboth CAP office. Her son, Jack, would also serve for a short time in this unit. Photo courtesy of Jack Beach.

ings, held at the DuPont Airfield, grew to include others in various aviation-related areas. At one of these meetings, Danish-born Sir. Holger Hoiriis was chosen to be Delaware Wing Commander.

Holger Hoiriis was a seasoned WWI pilot with a history that included many civil record flights. Hoiriis, with copilot Otto Hillig, on June 24, 1931, flew a Bellanca CH *Pathfinder* named *Liberty* nonstop across the Atlantic to Krefeld, Germany, the place of Hillig's birth. This was the fourth nonstop crossing of the Atlantic by a Bellanca airplane. At the time the Delaware CAP chapter was being organized, Hoiriis was employed by Richard du Pont's All American Aviation. Although Richard du Pont needed Hoiriis, he felt duty-bound to suggest Hoiriis for Delaware CAP Commander because of his experience.

On December 11, 1941, Hoiriis was accepted by the National Headquarters and assumed the rank of major. Major Hoiriis immediately appointed his Wing Staff and created Groups, Squadrons, and Flights throughout the State of Delaware. Within two months Hoiriis would send the following radio message to the National CAP Headquarters,

A very young Jack Beach, Rehoboth, followed in his father's footsteps by flying this AAF Fairchild, PT-19 Cornell. *He would later transition to B-17s and B-24 bombers. Photo courtesy of Jack Beach.*

> Have offered volunteer service to infantry. Conducting spotter flights. Made state inspection of roof-marking OCD. Graduating 30 with third-class radio license; enrolled 50 more. Finished chemical war class. Have conducted blackout tests by air.

At the time of Pearl Harbor, the Army Air Corps had only two operational bases along the east coast; one at Langley Field, Virginia, and the other at Mitchell Field, New York. The First Air Force, with a total strength of 337 fighters and bombers, of which 57 were unflyable, was charged with protecting the coast from Maine to Virginia. Rehoboth Beach, Delaware, and Atlantic City, New Jersey, were the nearest airfields to the highest submarine activity. Gill Robb Wilson was named Wing Commander of the Atlantic City site with its gravel runways and two hangars. Major Hoiriis established his Wing Command at Rehoboth Beach with Hugh Sharp in command. Rehoboth airfield had nothing more than sod runways and three aging wooden hangers with dirt floors. Captain Tom O'Dea, a Dover auto salesman, was a key figure in the Rehoboth expansion. He was a master at locating scrap lumber and installing heating and plumbing. The materials were donated, but all hands pitched in for the building. It did not take long for them to improvise a lean-to attached to one of the hangars which was used for operations. There were no lighted runways, therefore, operations were limited to daylight flying only.

The response to the CAP by the public was beyond everyone's wildest dreams. People came from all levels of society to volunteer, donating their airplanes, two-way radios, and anything else they thought

The original Delaware CAP Sikorsky S-39 was completely restored by volunteers from the New England Air Museum. The Delaware CAP official insignia is displayed proudly on the bow. Photo by Harry Van Den Huevel.

would be useful. The Rehoboth squadron consisted of Fairchild 24's, Stinson *Voyagers* and one Sikorsky S-39 amphibian. The privately owned Sikorsky S-39 was purchased by Major Hugh Sharp in 1942 from E. Paul du Pont, Jr. to serve as an aerial rescue ship.

Many older aircraft mechanics and ground support personnel joined the CAP to make sure the light, single-engine airplanes were in perfect flying condition. Flying over the ocean left pilots no alternative in the event of mechanical trouble. Captain Edward Marvin "Smitty" Smith, Rehoboth CAP, was in charge of aircraft maintenance. He would design a special "quick release" mechanism for the door so pilots could get out in a hurry if they had to ditch in the ocean. He also hand-crafted large airtight cylinders which he mounted in the rear of the airplane to keep it from sinking. The tail of the airplane sticking out of the water made an excellent marker for rescue. Pilots earned $8.00 and mechanics $7.00 a day. In most cases this was far less than they earned in the private sector.

Major Hugh Sharp, operations officer at Rehoboth, took the first CAP patrol flight over the Atlantic Ocean March 5, 1942. His first flight, slightly over an hour, revealed nothing but a couple of ships underway and an oil slick from a previous sinking. CAP pilots flew in all kinds of weather when heading east over the Atlantic Ocean. All flights were made in pairs for reasons of safety. If one plane went down, the other circled and radioed for rescue. In eighteen months the CAP logged over a million miles of over-water flying. Only two days of patrol duty were missed during this period because of weather. One CAP pilot remarked, "If we could see one-half the runway and our heads were not in the fog as we walked to our airplanes, we flew."

The CAP coastal patrol broke ground every morning at sunrise and more followed every two hours until night forced them to cease operations. The first submarine was spotted by the Rehoboth unit March 10, 1942, as it was about to launch a torpedo into the side of a tanker near the shallow water shoals off

Cape May. Pilots Eddie Edwards and Howard Carter spotted two vessels proceeding past Five Fathom Bank Buoy. One of the vessels appeared to have its decks awash and was floundering. As they flew closer they identified the vessel as a submarine. They dived their two unarmed *Fairchilds* toward the sub as the sub commander gave the order to "crash-dive." The sub slid beneath the waves and out of sight. They circled the area for an hour but the sub was frightened off and a tanker was saved. The Merchant Marines grew to love the CAP because it gave them a sense of security, no matter how small.

In spite of the intense CAP patrol, in March 1942 the enemy sent seventy-four ships to the bottom, all within three hundred miles of the coast. In May, the War Department refused to release the number of ships sunk by enemy submarines because of possible public reaction. It was learned later that ninety-one allied ships were lost in May. In spite of the sinkings, the CAP was credited with many saves. One CAP pilot saw a submerged submarine advancing behind a freighter and dove his airplane straight toward the periscope. The submarine captain immediately took evasive action and dove, thinking depth charges were going to explode any second. The next day however, the same pilot on patrol saw three Merchant Marine ships sinking or burning. This time the unfortunate submarine captain had surfaced to allow his crew to observe their kills sink to the bottom. The CAP pilot radioed for the Coast Guard which was fortunately nearby. The submarine crew scrambled for a crash-dive for the last time when Coast Guard depth charges hit their target.

This 1937, Fairchild 24W, restored to original CAP condition by George Hindall, Rehoboth, was chosen "Grand Champion Antique" at the 1997 EAA, Annual Convention held at Oshkosh, Wisconsin. Photo courtesy of George Hindall.

February 1943, President Roosevelt congratulates Edmond (Eddie) Edwards of Newark, Delaware, and Major Hugh Sharp of the Delaware CAP after awarding them the Army Air Medal. They would be the first civilians to receive this honor. Photo from Eddie Edwards.

The CAP patrol pilots became more than angry when they spotted submarines and had no bombs or depth charges to drop on them. Unfortunately, the Army would not agree to arm the small planes until after the following event took place.

Off the coast of Florida, a CAP plane spotted a sub in the mouth of the Banana River. In its haste to escape, the sub ran aground on a mud bank. The CAP pilot radioed frantically for help as he saw the whirling propellers try desperately to free the sub. For forty minutes the pilot circled his prey but help never came. The sub finally broke free and headed out to sea. This incident would have been a sure kill for the CAP plane if it had only been armed with demolition bombs or depth charges. An angry call to General Arnold finally brought action. The first CAP unit to be armed with 250 lb depth charges, and the first to score, was the Atlantic City unit on July 11, 1942. Major Farr and Captain Haggin were flying a Grumman amphibian when they spotted a submerged sub off Absecon, New Jersey. They circled the sub for four hours, before it finally began to surface. That is when the crew made their move and came in from astern to drop two depth charges from only one hundred feet. The submarine was blown completely out of the water with a direct hit.

On July 21, 1942, the Rehoboth operations radio picked up a distress call announcing that Lt. Henry Cross and Lt. Charles Shelfus had crashed their Fairchild (NC19144) into the ocean 20 miles north of Winter Quarter Light. The call was made from the Fairchild 24 sister ship flown by Lieutenants Carl Virdin and Shelly Edmonston, who were circling the area to mark the position. Major Sharp and Lt. Edwards responded to the call flying the aging 1930 Sikorsky S-39 amphibian. A strong northeast wind whipped up waves larger than they estimated and then nearly wrecked the Sikorsky when they set it down in the ocean. The port sponson was smashed in the landing, causing the airplane to list heavily to the left. Edwards clamored out on the hull and spotted Cross. Edwards successfully tossed Cross a line but there was no response. He shouted to Sharp that Cross looked paralyzed. Sharp skillfully maneuvered the lopsided amphibian directly alongside the injured man. Edwards reached out and caught him under the arms and hauled him onto the hull of the Sikorsky. With Sharp's help they carried Cross into the cabin. His back was broken in the crash. Shelfus, the observer, could not be found. He must have gone down with the shattered plane.

The broken sponson made it impossible to execute a takeoff, so they decided to taxi the airplane 20 miles to the nearest land. Unfortunately, the damaged wing that was dragging in the water made the airplane go around in circles. Lt. Eddie Edwards left the cabin and climbed out on the right wing strut to offset the imbalance. It worked, and he stayed there for several agonizing hours until they reached shore. Cross recovered from his injuries and was assigned to non-flying status for the duration. In response to their heroic actions, Sharp and Edwards were awarded the Air Medal at the White House by President Franklin D. Roosevelt.

The Sikorsky S-39 (NC803W) also survived. After leaving the service of the CAP after the war, it somehow found its way to Alaska. The airplane was wrecked by a subsequent owner in 1957 and left to die in the remote Alaska woods. In 1963, the Sikorsky was rescued by Philip Redden, of Anchorage, and shipped to the New England Air Museum. The airplane lay in storage until 1993 when volunteers from the Air Museum started restoration of the airplane. It is finished in its original colors of blue and yellow with a 20-inch CAP insignia on each side of the fuselage. The airplane also displays the official nose art from the Delaware Wing.

Major Holger Hoiriis passed away August 7, 1942, and Captain Hugh Sharp assumed command of the Delaware CAP Wing for the duration. Three other flyers from the Rehoboth CAP Squadron besides Shelfus were lost in the line of duty. All were lost during the frigid winter of 1942-1943. Lt. Delmont B. Garret of Rosetree, Pennsylvania, and his observer, Lt. Paul D. Towne, from Peoria, Illinois, crashed off Winter Quarter Shoal. Lt. Harold O. Swift of Wilmington froze to death on his life raft after he crashed in a blinding snowstorm off Fenwick Island while on dusk patrol.

Flying five-hundred feet over water for hours is not a scenic trip for pilots and spotters. When Major Sharp heard some grumbling from his men, he told his Rehoboth Squadron, "We are like insurance. As long as the enemy knows we are out there, they won't come in." The CAP left an impressive record of spotting 173 submarines and dropping bombs or

Another dangerous mission bestowed upon the CAP was towing aerial targets. The line was played out to about 1,500 feet behind the airplane before fighters attacked. Photo courtesy of Eleanor Wallin.

depth charges on fifty-seven, officially sinking two with four possibles. They assisted 91 ships in distress, saving 363 souls, at the cost of losing 90 airplanes, with 26 pilots killed, and 7 seriously injured.

After more than a year of patrols, submarine losses forced the enemy to retreat further out to sea. Now long-range, multi-engine military aircraft and Navy sub chasers took over. Admiral Horton, Atlantic Fleet Commander, decided to go on the offensive in spite of opposition from others. He purposely sent convoys into known German U-boat waters in hopes of drawing them out. This was a daring concept, but it worked. The introduction of sonar for tracking submarines became the curse of the submariner. Sonar made the ability to hide from foe virtually fruitless.

The German success of the U-boat activity in the beginning of the war is well-established. The end of the war holds a different story. Germany would lose 32,000 submariners out of the 39,000 who enlisted in the U-Boat service.

A letter found with Adolf Hitler after his suicide named Admiral Karl Doenitz as his successor. However, Karl Doenitz was arrested and tried as a war criminal for ordering his U-boat captains to sink unarmed merchant marine ships. Ironically, it was Admiral Nimitz who saved Doenitz from execution by admitting the United States applied the same tactics to Japan during the war. Admiral Doenitz was still given a sentence of ten years in prison. During this time he became a born-again Christian. He lived to publish several books on the war before he passed away on Christmas day, 1980.

The CAP was not disbanded after the war. The United States Air Force still has reign over these magnificent men and their flying machines. They are ready at a moment's notice to scramble into their small airplanes and respond to any call. They will search for a downed pilot, patrol from the skies after any kind of disaster, and even airlift someone in need.

Members still donate their time to encourage young people to join the CAP cadet program, where they are taught the basics of flying, navigation, aircraft maintenance, and weather. For those over sixteen, there is a pilot training program that takes students and prepares them for their FAA written test and flight examination. When they have completed the program requirements, they are issued an FAA private pilot certificate. It is a lot of work for these young cadets, but the ones who excel may go on to make flying their profession. As this is written, the CAP Delaware Headquarters are in New Castle County - with squadrons in Middletown, Dover, Georgetown, Brandywine, and Bear.

When the War ended, the Department of War was deactivated and a new agency, the War Assets Administration (WAA), began to deal with the huge surplus property owned or leased by the military. New Castle County Airport was evacuated by the military and returned to the County. Likewise, the Georgetown Airport was returned to Sussex County. Dover Army Air Force Base reduced its activities to custodial, but continued its lease with the City of Dover.

Delaware Aeronautical Commission

It was 1926 when the State Legislature first addressed the issue of Aviation in Delaware. Nineteen years later, on April 24, 1945, the State Legislature amended Chapter 301 of the Delaware Code to establish a Delaware Aeronautics Commission. During this period of Delaware assuming the role of "Rip Van Winkle," aviation had progressed from the wood and fabric airplane to the jet age. The Commission would consist of five members, all appointed by the Governor, to oversee all phases of Aeronautical activity in the state. The twenty-eight page legislation gave the Commission broad powers, which included the promotion of aviation, airports, navigational aids, eminent domain, etc., but no funding.

Governor Walter W. Bacon, on June 19, 1945, named the first members to the Delaware Aeronautics Commission: W. Samuel Carpenter, III, chairman; N. Max Terry, vice chairman; Stewart E. Poole, secretary; Richard H. McMullen, director; and James P. Scott, director. All five appointees served with no financial compensation. Any capital spent on aviation would have to be approved by the Legislature. The appointments were, for the most part, limited to ceremonial gatherings, and as a result Delaware aviation infrastructure fell behind the times.

Over the years, each governor would dutifully name his chosen five to oversee Delaware Aviation. Although neighboring states created aviation programs to further the development of civil aviation facilities, the Delaware Aeronautics Commission failed to introduce one piece of aviation legislation or submit legislation for funding airport development. When the state abolished the Commission form of government in 1971, for a more streamlined cabinet form of government, the Aeronautics Commission was transferred to the Department of Transportation (DELDOT). The Aeronautics Administrator now became a paid position to satisfy political favors.

Although most states have a Department of Aeronautics plan that offers aid to support privately owned, public-use airports, DELDOT management has failed to consider such a plan. This failure is magnified by the fact that the state aviation fuel tax generates approximately $50,000 annually. These funds are earmarked for the Legislative General Fund with no allocation to aviation. Many privately owned, public-use airports could be saved throughout the state if the powers in control would adopt aviation policies of neighboring states.

Three airports, one in each county, were built by the War Department in Delaware during WWII. Dover Air Force Base was the only one retained for military operations. This left Kent County and Dover, the capitol city, with privately owned, public-use grass airfields to serve aviation. The state failed to act on an airport to serve Dover until 1981 when a joint-use agreement was reached between DELDOT and the USAF. In spite of the many restrictions and high landing fees imposed by the military, DELDOT built a multi-million dollar Civil Air Terminal adjacent to DAFB. The emphasis was placed on obtaining a commuter airline to serve Baltimore and Philadelphia. Thus the building was designed with a passenger waiting lounge, baggage-handling facilities, and ticket counter. Jack Wallace, for whom the facility was named, had a marketing plan that was limited to "build it and they will come." Three attempts were made to establish a commuter service and each failed. Dover's location is simply too close by car, which was not taken into consideration. With the limited access and restrictions, DELDOT has failed to find use for the facility; DELDOT moved the two employees of the Aeronautics Commission into the building. Normal operations are 8:30 a.m. to 4:30 p.m. daily. The facility is closed on holidays and weekends.

The John R. Wallace Civil Air Terminal, dedicated October 1982. A 24-hour advance notice is required for landing. There is a $25.00 landing fee. (After hours, weekends, or holidays $35.00.) No fuel or other aviation services are available. Photo by the author.

CHAPTER 5

Bellanca

There would never have been a Bellanca Airfield in New Castle County if it were not for Henry Belin du Pont and the vision of the county citizens. Bellanca built over 3000 airplanes in Delaware and almost every model is entered in the record books. (His contribution to aviation is so profound his life story is worthy of being told.)

Giuseppe Mario Bellanca (1886-1960). Photo courtesy of Frances Rees Hayford.

Giuseppe Mario Bellanca was born in Sciacca, on the southwestern coast of Sicily in 1886. As a young boy he spent hours on the seacoast watching birds flying. He was captivated with the bird's ability to sustain flight through the invisible air. An obviously gifted young man, his parents sent him to Milan for higher education. By 1908, at the young age of twenty-two, he had earned a teaching certificate in mathematics. He furthered his education by enrolling at the Politecnico where he earned additional degrees in mathematics and engineering. It was at this institute that he concentrated on the efficiency of marine propellers.

The Wright brothers in America had already achieved powered flight in 1904, and Wilbur Wright had demonstrated the Wright *Flyer* in Europe as early as 1907. The publicity of the Wright flights created considerable interest in flying machines throughout Europe. European designers and pilots were flying their aeroplanes and carrying passengers by the time Giuseppe Bellanca had reached the age of twenty.

After hearing of a flying exposition in Turin, Italy, Bellanca's interest in aeronautics sent him on his road to destiny. At Turin he observed Leon Delagrange, a renowned French aviator, carry aloft Madam Therese Peltier, the first woman in history to fly. Bellanca was caught up in the moment. At the age of 22, he now knew where his energy would be

This is the parasol airplane built and flown by G. M. Bellanca in New York. Power was from a 30-hp three cylinder Anzani engine. Photo courtesy of Frances Rees Hayford.

directed. He quickly convinced two fellow students, Enea Bossi and Palo Invernizzi, to help finance and build a flying machine. They set to work on the design and construction with visions of flying like other aviators of the time.

Their airplane was completed and ready to fly from Baggio Airfield, Milan, September 8, 1909. Bossi was at the controls for the first flight. Although the flight ended in disaster, he did survive the crash and never faltered in his desire to continue in aviation. He went on to become a leading seaplane designer in Italy. Bellanca, although his dreams of flying had been temporally shattered with the crash, went on to design and build another airplane. However, lack of funds kept him from buying a suitable engine, and it never flew.

Giuseppe Bellanca had an older brother who, several years earlier, decided to leave Italy and travel to America, which was known throughout Europe as "the land of opportunity." His brother's letters painted a beautiful picture of this unspoiled land and the opportunities it offered. The name "United States of America" stimulated the feeling of "unity," which was in sharp contrast to the small, segregated countries of Europe. Giuseppe decided to leave home, as had thousands of others, and begin anew in this land of opportunity. He arrived in America, in September of 1912, to join his brother who lived in Brooklyn, New York.

With the financial assistance of other Italians in the community, Giuseppe started to build a parasol monoplane in the back room of a local grocery store. He worked long hours to complete the airplane by the spring of 1913. The airplane was removed from its Brooklyn home and transported to a nearby field, on Long Island, where Bellanca taught himself to fly. He accomplished this remarkable feat by flying only several feet then landing. Little by little, he gained more confidence in his ability to control the machine until one day he made a complete circuit of the airfield.

By 1914, Bellanca had opened a flying school at Garden City, New York. By this time he had built an improved version of his parasol airplane to use for teaching student pilots. One of his first students was a young lawyer, Fiorello LaGuardia, who would later become a WWI bomber pilot and the mayor of New York City. Times were very lean, and bartering was a major source of exchange at the time. Bellanca agreed to teach LaGuardia to fly in exchange for LaGuardia teaching him to drive his model "T" Ford. Both men arrived at the airport convinced that learning to fly would be simple in comparison to mastering the Ford. Bellanca taught his students to fly, first at Hempstead and later at Garden City, in the same manner that he taught himself. They made brief short hops of only a couple of feet in the air and then landed. When he felt they had control over the airplane, he would allow them to complete a cir-

This Bellanca model CD was built in 1916 by G. M. Bellanca when he worked for the Maryland Pressed Steel Company. Power was from a 30-hp Anzani engine. Photo courtesy of Frances Rees Hayford.

The Bellanca CE was the second model built for the Maryland Pressed Steel Company in 1918. Note the cable-operated ailerons. Photo courtesy of Frances Rees Hayford.

cuit of the airport. He taught flying to approximately a dozen students per year through 1915.

Clarence Chamberlin, famous barnstormer, flew the original Bellanca parasol in 1921 from Roosevelt field, New York. He wrote in his book, *Record Flights,* in 1928, the following conclusion;

> The pilot sat on a sort of a bird perch where he could look straight down and see nothing but thin air between him and the ground he had been rash enough to leave. The Bellanca Parasol had a three cylinder Anzani motor right in front of the pilot's face that could always be counted on to splatter the intrepid aviator with hot oil even if it did no worse to him — I flew the thing myself at Roosevelt field in 1921 after spending a lot of time replacing the original three-cylinder engine with a six-cylinder Anzani. It flew alright but I thought I would never get it down, the wind was so high and it was so light. My curiosity was quite satisfied; I was thoroughly convinced that Bellanca not only was a genius but a hero of a rare sort.

In 1916 Bellanca was offered a unique opportunity by the Maryland Pressed Steel Company, located in Hagerstown, Maryland. They offered him the position of "aeronautical engineer" for their newly formed aviation division. The company was producing shell casings for the war and felt that they should diversify into airplane production. During his tenure, Bellanca did design and build two biplanes known as model CD and CE, which were very conventional open-cockpit biplanes of the era. The airplanes were light and agile with a wing span of twenty-six feet and an empty weight of only 400 lbs. Test pilot Arthur Heinrich considered the CD a fine airplane in spite of the fact it had only a three-cylinder, 30-hp Anzani engine and the wing-warping lateral control.

By 1917, Giuseppe Bellanca introduced for the company the model CE which used the larger six-cylinder Anzani engine and ailerons for lateral control. The Anzani engine would only produce 90 hp, but it would still allow the Bellanca to cruise at 100 mph and land at 40 mph. Because of the war, although the airplane was now ready for production, difficulties in obtaining the Anzani engine from France created production delays.

The first CE was ordered by Clarence Chamberlin, a former Army flight instructor and notable barnstormer. It took fourteen months before he finally accepted delivery of the airplane. When he picked it up it had the smaller 70-hp Anzani engine in place of the 90-hp one that was specified. In spite of the lesser engine, the CE still had a top speed of 90 mph. These figures, in comparison to the Curtiss N *Jenny* that had a 90-hp engine and flew at 65 mph, were nothing short of outstanding for the era.

Chamberlin used the airplane for barnstorming and sold rides to the passengers who dared to fly. If a passenger was really brave, Chamberlin would do aerobatics for a few bucks extra. Chamberlin thought so much of his Bellanca-built CE that when Maryland Pressed Steel closed down in 1918, he purchased all the remaining tools and inventory. (It is unfortunate that not one of these splendid biplanes survives today.)

Bellanca's design criteria were endurance, speed, and safety. Drag reduction and increased lift were needed to meet his goals. By late 1920 he had finished plans for a high-wing, four-passenger monoplane that he designated the CF. He signed an agreement with the North Platte Aircraft Company in Nebraska on April 29, 1921, to manufacture the airplane. In this agreement Bellanca was to receive $390 per month salary and oversee the construction. The contract included the purchase of fifty new 90-hp Anzani engines for $375 each in anticipation of production volume. Unfortunately, after six months the company went bankrupt before the first airplane was completed. Bellanca had not even received the salary promised in the contract.

Victor Roos, a local motorcycle dealer in Omaha, rescued the project by providing funds to complete the airplane. Roos, A. H. Fetters of the Union Pacific Railroad, and several others formed the Roos-Bellanca Aircraft Company to complete the project. Instead of new Anzani engines, a surplus military 90 hp Anzani engine was purchased for $35 to save money. The airplane was moved to the rear of the local firehouse where a small team assembled to work with Bellanca. One of the team was a young woman named Dorothy Brown. Dorothy was the daughter of an Omaha businessman who also owned the boarding house where Bellanca had taken up residence. Dorothy was the "Girl Friday" who acted as

The Bellanca CF monoplane was sleek and fast. It broke thirteen flying records in 1922-23. This photo was taken in Omaha, Nebraska, June 1922. Photo courtesy of Frances Rees Hayford.

Harry G. Smith, test pilot, and Giuseppe Bellanca with the record-breaking Bellanca CF monoplane built in 1922. Photo courtesy of Frances Rees Hayford.

secretary, bookkeeper, fabric stitcher, and helped Giuseppe with aircraft drawings. In time they fell in love and married. Roos, hell bent on completing the project, gave them the afternoon off to get married.

Dorothy would spend her entire life with Giuseppe. They would have one son, August, who would also spend his life in aeronautical engineering. A lady of considerable class, she became known as the "Grand Dame of Shorewood Estate," where they eventually retired. The 250-acre Shorewood Estate, purchased in 1938, overlooked the beautiful Sassafras River on Maryland's Eastern Shore. The estate offered boating as well as a private landing strip for visitors. Dorothy was a close friend of Amelia Earhart, Elinor Smith, and Jacquelin Cochran, three of history's most renowned women pilots. Dorothy passed away September 4, 1979, at the age of eighty-two.

The Bellanca CF finally flew for the first time on June 8, 1922. Harry G. Smith, a thirty-one-year-old airmail pilot, had the honor of the first test flight. Smith recalls that after twenty hours of test flying, the Bellanca was so perfectly balanced that no adjustments were ever necessary to the rigging. Smith stated frankly, "This is the most remarkable airplane I have ever flown."

The CF design was very unique for its time. It carried four passengers in an enclosed cabin while the pilot sat aft in an open cockpit. In those early days of flying, many thought the pilot would lose control of the aircraft if he were enclosed in a cabin where he could not feel the effects of the wind. The CF lifted 14 pounds for each pound of thrust. Other designers were happy to achieve seven to one. The 90-hp Anzani engine was able to propel the CF at 100 mph, while carrying 950 lbs for 600 miles. This remarkable airplane also had a 12:1 glide ratio where others of the day would be hard pressed to obtain a 7:1 glide.

The Bellanca design incorporated a low-drag airfoil and eliminated all flying wires. The wing struts were unique in design to reduce drag and increase lift. This feature would become a trademark of the Bellanca airplanes for a number of years to come. The CF airplane was aerodynamically sleek and uncluttered and became the envy of others who tried to achieve the same goals.

Two days after the first flight, Smith flew the Bellanca to Crook, Nebraska, where it was officially clocked at 109 mph by the U.S. Army. On June 15, Smith flew the airplane in the Midwestern Flying Meet at Monmouth, Illinois. Smith ran away from every entry in each of the four events. The speed event was a five-lap, 15-mile triangular course in which the second place Curtiss *Oriole* biplane crossed the finish line 11 minutes 30 seconds behind the Bellanca. The purpose of the next event was to climb the fastest to 7000 feet. The Bellanca arrived at the assigned altitude in 11 minutes. The 160-hp *Oriole* conceded after arriving at 5600 feet in 12 minutes and 30 seconds. A gliding event required the pilot to shut off the engine 2000 feet over the airfield and land. Once again, the Bellanca stayed aloft for 4 minutes and 43 seconds to beat the second place Laird *Swallow* which landed in 3 minutes and 44 seconds. The three-day program netted Smith and Bellanca five trophies. The fifth trophy was awarded for winning all of the events.

William C. Hopson, the third-ranking airmail pilot of the day, was a skeptic. However, after his first forty-five minute flight of wringing the airplane out, he returned to the airport with a change of attitude. He remarked, "This is the most remarkable airplane I have ever flown or dreamed of flying."

Hopson later flew the Bellanca to Tarkio, Missouri, for the July 27-29 meet. Once again the Bellanca cleaned the competition's clock by winning every event. A promotional stunt that the newspapers played up was an aerial wedding. The pastor, the best man, the bride, and groom climbed aboard the roomy cabin of the Bellanca. Hopson flew from the open cockpit and was airborne in a mere 200 feet. The preacher married the couple while flying. Has there ever been another airplane built with 90 hp that could carry five people and perform at this level?

The Bellanca was entered next in the prestigious National Air Races of 1923 at Lambert Field, St. Louis. There were nine events scheduled for the meet, and the newest, fastest, and best airplanes in the world were entered. Seven were entered in the coveted *Aviation Country Club of Detroit* event. The 31-mile course required five laps for a total of 155 miles. Jack Atkinson took the checkered flag in the Bellanca CF, followed by Perry Hutton flying a Laird *Swallow,* and third was Casey Jones flying a Curtiss *Oriole.*

Over the years of documented aviation history, it is unchallenged that the Bellanca CF set the standard for significant change in airplane design. *Jane's All the World's Aircraft* stated in 1923 that, "The Bellanca cabin monoplane was designed and built as the first up-to-date transport airplane that was flown in the United States."

The Bellanca CF flew on to win thirteen performance and efficiency awards at various meets in 1922-23. Giuseppe was nominated for the prestigious "Collier Trophy" for outstanding achievement. The trophy, however, was awarded to the U.S. Post Office for outstanding air mail service to the public.

In spite of all the records and publicity, Bellanca did not receive a single order for this remarkable airplane although it was priced at only $5000, for the airplane market was still flooded with surplus WWI biplanes which could be purchased for as little as $300. Although the closed-cabin CF monoplane achieved outstanding performance, many pilots also held steadfast to the proven, open-cockpit, biplane concept.

The fuselage of Bellanca CF under restoration. Note the wooden tailskid. The main wheels did not incorporate any brakes. Photo from the Daniel E. Koble collection.

The Bellanca CF somehow survived the treacherous times in spite of the neglect. The CF and all the production rights were sold to Yellow Aircab Company of New York in 1924. When Yellow Aircab failed, the airplane was sold to Continental Aircraft Corporation. It too failed, and the airplane was bought at auction by P. Gilberti, of Hempstead, New York, on August 20, 1928. Gilberti sold it, less engine and instruments, for $150 to William Stanley Smith. Smith brought the airplane back to flying condition October 4, 1931. The first flight was to be made by Bert Acosta, well-known air mail pilot who had flown the WB-II for the Columbia Aircraft Company. Unfortunately Smith would never enjoy the fruits of his dream, as he was killed propping another airplane October 25, 1931. At this point in its history it became part of the Roosevelt Field Museum. Over the years, the museum fell into decay, and the Bellanca was transferred to the home of Giuseppe Bellanca in Galena, Maryland, where it was stored in a barn. After his father's death in 1970, Giuseppe Bellanca's son, August, donated the airplane to The National Air and Space Museum. Head Curator, Paul E. Garber, wrote to Giuseppe's widow, Dorothy,

> This airplane is a wonderful example of the genius of your husband, and we will be proud to maintain it as a memorial to his eminent place in aeronautical history.

This one-and-only unique airplane can be seen at the Paul E. Garber Facility, Silver Hill, Maryland. Restoration started January 25, 1979, and was completed May 13, 1980. It took Karl Heinzel and Rich Horigan 3911 man-hours to complete the restoration.

THE ULTIMATE PRIZE OF THE TIME

The first transatlantic flight had been accomplished by Lt. Commander Albert C. Read, U.S.N. flying a Curtiss-built flying boat. Actually, three of the 14-ton Navy flying boats (NC-1, NC-3, and NC-4, nicknamed Nancy Boats) made the attempt in unison. All three departed Rockway Naval Air Station, New York, on May 16, 1919. NC-1 was piloted by Lt. Commander Bellinger, NC-3 was piloted by Lt. Commander John Towers, and NC-4 was piloted by Read. NC-1 was lost at sea while being towed to port by the Greek ship *Ionia.* NC-3 had mechanical trouble and taxied through rough seas, for 200 miles, to port in Horta in the Azores. Read, piloting the NC-4, was the only one to complete the 3,925-mile flight by stopping in Newfoundland and the Azores. Read, upon landing in Lisbon, Portugal, radioed to his base, "We are safely on the other side of the pond. The job is finished."

On May, 22, 1919, prompted by the first crossing of the Atlantic, Raymond Orteig, a New York City hotel owner, on May 22, 1919, offered a $25,000 cash prize to the first *airplane* capable of crossing the Atlantic Ocean between New York and Paris nonstop. At the time of the announcement, there was no airplane capable of making the flight. However, the British flew their R-34 dirigible from England to New York City nonstop across the Atlantic on July 2, 1919. Before Lindbergh crossed the Atlantic in May 1927, 91 others, in 13 flights, had already conquered the Atlantic by making stops in the Azores and Newfoundland.

The 1920s continued to be lean years for everyone, especially those involved in aviation. Now that the Bellanca CF was completed, Bellanca began to focus on an idea he had for improving the performance of the de Havilland DH-4B used by the U.S. Post Office. New wings had been tried before with only limited improvement. The Post Office, therefore, was more than skeptical of Bellanca's request. However, Bellanca was so convinced his modifications would be successful that he offered the Post Office a deal they could not refuse. He guaranteed that he

Modified de Havilland/Bellanca DH-4B mail plane with new wings and lift struts. Photo from Francis Rees Hayford collection.

would forfeit the $7000 payment if his modifications failed to live up to his claims. Since the Post Office had nothing to lose, a contract was signed for the first modified DH-4 to be completed by June 15, 1923. Bellanca installed new wings and lifting struts (of the same design as his Bellanca CF). Ernest Allison, the pilot assigned to the first modified aircraft, reported,

> The Bellanca DH-4B outperforms the original DH-4 by carrying more load, climbing quicker, landing shorter and flying faster. I like this airplane in every respect more than the factory de Havilland.

Giuseppe, with his wife Dorothy, left Omaha to return to New York in the fall of 1923 where he formed the Bellanca Aircraft Company. In a Farmingdale, Long Island factory, Bellanca modified three more de Havilland mail planes for the Post Office. In spite of the success and praise from pilots, the Post Office was unable to increase the order due to lack of funds, and in March 1925, Bellanca had to shut down his business due to lack of work.

Giuseppe Bellanca and Clarence Chamberlin, as early as 1920, felt they had the ability to win the Orteig prize. Their only obstacle was financial backing. They met with officials of the Chicago *Tribune* and requested they sponsor them to the tune of $10,000 in the challenge for the prize. After much deliberation, the *Tribune* declined the offer. The *Tribune* felt that Bellanca and Chamberlin were crazy to attempt such a challenge with a single engine airplane. They also wanted no part of the bad publicity which would surely follow if their attempt failed. Bellanca and Chamberlin abandoned their plan when they realized that sponsorship during these lean times would not be possible.

With the help of Clarence Chamberlin, now a test pilot for Wright, Giuseppe Bellanca was hired by the Wright Aeronautical Company. Wright Aeronautical manufactured engines for the numerous airplane builders of the era in Paterson, New Jersey. Bellanca's assignment was to design a "showcase" airplane around the new 200-hp, Wright J-4 *Whirlwind* engine, for introduction to airplane manufacturers. Wright intimated to Bellanca that they had every intention of entering into actual construction of the aircraft if the design proved its worth.

The 1925 Wright/Bellanca WB-I was a true six-place airplane capable of flying 600 miles without landing. The number 15 is a National Air Race (NAR) number for competition. Photo from the Daniel Koble collection.

Bellanca went to New York University where he used its wind tunnel to help reduce drag on models he built for testing. He completed two airplanes for Wright that incorporated many of the features of his Bellanca CF. The all-wood WB-I (Wright-Bellanca) emerged as a six passenger aircraft, powered by the Wright 200-hp, J-4 *Whirlwind,* which could carry a payload of 1600 lbs at a speed of 120 mph. This airplane was lost in 1925 when test pilot Fred Becker crashed the airplane when it was being tested under maximum load conditions. Becker was injured but recovered to continue his flying career. Unfortunately the WB-I was a total loss.

The WB-II, a four-seat variant, was already under construction and would be powered by the improved Wright 220-hp, J-5 *Whirlwind.* The WB-II differed in construction when Bellanca changed to a welded steel-tube fuselage and empennage. The WB-II was issued NAR number 140 when entered in the efficiency contest at the 1925 National Air Races, Philadelphia. The Wright/Bellanca beat the second place Beech *Travel Air* by fifty-three percent.

The Wright Company, to Bellanca's dismay, decided to reverse its decision to take orders for producing the WB-II in fear it would alienate the airplane manufacturers to whom they sold engines. In an attempt to disassociate themselves from manufacturing airplanes, the Wright company entered into negotiations with the Huff-Daland Airplane Company in Bristol, Pennsylvania, for a buyout of the WB-II. All that Giuseppe Bellanca wanted out of life was to design and build airplanes. When Wright Aeronautical decided not to build airplanes, Giuseppe was again denied his dream.

THE ATLANTIC CROSSING

On September 26, 1926, Rene Fonck, French WWI ace, with a crew of three, attempted the crossing in an airplane built by Igor Sikorsky. A Russian aeronautical engineer, Sikorsky had successfully built and flown a mammoth four-ton, four-engine airplane in 1913. Within a year he constructed a similar five-ton airplane that carried 16 passengers and a crew of four. On June 29, 1914, he flew the behemoth from St. Petersburg to Kiev and back for a 1,590-mile round trip. The flight included on-board meals for the passengers who flew in armchair luxury. A total of 70 of the Sikorsky planes was built before the Bolshevik Revolution, and Igor fled penniless to America. The Sikorsky design for the Atlantic flight was a mammoth tri-motor airplane that weighed a staggering 28,000 pounds fully loaded. The engines selected for the task were French-built Gnome-Rhone, air cooled radials. Fuel tanks held 2500 gallons of aviation fuel to feed the three engines. Test flights were completed and covered extensively by the press. On the day of departure, four men climbed aboard with an ample supply of baggage and gifts for special friends abroad. Their visions of completing the trip were apparent.

This was, however, the first attempt to fly the Sikorsky with this amount of weight. The huge biplane gathered speed slowly as the engines belched out their sound of full power. The Sikorsky tried in vain to stagger into the air. The right wing dropped to the ground (the right landing gear failed because of overload) cartwheeling the plane into a mass of rubble. It came to rest inverted. Within seconds the 2500 gallons of fuel ignited into a huge ball of flame. Rene Fonck and copilot Louis Curtin escaped the inferno. Unfortunately Jacob Islamoff, the mechanic, and Charles Clivier, the radio operator, perished. (Jacob Islamoff was the brother of Ilia [Is] Islamoff, an aeronautical engineer who would work for Bellanca from 1919 to 1946.)

It was this failed attempt that sparked the idea in Charles Lindbergh that the single-engine airplane could make the crossing. The airplane uppermost in his mind was the Wright-Bellanca WB-II. Lindbergh was familiar with the airplane and saw it perform at several air shows. He also read many aviation articles praising the WB-II. There was little doubt the WB-II was far ahead of every other single-engine airplane of the day. Lindbergh talked to the president of Robertson Airlines, where he was chief pilot carrying mail, and told him of his idea. Bill Robertson was intrigued and agreed to set up meetings for Lindbergh with prominent St. Louis businessmen to request help in funding the purchase of the WB-II from the Wright Company.

Two months later on November 26, 1926, Charles Lindbergh arrived in Paterson, New Jersey, to talk to the Wright Company about purchasing the WB-II. At the meeting, Wright officials told Lindbergh that they were in negotiations with the Huff-Daland Company to sell the rights to produce the WB-II. If the Huff-Daland deal did not materialize, they would inform him of the outcome.

A tour of the factory and a meeting with Giuseppe Bellanca followed. Bellanca and Lindbergh had met before when Lindbergh marveled at the Bellanca WB-II at air shows. They discussed many aspects of the New York-Paris flight, including the use of multi-engine airplanes. Lindbergh and Bellanca agreed that a heavily laden, multi-engine airplane could no longer maintain altitude if one engine failed. The chance of an engine failure in a three-engine airplane was three times greater than a single-engine airplane. The amount of fuel required to feed two or three engines for the 3600-mile trip was simply too heavy to be practical. Therefore, multi-engine aircraft were worse off than a single-engine airplane in terms of safety.

Lindbergh asked Bellanca, "If I cannot buy the WB-II, how long will it take for you to build one for me?"

Bellanca became sullen and replied, "I work for Wright Aeronautical. I do not have a factory, I would need a factory to build the airplane."

Lindbergh asked another critical question. "How long do you think the WB-II could stay aloft without landing?"

Bellanca replied, "With extra fuel tanks, it will fly for 50 hours."

In their discussions, Bellanca and Lindbergh agreed that it would be wise to break the World Endurance Record as a test before attempting the flight across the Atlantic. The world record of 45 hours, 11 minutes, and 59 seconds had been established in August 1925 by two French Army officers, Drouhin and Landry. There was much praise to be gained by bringing the record back to America. Bellanca now had visions of Chamberlin and Lindberg flying the WB-II and becoming the first to cross the Atlantic nonstop. He also felt Chamberlin and Lindbergh could fly his airplane to victory for the World Endurance Record.

Lindbergh returned to St. Louis convinced that the WB-II was the airplane he wanted. After he arrived at Lambert Field, he sent Wright Aeronautical a telegram confirming that he wanted the WB-II and to notify him as soon as it would be available. It was not until the 4th of December that Lindbergh received an answer from the Wright Company. The message was not to his liking. Wright offered him a tri-motor airplane if the WB-II would not become available. Lindbergh had very valid reasons why he did not want a tri-motor airplane and had made them openly to the Wright officials. Lindbergh responded with a telegram indicating he was only interested in the WB-II.

The WB-II displaying NAR race number 140. Note the wooden propeller and laced covering over the wheels. This photo was taken in 1925, when it was still owned by Wright Aeronautical. Photo courtesy of Frances Rees Hayford.

On the afternoon of December 13, 1926, Lindbergh received the following telegram from Wright Aeronautical Company.

> REGRET THAT WE DO NOT DESIRE AT THIS TIME TO HAVE THE WRIGHT/BELLANCA USED FOR TRANS-ATLANTIC FLIGHT STOP SUGGEST FOKKER OR HUFF-DALAND THREE ENGINE PLANE WRIGHT AERONAUTICAL

The truth of the matter was that Wright Aeronautical did not want to take the chance of a possible failure which would create negative publicity for the company. Lindbergh's decision to fly alone was a major contributing factor in Wright not wanting to sell him the WB-II. Lindbergh's position of flying alone was not mentioned to Giuseppe Bellanca in their previous meeting.

Lindbergh grieved over the decision and decided he would have to locate a substitute. Unfortunately, he knew of no other airplane that fulfilled his requirements.

When the Wright Company decided not to sell the WB-II to Lindbergh, or to build the WB-II, Giuseppe Bellanca resigned and sought venture capital to establish an airplane factory of his own. He met Charles A. Levine, a New York millionaire of dubious reputation, who had sold surplus military gunpowder and equipment to South American countries for huge profits. Levine and Bellanca formed the Columbia Aircraft Company and bought the WB-II from the Wright Company in January 1927.

The Wright company had offers of $25,000 for the rights to the WB-II from other manufacturers, but decided to sell the airplane to Bellanca for $15,000. They felt honor-bound to sell the airplane to the person who created the design. Giuseppe Bellanca and the WB-II came to Columbia Aircraft with a contract that included fifty percent ownership and a stipulation that Bellanca had the final authority on who, and when, the WB-II flew. Clarence Chamberlin had more flying time in Bellanca airplanes than anyone in the country and also left Wright Aeronautical to become Columbia's chief pilot. The WB-II was now renamed the *Columbia,* which was proudly painted on the rudder.

Initially, Levine's vision of the Columbia Aircraft Company was to secure the lucrative New York to Chicago airmail contract using the WB-II. He also had visions of building the *Columbia* and selling it to others. Levine and Chamberlin made several trips to various places around the country in the *Columbia* in search of a site for the proposed factory. On these flights, Chamberlin brought up the subject of the Orteig prize that he and Giuseppe had secretly wanted to win.

Meanwhile Lindbergh was working hard to secure financial backing. He now had $13,000 from nine St. Louis businessmen plus $2000 of his own money. On February 3, 1927, Lindbergh sent a telegram to the following airplane manufacturers: Douglas, Travel Air, Boeing, Curtiss, Martin and Ryan Airlines.

> CAN YOU CONSTRUCT WHIRLWIND ENGINE PLANE CAPABLE OF FLYING NON-STOP BETWEEN NEW YORK PARIS STOP IF SO PLEASE STATE COST AND DELIVERY DATE
>
> ROBERTSON AIRCRAFT CORP

Robertson was an investor in Lindbergh's plan and allowed him to use the company name in achieving his goal. The only reply came from Ryan Airlines in San Diego, California, on February 4, 1926.

> CAN BUILD PLANE SIMILAR TO M-ONE BUT WITH LARGER WINGS CAPABLE OF MAKING FLIGHT FOR ABOUT SIX THOUSAND DOLLARS WITHOUT MOTOR AND INSTRUMENTS STOP DELIVERY IN ABOUT THREE MONTHS
>
> RYAN AIRLINES

On February 5, 1927, Lindbergh sent Ryan Airlines another telegram requesting a possible completion date shorter than three months. The telegram also requested general specifications of the proposed airplane.

By three p.m. the same day, Benjamin Franklin Mahoney, president of Ryan Airlines, sent the following telegram to Lindbergh.

> GAS CAPACITY THREE HUNDRED GALLONS CRUISING SPEED ONE HUNDRED MILES PER HOUR STOP LOADING ONLY TWELVE AND ONE-HALF POUNDS PER FOOT AND

TWENTY POUNDS PER HORSEPOWER STOP CAN COMPLETE AIRCRAFT IN TWO MONTHS FROM DATE OF ORDER IF NECESSARY STOP WILL REQUIRE FIFTY PERCENT DEPOSIT

RYAN AIRLINES

Once the Wright-Columbia deal was complete, Bellanca sent the following telegram to Lindbergh on February 6, 1927.

SORRY DELAY AS HAVE BEEN OUT OF TOWN STOP WILLING TO MAKE ATTRACTIVE PROPOSITION ON BELLANCA AIRPLANE FOR PARIS FLIGHT STOP SUGGEST YOU COME TO NEW YORK AS SOON AS POSSIBLE SO WE CAN GET TOGETHER IN QUICKEST MANNER STOP WIRE ME CARE OF COLUMBIA AIRCRAFT CORPORATION 5104 WOOLWORTH BUILDING NEW YORK

BELLANCA

Lindbergh left on the next train for New York in high hopes of buying the Bellanca WB-II. On the train to New York, Lindbergh had already decided to tell Bellanca to start installation of the long-range fuel tanks. He was going to establish a new World Endurance Record as a flight test.

In the offices of Columbia Aircraft, Lindbergh was introduced to Charles Levine, chairman of the board of directors, Clarence Chamberlin, test pilot, and Giuseppe Bellanca. In the conversation that followed Levine stated flatly that, "the *Columbia* is the best plane you can get for the Paris flight." Lindbergh confirmed that he had the money to buy. He also stated that he, and his backers, felt the manufacturer should contribute something since they would reap major publicity. Levine agreed, and told Lindbergh that he would sell the Bellanca for $25,000 and contribute $10,000 toward the project. Levine then told Lindbergh that he would not allow him to fly the *Columbia* alone. Actually, no one other than Lindbergh gave any consideration to flying the Atlantic alone. Levine insisted Lindbergh would have to name a copilot who met the approval of Columbia Aircraft Company. Lindbergh balked but finally gave in. Levine gave Lindbergh 24 hours to produce an acceptable name. The next day Lindbergh named a fellow airmail pilot employed by Robertson in St. Louis. Levine flatly refused saying, "I will not allow an unknown pilot to fly the *Columbia.*" He then stipulated that Columbia Aircraft would name the pilot and Lindbergh could be copilot.

Naturally, Giuseppe Bellanca thought that Levine was referring to Chamberlin with Lindbergh the copilot. Since Clarence Chamberlin had more experience flying the *Columbia* than anyone, it made sense. Lindbergh was shocked. There was no way he would accept the offer. He was going to be the pilot and that was final. Levine stood firm on his offer and would not budge. Giuseppe Bellanca felt that Clarence Chamberlin should be the pilot and did not care who went along so he did not comment.

Charles Lindbergh left town a very angry man. Had he known this in the beginning, he could have saved himself a trip to New York. It was time wasted that could not be regained. At least six others were well on their way to preparing for the Raymond Orteig prize of $25,000. On the train back to St. Louis, Lindbergh's mind was in a turmoil. He had been turned down by Fokker, Wright, Travel Air, and now Columbia. He also knew that the *Columbia* could depart any time Levine decided to name a pilot.

Lindbergh discussed the problem with his St. Louis backers. He felt the window of opportunity was closed and suggested they try a trans-Pacific flight. Knight and Bixby, two of the investors, felt that although time was critical, the window of opportunity was still open. Although it was late February, no one would attempt the crossing until May. There was still time available to build and test a new airplane. Lindbergh was astonished that the backers did not withdraw their support. His supporters were still rock-solid behind him and the New York-Paris flight. The Ryan offer was still valid, and Lindbergh made plans to leave for San Diego on February 23, 1927, to visit the factory.

Giuseppe Bellanca found out later that Levine had no intention of allowing Lindbergh to fly the *Columbia* as a pilot or copilot. Lindbergh was an introvert and Levine wanted someone more flamboyant. Levine also wanted both pilots to have a national reputation in order to cash in on the endorsements and promotions that were sure to follow. Levine had

The WB-II Columbia *being fueled at Roosevelt Airfield before its record endurance flight. Note the name change on the tail. It now reads "Columbia Aircraft Corporation." Photo from the Daniel Koble collection.*

the golden goose and planned to squeeze every golden egg out that was possible.

In spite of Bellanca's contract with Levine specifying that Bellanca would be the final authority on all matters relating to flying the *Columbia,* Levine had other ideas. Bellanca remained silent for the most part, not wanting to antagonize the man whose money was behind the entire venture. It was Levine's idea that a pilot with national fame be hired to fly the endurance flight to add to the fame and glory of the *Columbia.* Levine entered into negotiations with former Army Air Service officer Leigh Wade, who was a pilot on the Round-the-World flight in 1924. Bellanca however thought it was more important to have someone pilot the airplane who was experienced; that the flight would be successful on its own merits if the record was broken. If it flopped, it would not matter who was at the controls. In the end, Bellanca agreed to Wade, but only if Chamberlin went along.

Leigh Wade had ideas of his own. As soon as he felt secure in the position, he attempted to replace Clarence Chamberlin with one of his personal friends. Levine, eager to gain more publicity, fell into the plan. After Wade made this move, he then attempted to bring in another one of his friends as "General Manager." He also wanted to bring in his own man to handle his publicity and all announcements to the press. This aggressiveness proved to be his undoing, and on the 8th of April, Leigh Wade withdrew by mutual consent.

Bert Acosta was now hired to replace Wade. Acosta had given Levine flying lessons in a Curtiss *Jenny* several years earlier. This previous friendship gave Acosta the idea that he was the person to accompany Chamberlin on the trans-Atlantic flight.

Bellanca suggested that Columbia Aircraft use the *Columbia* to break the World Endurance Record. It would be an important test of endurance on the men and machine before making the attempt to cross the

Atlantic. Much would be learned from this flight that would be critical to a successful New York-Paris flight. Both Chamberlin and Acosta assumed they would be the chosen ones to fly the *Columbia* from New York to Paris when the time came.

The *Columbia* was fitted with extra fuel tanks and other safety modifications in order to accomplish the first phase of the operation by attempting to break the World Endurance Record. It would take almost a month to complete the modifications. The record would also be a test flight to shake out any problems that might develop prior to crossing the Atlantic. Chamberlin made fuel consumption tests on the *Columbia,* and it was ready to fly on April 12, 1927.

RYAN AIRLINES

Benjamin Franklin Mahoney, president of Ryan Airlines, was born in Wilkes-Barre, Pennsylvania, on February 8, 1901. Claude Ryan was an aeronautical engineer who had started Ryan Airlines several years prior to their meeting. Mahoney acquired half interest in Ryan's operation on April 25, 1925, for $7500, when Claude Ryan desperately needed funds to continue the operation. The business included a flight school, charter service, and sight-seeing service. By the summer of 1925, the pair purchased the Douglas *Cloudster,* a mammoth 56-foot wingspan, open cockpit biplane that had failed in an attempt to make the first nonstop transcontinental flight in 1921. They converted the airplane to hold five persons in a plush, closed cabin and initiated a daily run from San Diego to Los Angeles and return. In spite of a perfect safety record, Ryan discontinued the San Diego-Los Angeles run with the Douglas in 1926 due to a decline in business.

In 1925, Ryan Airlines also built, and sold, the Ryan M-1 monoplane which was used for flying airmail routes. The M-1 had an excellent reputation for being a dependable and rugged airplane. The "M" designation was for *monoplane* and was fabric covered. Initially powered by a Hispano-Suiza engine of 150 hp, it was later converted to the 200-hp Wright *Whirlwind,* and six were sold to Pacific Air Transport. They used the airplanes to carry mail between Los Angeles and Seattle.

By November 23, 1926, Ryan and Mahoney were unable to agree on a plan to raise needed capital to continue operations. Mahoney agreed to buy out Claude Ryan for $25,000. In the agreement, Mahoney also retained the right to use the Ryan Airlines name.

To this day, most people believe that Claude Ryan was the builder of the *Spirit of St. Louis.* He was not. Claude Ryan created a new company named "Ryan Aeronautical Company" (now Teledyne-Ryan Aeronautical). Claude Ryan, in the 1930s produced the Ryan ST *(Sport Trainer),* PT-16, PT-20, 21, and 22 *(Primary Trainers),* and the coveted Ryan SC *(Sport Cabin)* monoplane.

When Lindbergh met Mahoney, he told his story of the Bellanca-built *Columbia,* the Fokker tri-motor and the Keystone *Pathfinder.* By now the backers of these airplanes, who were seeking the same goal, were far ahead by having completed and proven aircraft. Lindbergh was convinced the New York to Paris flight would be made in the spring when the weather over the north Atlantic improved. If Mahoney agreed to build the airplane, it must be done with the utmost speed. Mahoney, after hearing Lindbergh's story, reconfirmed his commitment to Lindbergh that he would finish the airplane in two months.

A major obstacle for Mahoney, and his staff, was that Lindbergh had no design plan. Lindbergh's only comment for the design was that it be on the same lines as the Bellanca. Lindbergh knew beyond any doubt that the Bellanca-built *Columbia* was capable of completing the flight. He also knew that an attempt could not be made until May, when the North Atlantic weather would break. The window of opportunity was there, but it was closing rapidly.

It was February 25, 1927, when the challenge was accepted by Mahoney to build the airplane in two months time, with engine and instruments, for $10,580. Lindbergh notified his backers in St. Louis by telegram that he had signed the Ryan contract.

The design fell into the hands of twenty-eight-year-old chief engineer Donald Hall. This was Hall's first

The Spirit of St. Louis *ready for its first test flight near San Diego. Photo from the National Air & Space Museum.*

attempt to design a total airplane for such a purpose. However, he was determined to get the job done. Hall's first conclusion was that they could not modify the Ryan M-1 mail plane. The airplane would have to be designed and built from scratch for its single purpose. It must be capable of flying for forty hours nonstop. Aviation fuel weight is 6¾ lbs per gallon and the Ryan would have to carry 425 gallons. The 2601 lbs of fuel alone would weigh far more than the entire airplane when completed.

As rapidly as Hall drew plans, workers started construction. In the end, it would take Hall 775 hours to finish the design. The 220-hp, Wright J-5C, *Whirlwind* (serial no. 7331) engine arrived in April and was mounted to the fuselage. This was the same type of engine used by all the others including the Bellanca *Columbia*. A tried and proven air-cooled, nine-cylinder, radial engine, with dual magnetos for reliable ignition, was every pilot's dream.

The huge fuselage tank in front of the pilot would make forward visibility impossible. A periscope was installed as an alternative to a windshield. In an effort to reduce the airplane's weight, the electric starter, generator, and lights were not installed.

The following comparative specifications of the Ryan and the Bellanca show the striking similarities between the two competing aircraft.

RYAN-*Spirit of St. Louis* (built April 1927)

Wingspan	46'
Length	27' 8"
Height	9' 10"
Wing cord	7'
Engine	Wright J-5C
Horsepower	220 @ 1800 rpm
Fuel capacity	425 gallons
Oil capacity	25 gallons
Empty weight	2150 lbs
Loading weight	2985 lbs
Gross weight (Atlantic crossing)	5135 lbs
Maximum speed	124 mph
Range	4210 miles
Cruise speed	95 mph
Minimum speed (loaded)	72 mph
Minimum speed (light load)	49 mph

BELLANCA WB-II *Columbia* (built Sept. 1925)

Wingspan	46'6"
Length	26'9"
Height	8'9"
Wing cord	6'7"
Engine	Wright J-5C
Horsepower	220 @ 1800 rpm
Fuel capacity (Atlantic crossing)	455 gallons
Additional fuel for Trans-Atlantic flight	65 gallons
Oil capacity	25 gallons
Empty weight	1850 lbs
Loading weight	3600 lbs
Gross weight (Atlantic crossing)	5450 lbs
Maximum speed	130 mph
Cruise speed	110 mph
Range	4600 miles
Minimum speed (loaded)	55 mph
Minimum speed (light load)	46 mph

Lindbergh scanned the newspapers daily for information on the other contenders. John Wanamaker announced on March 4, 1927, that he would contribute $100,000 toward Admiral Byrd's attempt to win the Orteig prize with the Fokker tri-motor plane. Ten days later the American Legion members announced they would contribute the same amount to a new entry, Noel Davis and the Keystone *Pathfinder* (formerly the Huff-Daland Company). The Brooklyn Chamber of Commerce then contributed $10,000 to Levine to sponsor the *Columbia.* Lindbergh struggled desperately to find sponsors. Now they seemed to appear out of nowhere supporting the competition.

On March 26, the newspapers reported that Captain Charles Nungesser and Lt. Coli would pilot the French-built LeVasseur *White Bird,* and attempt an east-west crossing across the Atlantic soon. By April 9, Davis and copilot Wooster were test flying the latest entry, the Keystone *Pathfinder.*

When Acosta and Chamberlin were ready to depart for the endurance flight, skeptics felt that the WB-II would never leave the ground with 370 gallons (2295 lbs) of fuel. Many "experts" said it could not be done. They predicted the *Columbia* would share the same fate as Fonck's attempt. One of the Curtiss engineers estimated before takeoff that the *Columbia's* ceiling with the weight she carried was *600 feet below sea level!*

To everyone's astonishment, the *Columbia* left the ground in 1200 feet and climbed out smartly. The Columbia Aircraft Company, headed by Levine and Bellanca, made headline news on April 14, 1927. The *Columbia,* piloted by Clarence Chamberlin and Bert Acosta, broke the World Endurance Record by flying 51 hours and 30 minutes without landing. The men slept in shifts on a mattress placed upon the interior fuel tanks. Their only problem was thirst when the water supply ran out. To maximize the flight time, they climbed to 4000 feet over Roosevelt Field and allowed the engine to run out of fuel then glided down for a landing. In the end it was calculated that they had flown 4100 miles nonstop! This was 500 more miles than the calculated trip from New York to Paris. An enthusiastic crowd of over three thousand people was on the field awaiting their arrival. They hoisted the pilots upon their shoulders as movie cameras and photographers struggled for an opportunity to snap photographs of the historic event.

Bellanca and his chief mechanic, John Carisi, immediately started modifications to the *Columbia* for the trans-Atlantic crossing. A dump valve was installed in the main fuel tank to jettison the fuel in the event a forced landing was necessary. The valve could be opened which allowed the fuel to be jettisoned and the tank to act as a flotation device. Dual fuel lines were run to the carburetor in case one became clogged. Heavy rubber hoses were installed around the lines to prevent damage and freezing. A low-point water/dirt trap was also installed to prevent any foreign matter from reaching the carburetor. This trap could be drained while the airplane was in flight. Two compasses were installed in case one became faulty. Everything that was to be aboard the airplane was carefully weighed to be sure the airplane could carry the load.

In spite of the World Endurance Record being broken by Chamberlin and Acosta, Levine insisted he wanted a pilot who had national recognition to fly the *Columbia* to Paris. Almost immediately Levine announced that Lloyd Bertaud, a nationally known airmail pilot, would be hired for the historic flight. The real reason behind the selection of Bertaud was that Levine wanted to align the Eastern Division of the Air Mail Pilots Association (AMPA) to accept the *Columbia* as the airplane for airmail service. Levine was a member of the AMPA and wanted the New York to Chicago contract desperately. Another reason Levine did not want Chamberlin was because his slender build would not photograph well in movie offers that were sure to follow. Levine, in his never-ending pursuit of personal gain and publicity, announced to the press that the final choice of the copilot (presumably Chamberlin or Acosta) would be made at the last moment by the toss of a coin.

Levine's announcement created much turmoil between Bertaud, Acosta, and Chamberlin. The installation of heavy radios in the airplane was a real bone of contention. Bertaud wanted them installed. Chamberlin and Acosta felt they should be left out. Radios interfered with the compass, were short on operational range, weighed too much, and were unreliable. They also suggested that they were a fire hazard with all their electric wiring. Radios of this era only had a range of about 100 miles and would provide little, if any, help over the vast Atlantic. Both Chamberlin and Acosta would trade gasoline for the weight of the radios in a minute! The mechanics were ready to quit when Bertaud told them for the third time to re-install the radios. Levine got angry because every time the radios were installed or removed, the compasses had to be recalibrated to the tune of $75.00. The trio also fought over the route. Bertaud wanted to take the southern route, which was considerably longer, because of his fear of ice. Chamberlin and Acosta wanted the northern route because it was shorter and there were 1000-plus fewer miles of ocean to contend with. This bickering created a storm of criticism in the press. It became apparent that Levine promised all three the opportunity to fly when only two could go. Much to Levine's surprise, a fourth surfaced when Bertaud announced he wanted to name his own copilot for the flight.

Meanwhile the Ryan Company had been working around the clock in a desperate attempt to finish the *Spirit of St. Louis* in 60 days. The pressure on Lindbergh mounted when he read about Acosta and Chamberlin breaking the World Endurance Record. Lindbergh, once again, began to contemplate a Pacific flight as an alternative. He still concluded that the *Columbia* was his worst threat. It was a proven design, and Chamberlin and Acosta were seasoned pilots.

FOKKER CLOSES THE GAP

Another consideration was Anthony Fokker, who had a manufacturing plant in Hasbrook Heights, New Jersey. A year earlier (May 9, 1926) Admiral Richard Byrd and Floyd Bennett, with a crew of four, made aviation history by flying a Fokker tri-motor across the North Pole. Levine and the others felt they were far ahead of Anthony Fokker and his crew, which included Admiral Byrd, Floyd Bennett and George Noville. Much to everyone's surprise, Fokker rolled out its big tri-motor airplane *America* one month ahead of schedule.

Anthony Fokker, with the full crew, departed Teterboro Airfield on the afternoon of April 16, 1927, for the Fokker's first test flight. This was a mere four days after *Columbia* broke the World Endurance Record. The Fokker was built with extra fuel tanks in the fuselage to fly the Atlantic nonstop and win the elusive Orteig prize. What would normally be the passenger compartment was now fitted with huge fuel tanks to feed the three engines. Everything appeared to go well for Anthony Fokker, the pilot, until he landed. As soon as the airplane touched down it flipped over onto its back. After the accident it was found the CG (center of gravity) was calculated with the fuel tanks full of gasoline. The crew failed to place ballast in the rear of the airplane to compensate for the empty tanks. This caused the airplane to be nose heavy and the tail of the plane was too small to overcome the shift. The only one injured was Admiral Byrd, who suffered a broken wrist. The damage to the Fokker was not major, but it did eliminate the time edge they had worked so hard for. Levine and the others felt that if the Fokker had not crashed, it would have been a serious threat to their attempt.

On April 19, 1927, Captain Nungesser and Lt. Coli announced they would soon be ready to depart LeBourget, France, in their 450-hp LeVasseur, *White Bird.* All of the airplanes attempting the crossing, except the *White Bird,* used air-cooled engines. The LeVasseur used a water-cooled engine. There were concerns that a simple water leak in the radiator would overheat the engine and it would seize. All the water in the Atlantic would be of no use to a pilot with a leaky radiator.

While the *White Bird* was being readied, Bellanca and Carisi completed the modifications to the *Columbia* for the long trip across the Atlantic. Levine's nine-year-old daughter, Eloyse, christened the *Columbia* on Roosevelt Field, New York, Sunday, April 24, 1927. The *Columbia* was ready to go with the first favorable weather report (provided the arguments between Levine and Bertaud could be solved). To add to the festivities of the event, Chamberlin took Levine's daughter and her girl friend, Grace Jonas, for a short ride. Grace was the daughter of the president of the Brooklyn Chamber of Commerce.

On takeoff, unknown to Chamberlin, the left landing gear failed. The landing gear struts were set for the full load of fuel the airplane would require for the flight. When Chamberlin took off, the lightly loaded *Columbia* strut extended with full force when the airplane became airborne. The rapid extension of the strut sheared the limit pin and the wheel dangled beneath the airplane. Chamberlin continued the flight unaware that anything was wrong with the airplane.

Jean Smith, a veteran airmail pilot, shouted at a mechanic to grab a spare wheel to use as a signal. They jumped into the open cockpits of a Curtiss *Oriole* and flew up to signal Chamberlin of his problem. When they flew alongside waving frantically and pointing at the wheel, Chamberlin got the message. There were sand bags aboard the *Columbia* which were used as ballast because of the empty main fuel tank. John Carisi promptly started throwing them

Keystone Pathfinder *sponsored by the American Legion. Photo from the National Air & Space Museum.*

out the window over the airport to lighten the load. A policeman on a motorcycle, acting like a Keystone Cop, raced around checking every bag. He finally found one with an "OK" sticker and proudly rushed back to tell everyone that the pilot indicated everything was OK.

Chamberlin, with utmost skill, landed the *Columbia* on the right wheel with the right wing almost touching the ground and the engine shut down. The left wing finally dropped at the last possible moment and the *Columbia* sustained minimal damage. He received much praise from the press for his excellent flying skill under pressure. However, it would still take a week or more to repair the damage.

Two days after the *Columbia* incident another tragedy took place. Davis and Wooster were killed Tuesday, April 26, in a crash of the Keystone *Pathfinder.* The crash occurred near Langley Field, Virginia, when the airplane was being tested for full load capacity. The plane failed to gain altitude and crashed in a marsh and overturned. Fortunately, it did not burn. Every one of the multi-engine airplanes had crashed. It was becoming more evident that Bellanca and Lindbergh were right. The multi-engine airplanes must carry so much fuel they were unsafe for this flight.

The day after the *Columbia* christening, Bert Acosta resigned. His letter to Levine stated that he felt his heavier weight would detract from the performance of the flight. In reality, he simply got fed up with Levine's indecision and Bertaud's bickering. Bertaud meanwhile collaborated with a free-lance writer to buy the exclusive story of the flight. It so happened that the writer went to the North American newspaper *Alliance* and tried to resell the story to them. Unfortunately, Levine was already in negotiations with the *Alliance.* When Levine heard this news he was furious. He demanded that Bertaud withdraw from his agreement with the writer. Levine was so livid that he threatened to cancel the entire flight if Bertaud did not call and cancel the agreement right then and there. Bertaud called the writer and canceled the agreement in front of Levine.

The next day Levine came down to the hangar where Bertaud and Chamberlin were making last-minute preparations for the flight. Levine thrust out a contract, made up by his lawyer, and stated flatly, "Sign it, or I am cancelling the flight." The two pilots reviewed the contract which gave them each $150 a week for one year if they completed the flight. Levine was named "manager" and would direct all vaudeville, motion-picture and other contracts he could arrange. If the earnings exceeded their $150 weekly salary by a large margin, Levine would determine a bonus as he saw fit. In addition, Levine would get one-half of the prize money. With the endorsement prizes, this was nearing $100,000. There were no life insurance policies in the event they failed. Levine held the hammer. He could stop the flight even if Bellanca protested. Reluctantly, Acosta signed. Clarence Chamberlin could care less about the contract. He muttered, "Let's fly and then worry about the money." Levine could never take away their victory once it was won. Reluctantly, he also signed to keep the peace.

It appears that Levine and the others knew little of the Lindbergh-Ryan progress. If they did, they did not consider it a factor, especially since Ryan had to design, build and test-fly the finished product. Besides, Ryan Airlines had not accomplished one aviation record to create credibility. It takes a great deal of time and effort to be absolutely sure an airplane is free of problems before considering a flight that offers no second chance.

Everyone at Ryan was working on *The Spirit* around the clock. The first test flight was ahead of schedule and they were flying on April 28, three days after the *Columbia* landing gear failed. Lindbergh landed and reported that the airplane was unstable in flight. It would not fly "hands off" for even a moment. Hall decided it would require designing a larger tail assembly to cure the problem. Hall had used the tail from the M-1 to save time in construction and it now proved too small. Lindbergh felt there was not enough time to make the change and did not want the added weight. He decided the airplane was manageable and continued the test flying. (Lindbergh later recalled in his book that the instability of *The Spirit* probably saved his life as he dared not doze off for even seconds.)

THE FRENCH ATTEMPT

On May 8, 1927, the *White Bird* departed LeBourget Aerodrome, France. Nungesser and Coli were reported being sighted by ships at sea, off Cape Race, Nova Scotia, and even Boston. Crowds gathered on both sides of the Atlantic to hear the news that they had landed safely on Roosevelt Airfield. As time passed, with no arrival, it became evident their fuel was exhausted. They were never found.

Before Lindbergh left San Diego, he heard the news that Nungesser and Coli had departed LeBourget, France, and were on their way to New York. Lindbergh feared the contest was over and decided he would wait until he reached St. Louis to discuss what to do next with the investors. He also had to pick up his Department of Commerce pilot license (no. 69) there from Robertson Airlines. It was never easy to hear news that another pilot was lost. When Lindbergh arrived in St. Louis he heard that the Frenchmen were overdue, which cast a dark shadow over his attempt.

After the short stop in St. Louis, Lindbergh flew on to New York. To meet the requirements of the flight, Lindbergh immediately applied to the Department of Commerce for a permit to make the flight. Lindbergh was devastated when the Department of Commerce refused to issue him a permit because his airplane had no lights. It was illegal to fly an airplane at night without navigation lights. It was impossible to install them on the *Spirit* since it had no electrical system at all. In order to save weight, Lindbergh did not want an electric system. The engine used dual magnetos and did not rely on batteries. The engine had no starter, or generator, and was propped by hand. Because of pressure from the investors and Lindbergh, the Department agreed to a waiver since it was highly unlikely that Lindbergh would encounter another airplane flying across the Atlantic.

Although the *Columbia* was ready to depart Monday, May 9, 1927, and the weather reports were favorable, a delay was issued by the Department of Air Commerce. It appears that the U.S. Ambassador to France had cabled the State Department requesting that all flights be delayed in respect for the loss of the French pilots. It was simply too close to the French pilots' demise to celebrate a victory in France if an attempt were successful. The political implications were not worth the risk. The next date approved by the Department of Air Commerce was Friday, May 13. Although the *Columbia* was ready, and the weather was favorable, Levine, being superstitious, refused to allow the *Columbia* to fly on the 13th.

The more Bertaud thought about the contract, the angrier he became. His fury was unleashed when Levine announced to the press that "every nickel of the prize money was going to the pilots."

Bertaud then obtained a lawyer and demanded a suitable contract be negotiated before he would fly.

The dissension leaked to the press and Levine was cast in a bad light among supporters. A hasty meeting was called at the Garden City Hotel, which acted as headquarters for the Lindbergh and Fokker flights as well. The meeting included Ralph Jonas, Brooklyn Chamber president, former State Senator Charles Lockwood, Bellanca, Levine, his lawyer, Bertaud and his lawyer, and Clarence Chamberlin. Levine was raked over the coals for the hard-nosed contract he forced upon the pilots. Bertaud, who now felt he had the hammer, decided once again that he would name who would fly with him. Bellanca immediately put an end to that demand by announcing that "Chamberlin flies or the flight is off." Finally, with much reluctance, Levine promised to sign a contract May 11, which included giving all the prize money to the pilots, a $50,000 life insurance policy for each, and a percentage of the publicity contracts.

Everything appeared to be settled, but Levine failed to produce the signed contract on the specified date. Bertaud, not trusting Levine, promptly notified his lawyer to file an injunction in the Superior Court of New York. It was filed on May 18, signed by Justice Mitchell May, and was scheduled for argument Friday, May 20, at one o'clock in the afternoon. The injunction caused the *Columbia* to be grounded. This included a padlock on the hangar door, with two police guards, to thwart any attempt to remove the airplane.

All three airplanes were virtually ready to depart from Roosevelt Field. Byrd's Fokker, *America,* Columbia Aircraft's *Columbia* and Lindbergh's *Spirit of St. Louis.* Actually, *The Spirit of St. Louis* and *Columbia* were in separate hangars on Curtiss Field and would have to be hand-pulled down a short roadway that connected the two airfields. Roosevelt Field had the longer runway, and all pilots wanted the extra length for added safety.

Dean Blake, Weather Bureau chief, reported that the weather would not be favorable for several days. The thrill of competition among the three contestants made international headlines. A favorable weather report would ignite a beehive of activity. The media favored the *Columbia* which would carry two pilots and had already claimed the World Endurance Record. *The Spirit of St. Louis* was considered the dark horse with its lone pilot and lack of historical background. Thousands of spectators camped out on Roosevelt Field in the cold rain for days to witness the historic event. Many felt that when the weather was favorable, all three airplanes would depart within minutes of each other. It appeared that a race like no other in history was in the making.

There is much evidence pointing to many holding firm in their convictions that Lindbergh would not win. The press was not kindly to Lindbergh's rock-solid position of flying alone. And there were other factors. No other pilot would give any consideration to flying alone for over 30 hours over open water. *The Spirit of St. Louis* was hastily built and unproven. Prior to departure, it had only flown 27 hours and 25 minutes. The airplane was designed by 28-year-old Donald Hall, who had no previous experience in designing an airplane capable of carrying 2985 lbs and flying such a distance nonstop. The airplane could hardly be considered a "proven design" for such an undertaking. Its pilot was young, and with little experience compared to the others. Lindbergh had admitted the Ryan was unstable in flight and would go out of control without the pilot's full attention. Flying alone for over 30 hours without sleep was considered near impossible. Lindbergh refused to install a radio or even carry a sextant for navigation. He announced to the press that he would fly by "dead reckoning." The only electric power aboard was a flashlight. Lindbergh had no record of flying over vast waters without navigational checkpoints. Even the prestigious Lloyds of London publicly announced they would not insure the Lindbergh flight. The press openly stated that the Lindbergh flight was suicidal. Airplane designers agreed that the Bellanca *Columbia* was a far more technologically advanced airplane than the hastily-built Ryan. The press had hounded Lindbergh with questions about all the failures of pilots with more experience. One reporter even had the gall to ask his mother as she stood proudly by her son, "How do you feel when you know your son might be dead in 24 hours?"

She was so shocked she could not answer. Lindbergh made her leave New York on the next train. Lindbergh, a very modest person, was never comfortable with the press.

Because of continued unfavorable weather reports, Lindbergh was invited to attend the Broadway play, "Rio Rita" with his investors on Thursday evening, May 19. At this point Lindbergh knew that the first one to depart would have the advantage. The newspapers headlined Bertaud's injunction. One said that the injunction was lifted and another said the opposite. Lindbergh knew he must depart before the *Columbia* or face losing. Prior to leaving for the Broadway play, Lindbergh checked the weather one more time. Although the local weather was poor, the report indicated the weather was clearing over the North Atlantic. Lindbergh cancelled the trip to the Broadway play and gave the order to ready the *Spirit* for takeoff at the break of dawn. Although the clouds hung at a low 500 feet and produced light rain, Lindbergh felt it was time to go. He had many hours flying in this kind of weather and he would simply fly low up the coast to Nova Scotia where he should break out into clear skies. The mechanics removed the airplane from the hangar and pulled it down the road by the tail to Roosevelt Field before dawn. Now came the laborious task of hoisting fuel from the ground to the wing and fuselage tanks. Lindbergh retired to his Garden City Hotel room for what would be only a couple of hours of restless sleep because of constant interruptions.

Two sleepy-eyed policemen stood vigil over the locked hangar doors of the *Columbia* in the rainy early dawn. The Fokker crew was also waiting for a favorable weather report before loading the airplane with fuel. Clarence Chamberlin watched with awe as Lindbergh took off for Paris at 7:54 a.m. on Friday, May 20. There was no prior publicity and almost everyone was caught dumbfounded when they heard that Lindbergh had departed in poor weather.

The court hearing was resolved in Levine's favor by 2:30 p.m., May 20. Levine ordered the *Columbia* to be hastily fueled, and to take off as soon as possible in an attempt to overtake Lindbergh. When Giuseppe Bellanca heard this, he became angry. He felt it was not ethical to depart while Lindbergh was still in the air. He told Levine that he would not authorize the flight. The *Columbia* was pushed to the far end of the airfield where 100 gallons of fuel were drained on the ground to lighten the ship. Chamberlin and the crew moved the plane and Chamberlin threw a match into the gasoline. The explosion and fire convinced the crowd the *Columbia* had exploded. The ground crew pushed the *Columbia* into the hangar and locked the doors.

FROM SUICIDAL TO SUPER-HERO

What all of the detractors failed to take into consideration was the courage and determination of this young man named Lindbergh. The obstacles and the fierce competition failed to break his will and determination to see his dream come true. He never sought fame or fortune. He never asked for a contract from the investors or any guarantees of monetary gain. He never gave any thoughts to books, sponsors or movie rights. He never quarreled with the investors. He would fly alone and his word was final. He became a man obsessed with proving to everyone, in his own quiet way, that the single-engine airplane was capable of pulling off what almost everyone deemed impossible. "Lucky Lindy" beat the odds and made it nonstop in 33 hours and 30 minutes. In 33 hours time Lindbergh had the press change its opinion of him from "suicidal" to "super-hero" (This may be a record in itself). Overnight, Lindbergh and Ryan became household words.

The day after the announcement that Lindbergh had landed in Paris, Giuseppe Bellanca summoned the press and made the following statement,

> All of this has convinced me that two characters such as Levine and myself cannot continue in business together. During the controversy of the past weeks between Mr. Charles Levine and Mr. Lloyd Bertaud, I made every possible effort to settle the points at issue. My sole concern was to prove that my plane, built in America and manned by Americans, could successfully make the New York to Paris flight, thus adding another stage to the experimental development of aviation in this country. I was helped by many friends, including the American press, but I found the personal obstacles were too great to surmount. Nevertheless, I momentarily placed this apart so as not to place further difficulties in the way of the flight. Now the flight is over for my part, due to the monumental success of Captain Lindbergh. I desire publicly to announce the severance of any kind and all business relations with Mr. Levine and myself so that I may continue in the way of contributing to the growth of American aviation.
>
> I take the opportunity to express my admiration for Captain Lindbergh, a flight which will remain in the history of aviation as one of the greatest accomplishments of the present day. Captain Lindbergh deserves the highest praise that goes to the man that has given the world the most thrilling example of sportsmanship and aeronautical ability.

Bellanca followed up his speech with a fiery registered letter of resignation stating frankly that he was through and would file suit against Columbia Aircraft if Levine used his name in any promotional activities. He rightly blamed Levine for focusing too much attention on publicity, vacillating on who would fly the airplane, balking at a contract and failing to pay attention to the objective at hand. A deeply hurt and frustrated Bellanca made his feelings known publicly as to why he had resigned from Columbia Aircraft. Giuseppe Bellanca went into seclusion and lost himself in his work building airplanes in a rented warehouse on Staten Island.

Benjamin Mahoney, Lindbergh's mother, and A. J. Edwards, the Ryan sales manager, immediately boarded the liner *Mauritania* and met Lindbergh in Paris. The parties with notables were endless. Lindbergh told Herrick, the U.S. Ambassador, that he had no intention of allowing the *Spirit* to be shipped back home in a crate. He intended to fly across Europe, Siberia, the Bering Straits, Alaska, Canada and end the flight in New York. He could care less about the parties and speeches, which he loathed. He was

Clarence Chamberlin and the Columbia *on Roosevelt Airfield June 1927, for the Trans-Atlantic flight. PARIS has been painted out, since they planned to fly until the fuel was exhausted to establish a new World Distance Record. Cottbus, the name of the town where they landed, was painted in after landing in Germany. Photo from the Daniel Koble collection.*

eager to get back in his airplane and get away from it all. The ambassador followed Lindbergh to England where he applied all the diplomacy he could muster to change Lindbergh's mind. Obviously, there was a great hero's welcome awaiting him back home, but Lindbergh never sought or approved of all this notoriety. In the end Lindbergh, with the utmost reluctance, agreed to return with his airplane to America on the *U.S.S. Memphis.*

After painting out "Bellanca" on the tail, and New York-Paris on the fuselage of the *Columbia,* Levine announced that Chamberlin and a copilot (to be named), would still fly the Atlantic nonstop to establish a new World Distance Record. Levine's wife threatened to set fire to the airplane if he decided to go with Chamberlin. Once again Levine used his deception to dispel any thoughts of his being the copilot. On the day of departure, he was dressed in a business suit standing alongside the passenger side of the *Columbia,* but at the last moment he jumped aboard as Chamberlin roared down the runway.

It was only two weeks after the Lindbergh flight when Chamberlin and Levine flew the *Columbia* from New York to Eisleben, Germany. They had been flying for 43 hours (10 hours longer than Lindbergh) when weather and low fuel forced them to land in a field near Berlin. The *Columbia* flight was 3911 miles and considerably farther than Lindbergh's in spite of fierce headwinds they encountered. (If the winds were favorable they could have flown to Moscow.) Clarence Chamberlin and Charles Levine did establish a new World Distance Record.

It is not easy in today's world to appreciate the impact the Wright *Whirlwind* engine had on aviation in 1927. Previous transatlantic flights presented difficulties in terms of engine reliability and efficiency for engineers. Suddenly the Wright *Whirlwind* engine became the engine of choice. However, Wright was producing water-cooled engines for dirigibles, Hispano-Suiza aircraft engines under license, as well as marine engines. In spite of the successes of the *Whirlwind,* the company directors felt they did not want to focus their attention on a single engine in fear of disrupting their other markets. This angered Frederic Rentschler, president of Wright Aeronautical and he resigned. George Meade, chief engineer, and much of Wright's design staff followed. They located a very capable machine shop in Hartford, Connecticut, that had a reputation for high quality tooling work. The engineers approached the owners with the idea of building aircraft engines. The owners, Pratt and Whitney, were agreeable to the idea, and a new engine was born, the 300-hp Pratt & Whitney *Wasp Junior.* Although Wright changed its position after Rentschler left, the Pratt & Whitney engines would, in many cases, become the preferred aircraft engine.

It would be prudent to think that Mahoney and Donald Hall would have started immediately on a commercial version of the *Spirit.* That did not happen. In spite of the flood of inquiries, and firm orders, on the airplane's availability, Mahoney took two months off celebrating around the country before he returned to the factory. Donald Hall, completely spent from his months of grueling work, headed for a mountain retreat. No one was available to run the store while potential customers were beating a path to the door. When Mahoney finally returned, he hired Art Mankey away from Douglas Aircraft to fill the void of Donald Hall's absence. (Mankey and Jack Northrup were the ones Mahoney had hired as subcontractors to do the structural analysis on the M-1 mail plane.) Mankey immediately started the design for a cabin monoplane called the B-2 *Brougham.* (The B-1 *Brougham* project had been set aside when Lindbergh arrived with his order to build the *Spirit of St. Louis* in sixty days.) When Donald Hall finally returned from his retreat, he was given the title of "Chief Experimental Engineer."

It is interesting to note that Mahoney did receive one order for the only exact *duplicate* of the *Spirit of St. Louis.* The order was placed by "Tonichi and Daimai" the names of two newspapers, one in Tokyo and the other in Osaka, Japan. The plane was built in San Diego, and Mahoney designated the airplane NYP-2 (New York-Paris), which was registered in Japan as J-BAAC. Piloted by Fumio Habuto, it established a new Japanese endurance record of 13 hours and 23 minutes. The pilot was hailed in Japan as the *Japanese Lindbergh.* However, by 1938, the airplane was in dire need of a complete overhaul and was scrapped.

Mahoney (Ryan) was on the verge of bankruptcy before Lindbergh, during Lindbergh, and after Lindbergh. With his back against the wall financially and unable to fill orders, he finally agreed to sell the company to the St. Louis backers for $1,000,000 and a token job of $10,000 per year and $5000 expenses for five years.

On October 5, 1928, a new plant in St. Louis was finished and the last items moved from San Diego to Lambert Airfield (now Lindbergh Airport). The B-2 *Brougham* was a four-place airplane, powered by the Wright J-5 *Whirlwind.* The B-2 airframe was also modified to a six-place airplane (B-3) powered by the 300-hp, J-6 engine. Meanwhile, back in New Castle, Delaware, Bellanca was also fully operational producing the four- and six-place Bellanca CH, also powered by the Wright J-5 and J-6 engines. Because of its proven endurance, virtually every airplane manufacturer in the world wanted a Wright engine on his airplane in spite of the hefty $5,000 price tag.

THE BATTLE OF THE TITANS

The pressure was put on everyone at the Mahoney-Ryan plant to complete one B-2 and one B-3 for the Chicago Air Show in the fall of 1928. Their chief rival, Bellanca, would be there to show off his new four- and six-seat airplanes. To miss the event would create tragic losses in exposure and advertisement. The continued competition between the two parties would set off a battle to outdo each other. Aviatrix Marvel Crosson established a new women's altitude record of 23,966 feet in a B-3 *Brougham.* The record was short-lived when teenager Elinor Smith, in a Bellanca *Pathfinder,* flew to a new record of 27,418 feet. Every time the boys from St. Louis would go after a new record, Giuseppe Bellanca had an eraser ready to replace the Mahoney/Ryan name with a Delaware-built Bellanca.

The French did not take it too kindly that Bellanca took away their World Distance Record in 1927. In 1929 they responded by flying their bi-wing airplane named *The Question Mark* from Paris to Manchuria, a distance of 4,911 miles nonstop. Giuseppe Bellanca liked nothing better than a good challenge. By July 29, 1931, the Delaware-built Bellanca J named *Cape Cod* (N761W) was ready to shatter the European record. Powered by a 300-hp Wright, J6 engine, and grossing 7500 lbs with 728 gallons of fuel, pilots Russell Boardman and John Polando flew from New York to Istanbul for a new World Distance Record of 5,001 miles nonstop.

In spite of the competition from Bellanca, the Mahoney-Ryan sold well nationally and internationally. Orders were placed by China, South America, Mexico, Canada, the U.S. Department of Commerce, Robinson Airlines and others. The plant was not idle by any means. Early in 1929, the St. Louis investors saw the Lindbergh/Ryan connection begin to wane and felt it was time to cash in on the venture. Other manufacturers, such as Lockheed and Douglas, were introducing state-of-the-art metal aircraft. A Detroit connection of investors purchased control of Mahoney-Ryan and renamed it "Detroit Aircraft Corporation." At the same time, these investors also picked up Lockheed, Parks Air College, and eight other small companies (Bellanca refused to join the conglomerate). Their $20,000,000 infusion into the aircraft industry, they claimed, would create an aviation industry unlike anyone's wildest dream. In the Detroit factory they would build an airliner called the *"Detroiter."* Aircraft came off the assembly lines like model "A" Fords. Unfortunately, all the grandiose ideas came tumbling down when the Great Depression hit in late 1929. By year's end, the top-heavy Detroit Aircraft Company, with huge inventories, failed.

THE DELAWARE BELLANCAS

A thorn in the side of many Delaware leaders was the lack of aviation facilities available to the most honored aviator, Charles Lindbergh. New Castle County only had two landing facilities available at this time; the DuPont Airfield, which was privately owned, and Biggs Airfield. In 1927, the only public-use airfield in Delaware was Biggs Airfield, and it offered only minimum services.

The first attempt to further the growth of aviation in Delaware was made by the du Ponts in early July 1927. It was Henry Belin du Pont who contacted Giuseppe Bellanca and entered into negotiations to move Bellanca to Delaware. Henry B. du Pont was so sure Giuseppe Bellanca would accept his offer to build a factory, that he convinced Francis V. du Pont, Katherine Clark du Pont, and Margaret Lewis du Pont to join him in the purchase of the 360-acre Spring Garden Farm in New Castle. The du Ponts formally purchased the property on August 9, 1927. The negotiations with Giuseppe Bellanca broke down, however, when he would not agree to the plan which included controlling interest of the company by the du Ponts. Giuseppe Bellanca had his fill of controllers with Wright Aeronautical and Charles Levine. Andrew Bellanca, Giuseppe Bellanca's nephew, was a New York lawyer who represented Giuseppe in all his legal matters. He convinced the du Ponts that this issue was not negotiable. The du Ponts felt that if they were going to finance the venture, they should have control. Both parties were inflexible on this point, and negotiations came to an end.

Henry Belin du Pont still felt that Bellanca would be an asset to Delaware. The timing was right. Bellanca had a more marketable airplane than the *Spirit of St. Louis.* Lindbergh and Chamberlin had just proven the single-engine airplane was indeed reliable. All of a sudden there was an avalanche of worldwide interest in this type of airplane. Henry hoped a deal could be worked out between the community and Bellanca that would satisfy both parties. The community wanted an adequate airfield, and Bellanca wanted a factory to build airplanes. Each would certainly complement the other.

A Delaware delegation was formed in August 1927, consisting of Henry Belin du Pont, Francis V. du Pont, Alfred Chandler, Wilmington Chamber of Commerce, New Castle Mayor Klingmeyer, and members of the City Council, to approach Giuseppe Bellanca about moving to Delaware. The group put a subscription agreement together that included the purchase of the Spring Garden Farm which included frontage on the Delaware River.

Since the project was politically inspired, press releases were made to the public to gain their support. The Wilmington newspaper, *Every Evening* praised the committee in their site selection, since the railroad bordered the proposed factory and bordered the Delaware Bay which could offer seaplane operations as well. In the end, Bellanca would also have access to 1100 feet of Delaware Bay waterfront to use as a seaplane base. As part of the package, the Wilmington Chamber of Commerce agreed to pay $20,000 for the airfield beacon light, as well as the airfield boundary lighting, over a period of two years. It was agreed that Spring Garden Farm would be sold to Bellanca for $37,215. A subscription agreement was entered into in which Bellanca would redeem the stock out of earnings.

Everyone was not totally convinced the Bellanca deal was a good one. Lammot du Pont wrote the following letter to H. Belin du Pont on October 22, 1927.

> The first impression I got on reading the Bellanca Subscription Agreement (which I turned over to Mr. Hoopes) is that Mr. Bellanca is asking a good deal: a large salary guarantee for a number of years and fifty percent of the stock apparently being put up in money. The fact that he does not get this stock until the preferred is paid off is neither here or there, for if the Company is successful he gets it and if it is unsuccessful he has lost nothing. Of course, the fairness of this proposition, in this respect, depends on the ability of Mr. Bellanca and the value of what he brings to the Company in the way of finance and business. It is hard for me to realize that in these days of scientific development any one man can be so far ahead of all the others as to warrant such a compensation.

Emphasis was placed on the necessity of the airfield and the economic impact it would have on the state and the community. Delaware lay directly below the lucrative Washington-New York air route. The airfield would offer two aviation-related businesses to New Castle County. Many had visions of the Bellanca Airfield growing into a major East Coast terminal for freight, mail, and passengers. It was also quite feasible that the Bellanca factory would employ up to 120 people. (During the war years, Bellanca employed an estimated 3000 people.) When all aspects of the deal were analyzed, there were enough financial supporters to make the agreement become a reality. Henry du Pont purchased 200 shares of common ($1.00 par value) and 200 shares of preferred ($100 par value). Henry would also serve on the board as a director. However, he served only a short time due to his involvement in other aviation interests. No stock was available to the public. County records disclose that the du Ponts sold the Spring Garden Farm, October 27, 1927. By the end of March 1928, the necessary funds had been secured to start construction of Bellanca Airfield.

A letter, located in the archives of the Hagley Museum and Library collection of Henry Belin du Pont, confirms that the Bellanca Aircraft Corporation of America on June 30, 1928, transferred all its accounts to the books of the Bellanca Corporation of Delaware. Bellanca would become the only aircraft manufacturer in Delaware history to build airplanes to sell in a global market.

While the new Bellanca factory was under construction in Delaware, Giuseppe Bellanca was busy building airplanes in the rented Richmond Hill, Staten Island, factory. At least one model J and several CH-200s were built in this factory before moving to Delaware. Another Bellanca-built airplane, the K *Roma* was also built at Staten Island. Several aircraft in various stages of production were also constructed and moved to Delaware for final assembly.

1927 BELLANCA K *Roma*

The Bellanca model K *Roma,* serial no.107, NX4864, was built after Bellanca signed a special order with Cesare Sabelli, Roger Q. Williams, and Piero Bonelli. The buyers planned a highly publicized nonstop, trans-Atlantic, flight from the United States to Rome. The *Roma* was Bellanca's first

The Bellanca Roma *under construction at the Richmond Hill, Staten Island, factory. Photo from the Daniel Koble collection.*

attempt at building an airplane of this size. Bellanca got a jump on the other airplane manufacturers by installing an inward-retracting landing gear on the *Roma* to reduce drag. The *Roma* was built in the Staten Island factory and flown from Roosevelt Airfield to Delaware on July 28, 1928, for final tuning.

H. B. du Pont rendezvoused with the giant Bellanca *Roma* in the sky over the Delaware River in his Buhl *Airster.* They landed together for a celebration at the Bellanca Airfield where a crowd awaited their arrival. The *Roma* was scheduled to undergo further testing in Delaware before the Rome flight. Testing was completed in Delaware on July 31, 1928, and the *Roma* was flown back to Mitchell Airfield, New York, for its final preparation for the crossing. Unfortunately, a legal battle between Dr. Leon Piscolli, who helped finance the construction of the plane, and the crew caused the flight to be cancelled. Dr. Piscolli filed a breach of contract suit against Cesare Sabelli in the Delaware Supreme Court. It appears the contract stipulated that Piscolli would be the medical observer on the historic flight. Piscolli filed suit after several departure dates were announced and were later cancelled. In the end, the airplane was repossessed by Bellanca and returned to Delaware.

The *Roma* was later sold to Thor Solberg, a Norwegian, who wanted to fly it to Oslo, Norway. He renamed the airplane *Enna Jettick* for the shoe company that sponsored the flight. The plane encountered foul weather on the way to Newfoundland and was ditched at sea. The crew was rescued, the plane was lost. No other model Ks were built.

Leon Piscolli, M.D., Piero Bonelli and Roger Q. Williams with the finished Roma. *Photo from the Daniel Koble collection.*

The finished 1928 Roma *on Bellanca Airfield. Photo from the Hagley Museum & Library.*

Aerial view of the Bellanca factory in 1928. The large airplane at the hangar doors is the Roma. *Photo courtesy of Frances Rees Hayford.*

SPECIFICATIONS

Engine	Pratt & Whitney, *Hornet*
Horsepower	550
Wingspan	65'
Landing gear	Fully retractable

The Bellanca factory

In less than a year from conception, Bellanca Airfield became a reality with two grass runways, one 2500 feet long and the other 1800 feet. Three tons of fertilizer and thirty-five-hundred pounds of grass seed would be required to seed the surface.

The Bellanca factory consisted of an office, two hangars, engineering and assembly shop, paint and fabric shop, machine shop, woodworking shop and an aviator's house. The Bellanca airport was also lighted for night flying. Everything was built state-of-the-art, using the latest in efficient design and technology.

The original group who put the Bellanca package together decided to promote a "Grand Opening of Bellanca Airfield." The group would invite pilots from across the country to fly in for the show. The public also responded positively and businesses donated prizes for the flying events.

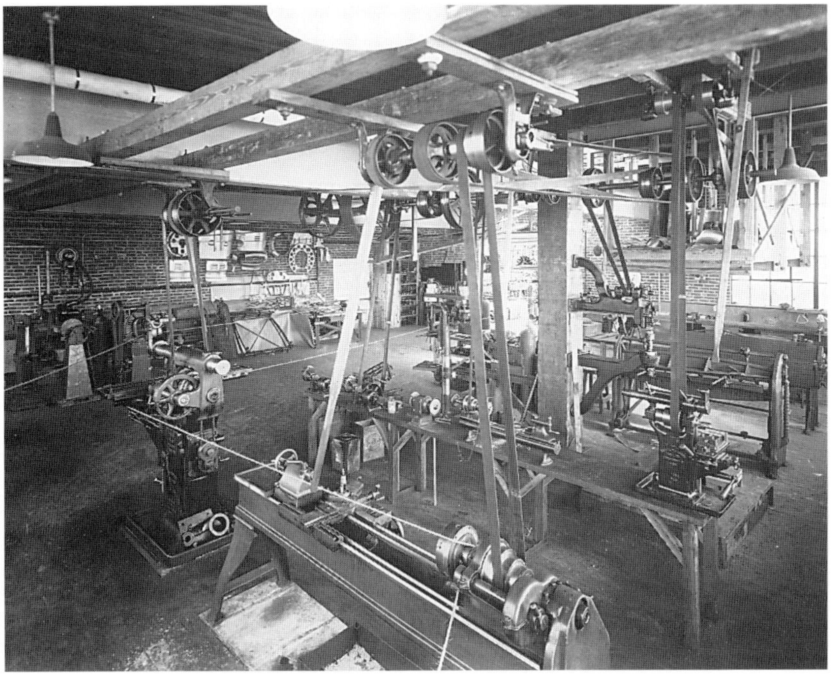

The Bellanca machine shop, New Castle, Delaware, 1928. Photo from the Daniel Koble Collection.

The Bellanca woodworking shop, New Castle, Delaware, 1928. Photo from the Daniel Koble collection.

The Bellanca wing assembly department. Note the heavy steel I-beams and the wood floors. Photo from the Daniel Koble collection.

The Bellanca CH fuselage and engine assembly department. Serial number 113 is in the foreground. Photo from the Daniel Koble collection.

The military services also agreed to participate. Local pilots offered to give free rides the week prior to, and the day of, the event scheduled for October 6, 1928. By October 5, 80 airplanes and the Goodyear *Baby Blimp* were listed on the program. The day before the event the newspaper *Every Evening* summed up the opening day.

> Tomorrow will mark the beginning of a new era for Wilmington and its metropolitan area. The dedication of the new Bellanca Airfield will definitely place this city, and the State of Delaware, on the National Air Map. The air meet will give us a hint of what the future has in store for us now that we have taken this advanced step.

The weather cooperated, and the early morning skies were crystal clear — perfect weather for flying. The program listed the events starting promptly at 11 a.m. Shortly before 11, guest speakers William McCraken, Assistant Secretary of Commerce for Aeronautics, Giuseppe Bellanca, the Governor, and the Mayor spoke briefly over a microphone. But the public did not attend the meet to hear speeches. They came for action. So promptly at 11 a.m. the oration ceased and the program of events was put into motion. A series of low-flying formation flights, consisting of 29 military aircraft, flew over the airfield. This low pass was the signal that the program was now underway.

Numerous airplanes of every description were parked on Bellanca Airfield for public viewing. Long lines of people waited eagerly for their first ride in an airplane. Pilots performed daring stunts with their airplanes to fascinate the crowds, and the military performed precision formation flybys. Delaware now had a proper airfield to serve a growing interest in aviation.

A NEW ERA FOR BELLANCA

Early in 1929, Luddington Airlines, operating out of Camden, New Jersey, included Bellanca Airfield as a

The Columbia *on Bellanca Airfield, circa 1931. Note that it now has the rear windows covered, a Townend speed ring around the engine, and a metal propeller to improve performance. Photo from the Daniel Koble collection.*

The remains of the world-famous Bellanca WB-II Columbia. *Unfortunately the airplane was scrapped. There was still enough left that it could have been restored to museum static display. Photo from the Hagley Museum & Library.*

stop on their Boston, New York, Philadelphia, Washington flights. Luddington used the Stinson Tri-motor for these flights. Unfortunately, the depression hit in late 1929, and Luddington Airlines failed to survive through 1930. Although short lived, Luddington became Delaware's first scheduled airline service.

It is not known who gave Giuseppe Mario Bellanca the nickname "GM." It started before he came to Delaware and he carried it through his entire life. When he came to Delaware, he also brought with him the world-famous WB-II, *Columbia.* The *Columbia* occupied a separate hangar on the Bellanca Airfield and was brought out to make more record flights. Unfortunately this historic airplane was lost on January 25, 1934, when a grass fire set the hangar on fire.

Where other manufacturers would build airplanes, then try to sell them, Bellanca airplanes were sold first, then built to meet the customer's requirements. Interior trim, engine options, color combinations, wheels, skis, floats, brakes, instrumentation were all agreed upon prior to the airplane's entering production. Many were ordered, or remanufactured, for the sole purpose of attempting another record flight. Although the customer would pick one of several models, the Bellanca airplanes were an extension of the customer's particular taste or intentions. To say that Giuseppe Bellanca was a master of manufacturing many variants from his designs is an understate-

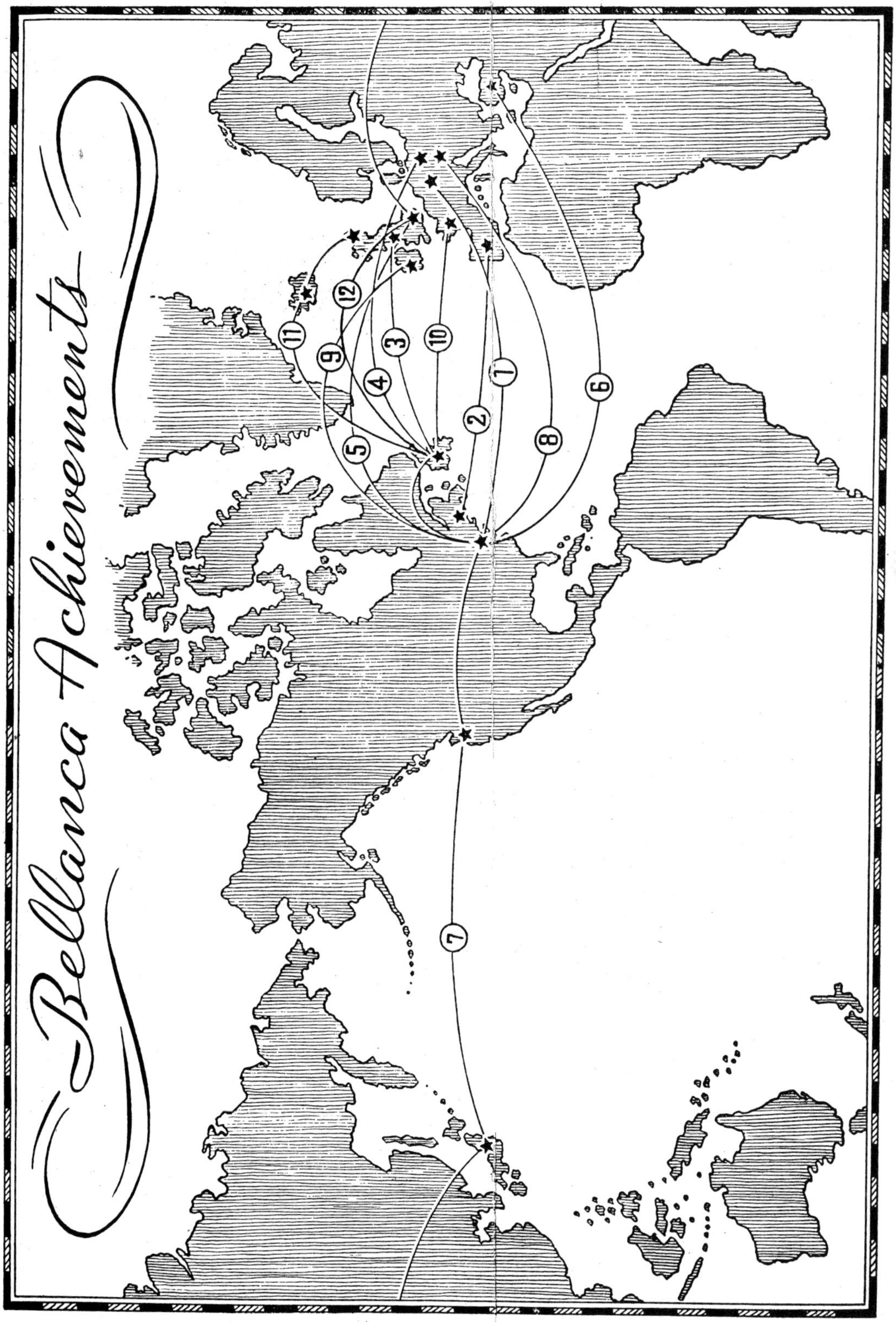

This map, reproduced from the Bellanca News Bulletin, *April, 1937, graphically displays the overseas records set by Bellanca airplanes.*

ment. This is especially true when reviewing the "CH" and "Airbus/Aircruiser" series. His ability to do this and still compete in the volatile market of the 1930s and 1940s would be copied by other manufacturers for many years.

A concerted effort was made to document the *exact number* of airplanes Bellanca produced of each model during his tenure in Delaware. This may not be possible since Bellanca made extensive modifications to airplanes and would, in some cases, change the model number when doing so. There is also evidence that Bellanca never released accurate production numbers on controversial aircraft, such as the 28-90 *Flash*. Giuseppe Bellanca's policy of *every airplane off the assembly line should be a little bit better than the one before it,* makes it virtually impossible to document each airplane's exact configuration. In many cases a Supplemental Type Certificate (STC) was approved by the Commerce Aviation Branch (CAB) to allow continued improvement on aircraft in production. Therefore, specification changes between the first and last unit produced is almost a certainty.

BELLANCA J *PATHFINDER*

The name *Pathfinder* was carefully chosen by Bellanca for the model J because of their missions. *Pathfinders* would be the record-breakers or extra-long-

Lewis Yancy, GM Bellanca and Roger Williams with Bellanca J Pathfinder *named "Cape Cod." Photo courtesy of Frances Rees Hayford.*

In 1928, William Brock (in photo) and Edward Schlee set a new World Endurance Record of 59 hours, 7 minutes in this Bellanca named Rosemarie. *Photo courtesy of Frances Rees Hayford.*

range airplanes. All were highly modified versions of the Bellanca CH. The model J *Pathfinder,* named *North Star* (no. 101 NX3789), was used in 1929 for the trans-Atlantic crossing from Maine to Spain. Serial no. 104, NX4484, is also listed as a J model, named *Reliance,* which was used by Lt. Royal V. Thomas to set a Solo Endurance Record of 35 hours and 25 minutes. Serial no. 106, NX5315, is also listed as a J model and is believed to have been used by Admiral Byrd for a North Pole expedition. Serial no. 109, NX7085, also a J model, named *Rosemarie,* was flown by William Brock and Edward Sheele to a new World Endurance Record in 1928. To make things more confusing, Bellanca would use the same model J and rename the airplane for additional flights. Bellanca J serial no. 3004 was *Miss Veedol,* flown by Clyde Pangborn and Hugh Herndon from Japan to the USA. It was later sold and renamed *The American Nurse,* which was lost at sea on another Atlantic crossing attempt. The *Olympia* was also renamed *Warszawa.* All the model J *Pathfinders* were issued government registration numbers with the prefix NX, or NR denoting experimental or restricted.

In today's world of high-speed jet travel, it is hard to imagine the obstacles and overwhelming odds the pilots of early aviation faced when attempting to break a record. The 4500-mile flight across the Pacific Ocean by Clyde Pangborn and New York socialite Hugh Herndon, in 1931 in a single-engine airplane defies the imagination.

CONQUERING THE PACIFIC

Clyde Pangborn lived the Roaring '20s as no other. He flew death-defying aerobatics for the Gates Flying Circus like a ballet dancer. At 3000 feet he would shut the engine off and complete a series of stunts that would have the crowd awestruck before coming to a silent landing before an applauding audience. Clyde Pangborn would become famous in his eleven years of barnstorming in every state of the union and would carry 150,000 passengers without a mishap. In these early days of aviation, this must be a record in itself. However Clyde Pangborn had another destiny with history.

Born in Wenatchee, Washington, in 1893, he earned a degree from the University of Idaho in civil engineering at age twenty-two. When the country entered into WWI, he enlisted as a flying cadet in the Army Signal Corps. An obviously gifted pilot, he served his entire time as a flight instructor at Elling-

Miss Veedol *being rolled out of the hangar in New York before attempting to beat the 'Round-the-World record of the* Winnie Mae. *Photo from the Daniel Koble collection.*

Bellanca NR796W named Miss Veedol. *Note the "Guppy" belly tank for extra fuel. Clyde Pangborn and Hugh Herndon Jr. spanned the Pacific nonstop in this airplane in 1931. Photo from the Daniel Koble collection*

ton Airfield, Texas. After his discharge in 1919, Clyde took up barnstorming to make a living. It was 1929, in Syracuse, when Clyde met Hugh Herndon, Jr., the grandson of John H. Carter, an oil magnate, and the son of wealthy socialite Mrs. Russell Boardman. Although a Princeton Graduate, Hugh teamed up to barnstorm across the United States with Clyde. By 1930 they came up with the idea to fly around the world together.

The world-renowned Bellanca was their choice of airplanes. They came to New Castle, Delaware, where they met with Giuseppe Bellanca and his Russian-born chief of special operations Iala "Is" Islamoff. Construction began immediately, and the airplane was near completion when Wiley Post and Harold Gatty, in the last week of June 1931, departed New York in the *Winnie Mae.* They flew the most northerly route possible near the Arctic Circle and completed their trip around the world in 8 days, 15 hours, and 51 minutes.

Undisturbed by the flight of the Lockheed-built *Winnie Mae,* Clyde and Hugh departed New York one month later on July 28 in an attempt to beat the record. The bright red Bellanca, named *Miss Veedol* after the sponsor Veedol Oil Company, carried 915 gallons of fuel in five separate gas tanks and 45 gallons of oil. Space was so limited the oil tank was also integrated into the two front seats. High above the fuselage 500-gallon gas tank, there was just enough room of one person to "stretch out" in an attempt to get some sleep. How anyone could sleep over the roar of the unmuffled, 420-hp Pratt & Whitney *Wasp* engine is a wonder in itself. The airplane was fitted with large tires to allow safe landings on unimproved fields, if needed. Other than maps, a magnetic compass was their only navigational instrument. To conserve weight, no radios, parachutes or life rafts were aboard. However, the 500-gallon fuselage fuel tank had two 8" dump valves that could jettison the fuel in seconds which, when closed, would allow the airplane to float. Although the Bellanca was not as fast as the *Winnie Mae,* they felt the extra range would allow fewer stops allowing an opportunity to beat Post's record.

Halfway across the Atlantic, they ran into a terrible storm which sent the compass spinning. Completely disorientated, they finally landed in Wales 32 hours later. After a brief stop at London's Croydon Aerodrome, they were off to Berlin. Calculating their time and comparing it with Wiley Post's confirmed they were 22 hours behind. In spite of the deficit, they took off immediately for Moscow, then across Siberia where they encountered horrible weather. When they finally reached Khabarovzk, they were now over 27 hours behind the *Winnie Mae*'s time. To make matters worse, they were forced to stay in Khabrovzk until the weather improved. They finally accepted the fact it was impossible to make up the time lost. To their surprise, several telegrams arrived from friends in Tokyo suggesting they try for the $25,000 prize offered by the Japanese newspaper *Asahi* to the first one to fly the Pacific nonstop from Japan to the United States.

They immediately wired the U.S. Embassy in Tokyo to obtain permission to fly over Japan. In their haste to take advantage of clearing weather, they did not wait for a reply and left Khabarovzk without a permit. Japan had invaded Manchuria and the Japanese military had set very strict rules concerning any flights over fortified areas. As innocent as a couple of babies in a crib, Pangborn and Herndon flew on taking pictures with their 16mm movie projector along the way. Unknown to them, they were being tracked by every Japanese warship en route. When they landed at Tokyo's Tachikawa Airport they were promptly arrested.

Clyde Pangborn and Hugh Herndon made international news when they were charged with no less than fifty-five violations and placed under house arrest in the Imperial Hotel. A week later they went to trial and were found guilty of all charges. Their camera was confiscated and they were fined $1050 each or 205 days in jail at hard labor. Surprisingly, they were allowed to retain ownership of their airplane. The news media had spread the news internationally before Mrs. Boardman cabled them the money for their release, and some extra for preparing for the trans-Pacific flight.

Once Pangborn and Herndon were released, they announced to the press that they were going to attempt to fly the Pacific Ocean and claim the $25,000 prize offered by the Japanese newspaper.

Another stumbling block surfaced when the Japanese government refused to issue a permit without their certification of the aircraft due to several other failed attempts. For weeks they tried in vain to acquire the necessary permit. It was not until the U.S. State Department intervened that the Japanese gave their permission. However, it was not without some very stringent conditions. Only one attempt would be allowed. If that failed, they would be required to load the airplane and themselves on the first available steamer and leave Japan.

Pangborn devised a bold scheme to lighten the airplane, by 350 lbs, by jettisoning the landing gear in flight. Once airborne, its speed would increase by 17 percent and simultaneously make the airplane safer if forced to ditch in the sea. With the help of friends, he removed the bolts in the landing gear attach points and inserted removable pins in their place. Once airborne, and out of sight of land, he would pull the pins and drop the entire landing gear into the ocean. Additional metal strips were added to the bottom of the fuselage to act as a skid during the U.S. landing.

Another problem surfaced when they calculated the runways were not long enough at the Tokyo airfield for a takeoff. The only suitable place they could find was Sabishiro Beach, 300 miles north. On September 29, 1931, they departed for the beach to ready the airplane for the attempt. By the 3rd of October the airplane was loaded with 915 gallons of fuel and 45 gallons of oil. The airplane weighed *three times its own weight* when Pangborn shoved the throttle wide open for takeoff. Slowly, but surely, the Pratt & Whitney *Wasp* pulled the heavy airplane down the beach. Forty, fifty, sixty, seventy, eighty and the Bellanca would still not leave the ground. Finally, at 100 mph it broke the bonds of earth and headed low out to sea.

Once settled down and convinced all systems were operational, Hugh Herndon pulled the pins on the landing gear and dropped it in the ocean. With the landing gear gone, the Bellanca picked up the much-needed, fifteen-miles-per-hour speed. However, their work was still not done. Two landing gear strut braces still dangled from the fuselage. Although they were lightweight braces and did not cause any flying problems, they could be potentially dangerous when landing, especially so if one should deflect into the cabin seating area. Another discussion ensued. The solution was simple. Climb out of the airplane (now cruising at 10,000 feet over the Pacific), detach the holding bolts, and climb back in. Since both pilots were bona fide barnstormers, they probably drew straws to see who would have all the fun. Pangborn won (or lost, depending on your viewpoint) and climbed out into the cold, 115-mph wind and released the struts.

At 10,000 feet over water, the clear sky and the water become boring for pilots. Neither offers any variety of scenic wonders compared to the patchwork of landscape. Although they were flying at 115 mph, the airplane appears suspended in space. Alter-

Clyde Pangborn and Hugh Herndon, with engine shut down make their final approach for landing at Wenatchee, Washington, in Miss Veedol. *Photo from the Daniel Koble collection.*

nately they took turns climbing on top of the fuselage gas tank for a nap. After many hours of sheer boredom, followed by more boredom, their first landmark, Mount Rainier, was spotted. The Spokane Airport was just 200 miles east, and the runways were long enough for a safe landing. Unfortunately, when they arrived, the airfield was fogged in and a landing was impossible. To add frosting to the cake, they gave fleeting thoughts to flying further inland to break the 5,001-mile nonstop distance record. They calculated they had flown 4558 miles. If they were to break the distance record they would have to fly to Dallas, Texas. They decided it was not worth the effort. They had now been in the air for over 41 hours. They turned back and decided to land at the Wenatchee Airfield, the town where Pangborn was born.

The townspeople had anticipated Pangborn's arrival and were waiting for him to land. He made one low pass across the airfield and the crowd gasped in horror when they saw the airplane had no landing gear. Before the final approach, Pangborn and Herndon jettisoned all but a few gallons of fuel in a wing tank to prevent any chance of fire and landed with no other damage than a broken propeller. One of the first people on the scene was a Japanese representative from the newspaper *Asahi* waving a check frantically.

A hero's welcome does brighten up ones spirits. However, after two weeks of banquets it soon became boring. The new landing gear and propeller had arrived from Delaware and were installed on the plane. The plane was loaded with fuel and several cases of prize Washington apples as they bid farewell to Wenatchee. Their record stood for sixteen years when a modified U.S. Air Force B-29 made the flight in 1947. However, as we know from the Lindbergh flight, coming in *first* is what really counts in history. Clyde Pangborn and Hugh Herndon, with their Delaware-built Bellanca, captured the honor for all time.

Clyde Pangborn's brother, Percy, organized the citizens of Wenatchee and Chelan County to erect a permanent monument on the site of the landing after the war. The basalt monument, with a concrete base was dedicated in May 1969, and is now listed in the National Register of Historic Places. Pangborn and Herndon spent their entire lives in aviation and died of natural causes. They were the true pioneers of early flying. Their historical Delaware-built Bellanca was purchased by Hugh Herndon and later sold to Dr. Pisculli who renamed it *The American Nurse.* Pisculli and the airplane were lost at sea in another Atlantic crossing attempt in 1932.

This is just one of the many historical flights of Delaware-built Bellanca airplanes. A complete list of the record flights is listed at the end of this chapter. The Bellanca employees who labored building these primitive aircraft are to be commended for their craftsmanship. The pilots' lives depended upon their dedication to hand-building the best airplane possible.

Bellanca CH-200 (Approved Type Certificate No. 47 dated June 1928). Photo courtesy of Frances Rees Hayford.

Bellanca CH-300 Pacemaker *with Bellanca pilots Wes Smith and George Haldeman. The trophies are from the Cleveland Air Races, 1929. Photo courtesy of Frances Rees Hayford.*

BELLANCA CH-200

The CH-200 was the first ATC approved production airplane built in Delaware. It was essentially an improved version of the famous *Columbia* in which Chamberlin and Levine flew from New York to Germany. All CH Bellancas were constructed as follows: the fuselage and tail were constructed of hand-welded, chrome molybdenum-steel tubing and the fuselage formers were made of wood to create the form. The wing spars and ribs were spruce. The airplanes were covered with Aircraft-grade linen and painted with DuPont nitrate dope. Any DuPont nitrate dope colors were available to the customer. The CH-200 could seat four to six in comfort.

CH-200 SPECIFICATIONS

Number of seats	4-6
Engine	Wright J5
Number of cylinders	9
Horsepower	220 hp
Fuel capacity	90 gallons
Range	750 miles
Length	27' 9"
Wingspan	46' 4"
Wing area	273 sq/ft
Landing speed	46 mph
Service ceiling	13,000 ft
Fuel consumption (prox)	11 gph
Rate of climb	850 ft/min
Payload, including pilot	1235 lbs
Empty weight	1890 lbs
Top speed	128 mph
Cruise speed	110 mph
Landing speed	46 mph

In the fall of 1929, Bellanca displayed the CH-300 *Pacemaker* for the first time at the National Aviation Show in Chicago. The CH-300 *Pacemaker* would carry six people in a luxurious, upholstered cabin. The engine was the newly improved, Wright 300-hp J6 *Whirlwind.* The extra power increased performance dramatically. H. B. du Pont purchased a CH-300 model (serial no. 105) to enter in the National Air Races in Los Angeles in 1928. Bellanca CH-300, serial no. 108, was test flown by Schlee-Brock Aviation before being sold to orchestra leader Roger Wolf Kahn. This airplane was later modified and used to demonstrate an aerial mail pickup system invented by Godfrey Cabot in 1929. (As this is written, this Bellanca is owned by Dan Cullman, Washington State.) By June 1930, Bellanca offered the 300-hp Pratt & Whitney *Wasp Jr.* in the airplane as an option. This engine change required a separate ATC (number 328), issued by the CAB (Department of Commerce, Aviation Branch). This model was designated the 300-W *Pacemaker.*

PACKARD DIESEL POWER

Engineers considered gasoline-powered airplanes expensive to operate and highly explosive in the event of a crash. For years they were working on ways to reduce the weight and the cost of manufacturing to compete with the gasoline engine. There was little doubt among engineers that a properly designed diesel engine could outperform the gasoline engine in terms of safety and economy. Captain Lionel Woolson, Chief Aeronautical Engineer for Packard Motors, and Herman Dorner, from Hanover, Germany, designed the Packard diesel engine. The first test flight was made by Walter Lees, September 19, 1928, Utica, Michigan, in a Stinson *Detroiter.* When the Packard Motor Car Company introduced its air-cooled, radial, diesel engine to manufacturers for further testing, Stinson, Buhl, Mahoney/Ryan, Waco, Verville, Lockheed and Bellanca responded to the challenge. On March 9, 1930, Walter Lees and Lionel Woolson flew the *Detroiter* nonstop from Detroit

The Packard Diesel-powered Bellanca. Photo courtesy of Jo Cooper, author, Pioneer Pilot.

to Miami, Florida, in 10 hours and 15 minutes. Fuel costs for the trip were reported at $8.50. Although the other manufacturers airplanes flew well with the Packard diesel engine, Bellanca was the only one to set a new record.

Packard sponsored Walter Lees and Fred Brossy's attempt to challenge the World Endurance Record of 67 hours and 13 minutes, held at the time by two Frenchmen, Maddelena and Creeorni. The choice of airplanes to be fitted with the specially prepared diesel engine, was a Bellanca J-2 *Special.* The Bellanca would be built, especially for this flight, with longer wings and many other features agreed upon by Bellanca and Packard engineers.

The planning of the flight by pilots Walter Lees and Fred Brossy required paying attention to even the smallest detail. The airplane weighed 2400 pounds empty. The weight of the two pilots and their equipment, which included food, water, parachutes, and other related items, brought the weight of the airplane to 3,042 pounds. Added to this was 558 gallons of fuel and lubricating oil which added another 3,350 lbs. Oversized tires were installed to keep the airplane from bogging down in the sand under the load.

The decision to depart from the miles-long, wide and smooth, sandy beach at Jacksonville, Florida, was made in the interest of safety. Once the airplane was airborne there were no obstructions to overcome, which allowed for a slow rate of climb to cruising altitude. Other factors taken into consideration before takeoff were, three days of forecasted clear weather, wind direction and velocity.

The record flight would take three separate attempts. The first of three attempts would be made on March 11, 1931. This attempt was aborted, after 30 hours and 13 minutes, when the oil temperature and oil pressure gauges failed. In addition to the failed gauges, a small fuel leak developed behind the firewall, which was not accessible while in flight.

During the month-long preparation for the second attempt, two Frenchmen, Rossi and Boussourot, established a new record of 75 hours and 23 minutes. Lees and Brossy calculated they would require additional fuel to fly longer, which changed the

Walter Lees and Fred Brossy inspect the load of fuel and oil they will be carrying aboard the Bellanca. The five-gallon fuel cans were thrown overboard when emptied. Photo courtesy of Jo Cooper, author, Pioneer Pilot.

Colonel Julian's jet-black Bellanca J2 Pathfinder Special *named Abyssinia. Photo from the Daniel Koble collection.*

takeoff weight of the airplane to 6,666 pounds. On April 12, 1931, Lees and Brossy departed the beach at Jacksonville once again. The rules stipulated that the airplane must be visible to the contest officials once every 30 minutes. After flying in circles for 74 hours and 12 minutes (slightly over two hours short of breaking the record), they were forced to land due to heavy rain and fog.

Another agonizing month-long delay was needed to prepare the airplane for a third attempt. This time, a sixty-gallon fuel tank was specially built to fit under the fuselage of the airplane. This increased the gross takeoff weight to 6,991 pounds. At 6:47 a.m. on May 25, the duo made their third successful takeoff. This time the weather was favorable, and the airplane performed flawlessly. Although Lees and Brossy planned to fly until they were down to the last remaining gallon of fuel, in the interest of safety they landed at sunset with four hours of fuel remaining. Walter Lees and Frederick Brossy set a new World Endurance Record for non-refueled flight of 84 hours and 32 minutes. This record would not be broken until Burt Rutan's *Voyager* circumnavigated the earth 55 years later in 1986.

Bellanca serial no. 1101 was built as a test bed for Packard Motor Company and built as a model J-2 *Pathfinder Special.* Since it was experimental, the Aviation Branch issued the number NR782W. The diesel-powered Bellanca was created solely for promotional and testing purposes and was never issued an ATC number. The major complaint of the diesel-powered airplane were the fumes. The pilot and passengers would reek of diesel fuel, and passengers complained of becoming ill. No matter how economical the engine operated, this was not acceptable. The diesel engine was removed after the record flight and a 300-hp gasoline engine installed to make the airplane saleable.

About this time a buyer appeared on the scene. Colonel Hubert Fauntleroy Julian came to Bellanca from Abyssinia, Africa, to learn to fly and buy an airplane. His intention was to have Bellanca construct an airplane for him to fly back to Africa. Bellanca offered to remanufacture the airplane to meet his needs.

Julian was directed across the airfield to Air Service, Inc. where Dick Morgan gave him flying lessons.

Left to right: Shirley Short, test pilot, holds the altitude recorder while Elinor Smith and R. M. Noorduyn, vice president, congratulate George Haldeman for his record flight. Photo courtesy of Frances Rees Hayford.

The black aviator was quite impatient and became very vocal about not wanting to spend all his funds on flying lessons. On July 29, 1931, Pop Hanson, the local Aviation Branch flight examiner, approved Julian's pilot license in spite of his barely meeting the minimum flying skills.

Meanwhile Bellanca was busy rebuilding serial no. 1101 for the proposed Atlantic crossing. The tail number was changed to N782W. Julian named the airplane *Abyssinia,* with a subtitle of *Emperor Hail Salassi I King of Kings.* Under the pilot's cabin window was inscribed "Pilot, Col. Hubert Julian, *Black Eagle.*" Displayed on the forward fuselage was, "HOLDER World's Non-Refueling Record 84 hours and 33 minutes."

Colonel Julian made an erratic takeoff, from Bellanca Airfield, bound for Roosevelt Airfield where he would start his trans-Atlantic flight. Unfortunately, when Julian tried to depart Roosevelt Airfield, the airplane's tail would not come up and the airplane veered off the runway, crashed and burned. The CAB investigation reported that Julian failed to set the trim setting to the "takeoff" position. Although Julian survived, his continued interest in aviation is unknown.

Bellanca Pacemaker

Giuseppe Bellanca capitalized on the added horsepower of the CH-300 *Pacemaker* by establishing two new altitude records. Elinor Smith broke the Women's Altitude Record by flying to 27,418 feet.

This was followed by Bellanca test pilot George Haldeman, who set the Men's Altitude Record of 30,453 feet.

Bellanca made many improvements during production of the popular CH over the years that are not immediately apparent. An optional landing gear did away with the narrow stance and the floating axle with rubber shock cords. The new landing gear was much wider and utilized oil-hydraulic shock absorbers. All the control systems were changed, starting with the *Pacemaker,* to smooth out their operation by changing their leverage and balancing the surfaces. The trim tab, which operated on a cable and drum, was also changed to a chain and sprocket. All these changes were made to the existing ATC by filing for an STC (Supplemental Type Certificate) with the CAB.

The four forward seats were of the bucket type and the aft seat was of the bench type. Sliding safety-glass windows were installed which allowed for better ventilation. A cabin heater was offered as an option after serial no. 128. The tail skid, which acted as a brake, gave way to a standard full-swivel tail wheel beginning with serial no. 151. Main landing gear wheels were then fitted with Goodyear brakes. A two-band LF radio was also offered as optional equipment.

If a passenger took the copilot seat, the control stick and the rudder pedals were easily removed. The pilot and copilot seats had an isle between them for easy entry. All Bellanca CH models were also offered as a freighter with only two front seats. Edo pontoons were also available at extra cost. An optional NACA-designed engine cowling was also offered. Standard equipment included an oil cooler, which improved cooling of the cylinders as well as the oil.

On April 3, 1932, Bellanca was issued approval under ATC number 476, for the installation of the 420-hp Wright R-975-E2 engine in the *Pacemaker.* This Bellanca is listed as an E model. Seven were built especially for discriminating buyers who wanted more performance. Three were built for the CAB, another was purchased by the *Des Moines Register & Tribune,* and the final three went to private parties unknown. The E was heavier than the standard *Pacemaker,* but the extra power allowed a staggering 4657-lb gross weight. Larger, 120-gallon fuel tanks were installed to accommodate the thirsty engine's 24-gallon per hour fuel consumption. Outwardly, the changes were so subtle, the E could not be distinguished from the 300 *Pacemaker.*

The final *Pacemaker* was the model 31-42 *Senior Pacemaker.* Using the larger fuselage from the *Skyrocket* and offering a wide variety of interior options, including seating for four, six, or eight, the 31-42 was in a class by itself. An extra belly storage compartment, with a capacity of 100 lbs and a Hamilton controllable-pitch propeller were also standard equipment. Additional configurations were also available for hauling cargo with only one pilot seat. With the Wright 450-hp engine and the performance of a scalded cat, the majority of the 31-42s were shipped out of the U.S.A. as bush planes. In spite of its traditional steel-tube, wood and fabric construction, the Bellanca 31-42 was loved by pilots who would load the airplane with anything that would fit through the door. It was built as a work horse and did its job well.

Bellanca's sturdy construction and ability to carry heavy loads from unimproved landing sites made the CH series Bellancas an ideal work plane. America, Canada, Alaska, Mexico, China, Central and South America were all established markets. Bellanca built some of the most dependable aircraft of the period, and his planes had an excellent reputation, especially in these areas of rugged terrain.

PACEMAKER 300 SPECIFICATIONS

Number of seats	4-6-8
Engine	Wright J-6*
Number of cylinders	9
Horsepower	300
Fuel capacity	86 or 105 gallons
Range	675 miles
Wingspan	46' 4"
Wing area	273 sq/ft
Length	27' 9"
Rate of climb	1,100 ft/min
Landing speed	50 mph
Service ceiling	18,000 ft
Empty weight	2253 lbs
Gross weight	4050 lbs
Cruise speed	120 mph
Top speed	140 mph

* Pratt & Whitney 300-hp. *Wasp Junior* also available. Wright R-975-E2, 420- hp, available in E model.

Victor Dallin purchased Bellanca CH-200 (no. 117) in 1928, which he used for aerial photography. Over 1800 negatives of his work are stored at the Hagley Museum & Library. Photo from the Hagley Museum & Library.

Bellanca (no. 138) Pacemaker *CH-300 on optional Edo pontoons built in 1929. Photo from the Daniel Koble collection.*

Luxurious interiors were a trademark of Bellanca aircraft. Note the tassels to pull down the window shades and elaborate wood framework around the windows. Photo courtesy of Frances Rees Hayford.

Skis were a very popular option for Bellancas that operated from Alaska and Canada. Photo courtesy of Frances Rees Hayford.

Ilia Islamoff, special projects engineer. He was responsible for overseeing all J models and other special Bellanca projects. Photo from the Daniel Koble collection.

[1] This is the original version of the TES with the dual Pratt & Whitney engine installation and the single vertical fin.

[2] This is the final version with the Curtiss, Conqueror *water-cooled engines, triple vertical fins and relocation of the pilot. The size of the TES dwarfs the two men standing beside it on Bellanca Airfield. Photos courtesy of Frances Rees Hayford.*

BELLANCA MODEL TES
Blue Streak (1929)

Without a doubt, the 1929 Bellanca TES (Tandem Engine Special) *Blue Streak* X855E, serial no. 118, was the most unusual airplane ever built by Bellanca. Although it is considered a Bellanca design, it was the brainchild of a Russian aeronautical engineer Ilia "Is" Islamoff, chief designer, who was employed by Bellanca long before he came to Delaware. Is Islamoff was instrumental in the designs of *all* Bellanca planes built in Delaware.

The installation of twin, center-line-thrust engines was a safety feature unobtainable from wing-mounted engines. The TES *Blue Streak* would not bring fame and glory to Bellanca in spite of the crowds of people it drew. A behemoth of an airplane, the first version had two 420-hp Pratt & Whitney *Wasp* air-cooled engines mounted in tandem. The second engine was mounted, behind the first, facing aft. A long shaft from the second engine passed through the center of the aircraft where it exited the fuselage between the two booms that carried the empennage. The front engine "pulled" and the aft engine "pushed."

The TES went through a major rebuilding after the test flights with the Pratt & Whitney air-cooled engines. It is interesting to note that the first variant had no forward visibility at this time (like the *Spirit of St. Louis*). Exhaust stains at various points around the closed cowling indicate the exhaust was discharged at these points. The noise level of these two engines must have been appalling. Inability to cool the rear engine was the reason that the Pratt & Whitney air-cooled engines were replaced with Curtiss *Conqueror* water-cooled engines. Other changes made during the engine replacement, which are apparent in the photographs, were the triple tail versus the single tail, and a three-blade propeller was also used both fore and aft in the final upgrade. The pilot was also moved to a conventional position with forward visibility.

The TES was sponsored by the *Chicago Daily News* as a high-speed, cross-country aircraft. No specifications of the airplane were available from NASM for review. All fuel was stored in wing tanks and a 2200-gallon fuselage tank. Bellanca test pilot, Mr. Shirley Short was offered the position of corporate pilot for the *Daily News* when the airplane was finished. Apparently Mr. Short felt comfortable with the *Blue Streak* and accepted the position. On a test flight in Chicago for maximum speed at full gross weight, May 26, 1931, the TES crashed killing pilot Shirley Short, copilot Richard K. Peck, mechanic Robert Gormley, and radio operator Lewis S. Rice. The TES crew were attempting to set a max-gross weight, 2,000 kilometer speed record under the supervision of the National Aeronautics Association. At the time of the accident, the TES, was carrying 4,400 lbs of lead, 1600 gallons of fuel, plus the pilot and crew, which totaled 5000 kilograms. Reports indicated

The skeleton frame shows the first version installation of the two 420-hp Pratt & Whitney Wasp *air-cooled engines in the TES. Photo is dated 4-6-29. Photo courtesy of Frances Rees Hayford.*

Test pilot Shirley Short, James Houghteling, vice president, Chicago Daily News *and GM Bellanca appear pleased with the first version TES* Blue Streak. *Photo courtesy of Frances Rees Hayford.*

the airplane appeared to have suffered structural failure while shuttling between Curtiss-Reynolds Airport and Sertna, Illinois. A farmer, who was first on the scene and witnessed the accident, reported,

> The right wing appeared to fail. The crew responded immediately by dumping the fuel and shutting down the engines in an attempt to glide in for a landing. The airplane appeared to dive straight for the ground.

The airplane did not burn as reported by much of the news media. The airplane was intentionally set on fire by the CAB in the interest of safety of the spectators and souvenir hunters.

When attempts are made to break existing records of any kind, it should be remembered that the risks are very high. The TES, although an unorthodox design, did fly successfully for two years before it crashed. The 1930s was an era of frenzied activity in aviation record-breaking. The economy was in a depression, and those who broke old records reaped a gigantic harvest of publicity and lucrative sponsor-supported financial rewards. In spite of the rewards gained by the victors, it was a hard and dangerous life. Men and machines were driven to their limits and, in this case, beyond. Giuseppe Bellanca was deeply despondent for a long time over the loss of Shirley Short and the crew who had flown and tested his airplanes for many years.

BELLANCA Skyrockets

The Bellanca CH-400 *Skyrocket* was a beefed-up version of the CH-300 *Pacemaker* to allow for the extra horsepower. The airplane was sold to a limited buyer group who looked for outstanding performance. The select group included Ann Harding, Wallace Beery, Shell Oil, Western Telephone, and Bendix Corp., to name a few. Although the speed was not that much more than the CH-300, the ability to get off the ground with a full load was astounding. Since the air-

This rare concept drawing of the TES is signed by aeronautical engineer, "Is" Islamoff and dated 3-9-31. It appears Islamoff intended further development of the TES as a double-deck airliner. When the original version crashed three months later, the project ended. Photo courtesy of Frances Rees Hayford.

plane was built to order, plush interiors and superb attention to detail were adhered to.

CH-400 SKYROCKET SPECIFICATIONS

Base price	$17,950*
Engine	P&W C-1 Wasp
Horsepower	420
Wingspan	46' 4"
Wing area	273 sq/ft
Length	27' 10"
Height	8' 4"
Number of seats	6
Empty weight	2008 lbs
Gross weight	4600 lbs
Fuel consumption	22 gal/hr
Fuel capacity	120 gallons
Propeller	9'
Oil capacity	12.5 gallons
Maximum speed	155 mph
Cruise speed @1700 rpm	134 mph
Landing speed	56 mph
Climb (sea level, normal load)	1250 ft/min
Service ceiling	20,000 ft
Range	670 miles

*$21,150 with Edo pontoons.

On April 26, 1935, Bellanca introduced the model 31-50 *Skyrocket* and the 31-55 *Senior Skyrocket.* The model 31-50 was issued ATC 565, which was amended to include the model 31-55. After this airplane, there was nowhere else for Bellanca engineers to go with the design which originated with the WB-I back in 1925. The fuselage was widened, the stick control was replaced with a throw-over wheel, and custom interior designs were unlimited. Interiors were elegantly appointed in fine leather or other materials chosen by the customer. A wide variety of paint colors were also offered. Designated the 31-50 and 31-55 respectively, it was an airplane built to customer specifications. The non-supercharged 525-hp Pratt & Whitney was used on the 31-50. The supercharged 550-hp Pratt & Whitney S3H1 *Wasp* engine and Hamilton Standard controllable pitch propeller were standard on the 31-55. All glass was shatterproof. Cabin walls were heavily insulated for passenger comfort. All seats were equipped with arm rests. A rear cabin door was installed for easy entry. An instrument panel was equipped for flying without visual reference to the ground. The *Senior Skyrocket* was equipped with wing flaps to lower landing

Two famous Bellanca Senior Skyrockets. *The top photo (no. 621) is Dr. Richard Light's, that he flew around the world in 1935. The bottom photo (no. 807) is Hollywood actor Wallace Beery's airplane. Photos courtesy of Frances Rees Hayford.*

speed and takeoff distance. Edo pontoons were available at extra cost. A two-band radio was also available along with many other aviation options one could want. A total of seven were built, serial nos. 806 thru 812.

SENIOR SKYROCKET SPECIFICATIONS

Price	Variable
Wingspan	50' 6"
Length	27' 11"
Height	8' 6"
Wing area	359 sq/ft
Empty weight	3440 lbs
Useful load	2160 lbs
Gross weight	5600 lbs
Fuel	200 gallons
Range	700 miles
Maximum speed	190 mph
Cruising speed	180 mph
Landing speed, with flaps	56 mph
Landing speed, no flaps	64 mph
Climb	1500 ft/min
Ceiling	25,000 ft
Cruising range	700-1000 miles
Fuel consumption at cruise	36 gph

BELLANCA P-100 *Airbus* (ATC-360, August 27, 1930)

The *Airbus* was an extension of the model K *Roma.* Only one P-100 *Airbus* with 12-seat capacity was built. The *Airbus* was designed primarily to haul passengers and airmail. George Haldeman, Bellanca test pilot, flew the P-100 on a countrywide tour of 4400 miles to promote interest in the airplane for the airlines. Keeping close track of expenses, the *Airbus* turned in an average speed of 125.5 mph and $0.08 per mile operating cost. In spite of these astonishing numbers, the airplane did not sell. This was due to several causes. The public air traveler felt the three-engine airliners provided a safety margin unavailable to the single-engine airplane. The Curtiss water-cooled engine, with its complex plumbing, was considered obsolete compared to the air-cooled radials. This lone aircraft is rumored to have been sold in Mexico and was later destroyed in a hangar fire.

Serial no. 701: This is the first Bellanca Airbus, *Model P-100. It is the only one built with the Curtiss Conqueror water-cooled engine. Photo from the Daniel Koble collection.*

AIRBUS P-200 (ATC-391, January 26, 1931)

When Bellanca introduced the P-200 *Airbus,* in 1931, with a new 575-hp Wright, air-cooled engine, it was capable of carrying 12 or 15 passengers. One pontoon-equipped P-200A, serial no. 702 was purchased by New York & Suburban Airlines who offered flights from outlying Long Island to the docks at the foot of Wall Street. The airplane was sold in Canada in 1937 and registered as BF-BBJ, where it was last operated by Wings Ltd. Winnipeg, Manitoba. The second P-200 was serial number 703, which was a special order for RCA president Fred King, and named *Miss Fidelity.* Bellanca dubbed this deluxe airplane a P-200 *Aircruiser* which seated nine passengers in corporate level luxury. Another P-200 was built for La Touraine Coffee Company as a P-200 *Airbus.* It was later returned to the factory to be upgraded as a model 66-67. Although the P-200 performed well, a larger engine would later increase the useful payload to a point it could carry more than its own empty weight. Airlines did not buy the P-200 as was expected, due to the stifled economy and passenger perception of multi-engine safety.

The Army Air Corps was impressed with the *Airbus* performance and ordered four as utility transports labeled YIC-27. The first two of these were unsold civilian-versions (no. 705, N10797 and no. 706, N10798) P-200s and were converted to military YIC-27 specifications. In 1934, the Army returned this initial order of four to the factory to be upgraded to 675-hp or 750-hp engines. These installations changed the model numbers to C-27B and/or C-27C, depending on configuration. After the first deliveries, the Army returned to Bellanca to order ten more, this time with the Pratt & Whitney *Hornet* engine, called the C-27A. The Army also used the C-27A as an ambulance with litters in two rows, three high; or it could seat 15 passengers. In the cargo configuration, seating was limited to the pilot and copilot. A trapezium-shaped door, 5 feet 8 inches high by 4 feet 4 inches wide, was installed on the cargo version to access the 372 cu/ft cargo area. The operating costs were reported to be ten cents per mile. The airplane was approved for installation of pontoons, skis, or wheels, thereby making it the most versatile airplane of the time.

Serial no. 703: Listed as a Deluxe P-200 Aircruiser, *NC786W was sold to Fred King, president of RCA. This Bellanca* Aircruiser *was designed to carry nine people in absolute luxury. Photo courtesy of Frances Rees Hayford.*

Serial no. 704: NC10796 is listed as an Airbus *and was sold to LaTouraine Coffee Co. Photo from Hagley Museum & Library.*

Serial nos. 705 thru 718: These are listed as YIC-27s or C-27As and were sold to the U.S. Army Air Corps. Photo courtesy of Augie Bellanca.

BELLANCA Model 66 *Aircruiser* (ATC-563, March 16, 1935)

In an effort to promote safety for passengers, the CAB was cool toward single-engine airplanes operating as airliners. By 1934, the CAB announced that single-engine airliners would be banned from operating in the United States. This was a golden gift to the companies committed to building all-metal, multi-engine airliners. Douglas, who was producing the DC-2, and Lockheed, who was building the model 10 *Electra,* could not keep up with the demand. It was not good news for Bellanca who had steadfastly built all of his single-engine airplanes from wood, steel, and fabric. Bellanca, shifted his market strategy to focus upon airplanes that could operate from short, unimproved airfields. His airplanes proved sensational in operating from airfields that airliners could not use.

Bellanca's strategy paid off with the introduction of the model 66 *Aircruiser* (ATC-563) in 1935. The *Aircruiser* was now fitted with a 670-hp Wright *Cyclone* engine. This single-engine airplane could outperform the tri-motor airplanes of the day and was used primarily in Canada and the Philippines. The notable exterior difference in production models was in the side window arrangement, which was now round. The *Aircruiser* was so popular, another engine was offered as the model 66-70. With the supercharged 715-hp Wright *Cyclone* engine, the *Aircruiser* was capable of lifting loads heavier than the airplane's empty weight.

A book could be written on the many outstanding stories documented on the performance of this remarkable airplane. One, however, bears mentioning. Elizalde & Company, Ltd. operated an ore mine in the mountains on the island of Mindanao in the Philippines. It would take mules three weeks to bring the ore out over the mountains on dangerous narrow trails to the city of Davao. In comparison, the Bellanca *Aircruiser* (registered NPC-41, serial no. 721) made the trip by air in 26 minutes. A mountain ridge soared 3000 feet into the air above the 1600 feet above sea level dirt airstrip carved into the valley below. The Bellanca had a special, large double-door hatch cut in the roof to load bulky machinery needed for mining. Loaded with over 4000 lbs of ore, the Bellanca broke ground after 1000 feet and climbed smartly in the hot air to clear the 3000 foot ridge with an ample margin of safety. Month after month, the Bellanca hauled over 300,000 pounds of freight. Even more incredible, Elizalde

1938 Bellanca Aircruiser, *model 66-67, CF-BKV, serial no. 722 sold to MacKenzie Air Service, Canada. Photo courtesy of W. P. Sansom.*

Photo dated July 10, 1935. MacKenzie Air Service, Ltd. (CF-AWR Canada). Photo from the Frances Rees Hayford collection.

Bellanca Aircruiser *mystery ship. The back of this photo has a Bellanca label marked, "Bellanca all-purpose* Aircruiser. *Geared 750 hp Pratt & Whitney S2EG* Hornet. *Plane in the service of Constitutional Army of the Republic of Cuba." Photo from the Daniel Koble collection.*

only operated the one airplane until the mine was spent. The airplane eventually found its way back to Canada, via the Charles Babb Company, Glendale, California, where it was registered CF-BTW.

The Canadian Historical Society, which has chapters in Toronto, Montreal, and Ottawa, produces a journal which covers the history of Canadian aviation. A feature article, printed in the Vol. 9, 1971 edition by Paddy Gardiner and Doug Anderson, outlines the complete history of each of the five *Aircruisers* that were registered in Canada. These remarkable airplanes were used by Canadian Airways, British North American Airways, Canadian Pacific Airways, MacKenzie Air Service, and Transair, servicing the otherwise inaccessible Canadian territories. Their ability to operate from pontoons, skis, or wheels allowed them to be operational year-round.

It appears there are two survivors of this remarkable airplane. Serial no. 721, Canadian registration CF-BTW, is now registered in the U.S. as N2191K. It is on display in the Blimp Hangar, Tillamook Naval Air Station Museum, Oregon. The airplane, owned by Jack Erickson, is maintained and flown occasionally. The second *Aircruiser,* also of Canadian registry,

(CF-AWR) has been retrieved by the Canadian Aviation Museum from the Western Canadian wilderness where it crashed in 1947. It is undergoing restoration for static display in their museum.

The main drawback for more sales was the CAB regulation which mandated all scheduled airlines carrying passengers must use multi-engined aircraft. This ruling limited the airplane to corporate, military, and freight use. The last *Aircruiser* built was serial number 723, (on pontoons) registered as CF-BKV. It was sold to MacKenzie Air Service, LTD., Edmonton, Canada, in 1935.

Once again, Bellanca records are in the hands of NASM and are not available for review. There is much controversy between historians as to the actual number of this airplane produced. After extensive research, the author can account for 23 confirmed serial numbers. However, the Cuban *Aircruiser* serial number has never been documented. It is possible more could have been built for export. Until NASM allows the Bellanca history to be made public, no one will know for sure.

SPECIFICATIONS MODEL 66-70

Base price	$32,900
Power plant (supercharged)	Wright *Cyclone***
Rated horsepower	715 hp*
Propeller	3-blade, adjustable
Fuel capacity	200 gallons
Normal range	700 miles
Fuel consumption	42 gph
Wingspan	65'
Length	43'4"
Height	11'9"
Tread	216"
Wing & strut area	664.2 sq/ft
Empty weight	5,983 lbs
Gross weight	10,853 lbs
Maximum speed	167 mph
Cruise speed	155 mph
Landing speed, with flaps	58 mph
Service ceiling	16,000 ft
Number of passengers	15

* Price raised to $35,800 with Wright 850 hp engine.
** Pratt & Whitney S2EG, 750 hp *Hornet* also available. Wright *Cyclone*, 850 hp available in model 66-75.

Navy XSE-1 Scout. *Serial no. 501. Photos from Daniel Koble collection.*

1932 Bellanca NAVY XSE-1 (No ATC number issued)

The U.S. Navy invited the aviation industry to submit bids for building a new scout plane in 1931. Bellanca won the contract and built the XSE-1 ("X" for experimental, "S" for scout and "E" for Bellanca). Power for this airplane was the 9-cylinder, 575-hp Wright *Cyclone F* engine. The Bellanca design was a nontraditional, high wing sesquiplane. It made its maiden flight December 15, 1932. Stewart Chadwick, test pilot, died on Bellanca Airfield when he failed to recover from a flat spin, February 28, 1933. The remains of this airplane were destroyed in the same hangar fire that consumed the Bellanca WB-II *Columbia.*

A second airplane was constructed XSE-2 (serial no. 502) in 1934. A comparison of the photos indicates this airplane was an entirely new design. One change that is particularly evident is a completely new empennage and heavier aft fuselage. The engine in this airplane was also changed to the 765-hp Wright twin-row, 14-cylinder, R-1510. The XSE-2 made its maiden flight on December 23, 1934. This airplane was delivered to the Navy in Washington, D.C., on April 26, 1935. It was the only one built. Its fate is unknown at this writing.

This is the only Navy XSE-2 built by Bellanca. Photo from National Air & Space Museum.

Navy XSE 1

1932, Marines RE-3. National Air and Space Museum photo.

U.S. Navy JE-1 Bellanca. Photo from Daniel Koble collection

1932-1937 Military RE-1,2,3

The three RE aircraft Bellanca built for the military were essentially variants of the Bellanca CH-400. Each varied from one another as required by their special missions. The RE-1 was used for high altitude radio research and included a Townend *speed ring* around the engine for improved cooling. RE-2 was used by the Navy brass for staff transportation. The RE-3 was fitted with two stretchers for ambulance work and used by the marines. Exact specifications and engines used are not available.

In 1937, the U.S. Navy ordered a JE-1 Bellanca for more high altitude radio testing. This one-and-only airplane was fitted with a Pratt & Whitney R-1340, 570-hp, supercharged engine. It was basically a *Senior Skyrocket* model CH 31-55 with Navy-requested alterations. The airplane was used for high altitude radio surveillance. The aircraft was stationed at Anacostia Navel Air Station, Maryland.

Apparently Bellanca had the airline market in mind when the front of this photo was air-brushed to simulate a possible change in the forward cockpit. None were known to be built in this manner. Photo from the National Air & Space Museum.

This photo shows the three model 77-140 Bellancas on floats built for Columbia. The planes were powered by two 715-hp, Wright Cyclone *engines. Photo courtesy of Mike Guididas.*

1934-35 BELLANCA 77-140 patrol planes (No ATC number issued)

SALES TO COLOMBIA

In spite of the depression years, Giuseppe Bellanca enjoyed a considerable domestic and export business. Along with the renowned Bellanca CH models, which were often used as bush airplanes, the 77-140 was born. The Government of Colombia agreed to buy four model 77-140s. Three were built as float planes and one as a land plane. It would be Bellanca's first attempt at building a multi-engine aircraft of this type. Power was supplied by two Wright *Cyclone* 715-hp, radial air-cooled engines with electric controllable-pitch propellers. Designed as armed patrol bombers, the four aircraft had serial numbers 1201 through 1204.

Mike Guididas, Kennett Square, was employed by Bellanca Aircraft as an aircraft assembler during the construction of these beautiful planes. He remembered Frank Bellanca (Giuseppe's brother) telling

Serial Number 1204, land-based, model 77-140 bomber or patrol airplane. Note the forward gunner position, which is a carryover from WWI. Photo from the Daniel Koble collection.

him that the Colombian airplanes were paid for with coffee beans. The beans, he said, were shipped from Colombia to a New York broker where they were sold. The proceeds of the sale were then transferred to Bellanca. Unfortunately, no specifications are available for this unique airplane.

BELLANCA 1934
Model 28-70 (Serial no. 902) *Irish Swoop*

The *Irish Swoop* was designed to compete in the 1934 MacRobertson Race from London to Australia. The announcement of a free-for-all race between London and Australia in 1934 brought a variety of aircraft and pilots to accept the challenge. The De Havilland *Comet,* Boeing 247, Douglas DC-2, Granville Brothers *Gee Bee* and, most interesting to Delaware, the Bellanca *Irish Swoop,* were a few of the contenders.

The *Irish Swoop,* as it came to be known, was built by Bellanca for The Irish Hospitals Trust and piloted by Captain Bonar, who hoped to win the coveted London to Australia race. This would be Bellanca's first attempt to design and build a low-wing, thin-airfoil, speed-wing, retractable landing gear monoplane. Bellanca had never attempted to build a cantilever wing from wood and decided to use streamlined flying wires to support the wing structure. To accomplish this on his low-wing monoplane, he created belly-mounted pylons to hold the lower wires. The upper wires were mounted to the fuselage in the conventional manner. The sleek airplane was designed to carry a pilot and navigator in tandem seating. The engine would be the latest Pratt & Whitney JR-23 *Twin Row Wasp,* radial air-cooled engine.

Test flights from the Bellanca Airfield proved successful, but few were flown due to a shipping deadline. The *Irish Swoop* was given race number 29 and shipped to London for assembly and further testing prior to the race. Unfortunately, cowling problems which could not be overcome before the race scratched the flight. The airplane was later returned to Delaware for further testing. Unfortunately Cap-

James A. Mollison, left, and Spalding Allen with the Bellanca model 28-70 Irish Swoop. *Photo courtesy of Frances Rees Hayford.*

tain Bonar came in for a landing too fast to stop the airplane and ran off the end of the Bellanca Airfield runway. The accident caused major damage and the airplane was stored. Jimmy Mollison rescued the airplane October 15, 1936, and rebuilt it completely. Dick Morgan test flew it once again from Bellanca Airfield in late 1936. After the test flights were completed, Mollison flew the *Irish Swoop* across the Atlantic, setting a new speed record in the process. Mollison later flew another record flight from England to South Africa. It is not known what happened to the airplane after this recorded flight.

1936-39 BELLANCA Model 28-90 *Flash*

The Bellanca 28-70 *Irish Swoop* design lived on in an improved version, the model 28-90 *Flash.* It is an airplane cloaked in mystery, intrigue, and deception. The deception came from Bellanca, who announced the airplanes were being built as high-speed mail planes for Air France. The State Department blocked their delivery when it learned they were destined for the Republicans in the Spanish Civil War. Another attempt was made by a fictitious Greek firm in South America to get them out of the country, but once again the permits were denied.

Nineteen were known to be shipped to Hankow, China, along with Ed "Smitty" Smith, a Bellanca mechanic, for assembly and maintenance. Smitty had a colorful career in Delaware aviation. He was the corporate pilot for the du Ponts, head mechanic for Bellanca, chief mechanic for the Rehoboth CAP, and Chief mechanic for Summit Aviation.

The Chinese were at war with Japan, and two 30 caliper machine guns were mounted in the wings and one flexible machine gun in the rear. In addition, four bomb racks under each wing would carry eight

120-lb bombs. The first six were destroyed in a hangar by the Japanese during a bombing raid when they uncovered the operation. In researching material on the airplane, production figures appear to be at forty-two. It also appears the first twenty were built without wing flaps. The remaining planes were designated 28-90B models which included split wing flaps to lower the landing speed. These were shipped to Vera-Cruz, Mexico, where they were purchased by the Mexican Air Force. Power was supplied by a 900-hp, twin row, Pratt & Whitney *Wasp Senior* engine.

The *Flash* was very similar to the *Irish Swoop* in design concept. The retractable landing gear, however, was changed to fully enclose the wheels when retracted. This change improved top speed to a reported 280 mph. Another unique innovation was a retractable oil-cooler which could be deployed into the windstream if the engine oil temperature began to rise when using full power. An oddity for this type of airplane were the wire-braced wings, which required struts mounted below the fuselage.

Production came to an end when Congress passed the Neutrality Act in 1939 which forbade any U.S. business from supplying war materials to any belligerent nation. No known units are believed to have survived.

28-90 SPECIFICATIONS

Engine	Pratt & Whitney
Horsepower	900
Wingspan	46'2"
Length	25'11"
Height	8'8"
Wing area	279 sq/ft
Empty weight	4330 lb
Gross weight	6755 lb
Rate of climb (at gross weight)	3760 ft/min
Ceiling	30,500 ft
Maximum speed	280 mph*
Cruise speed	240 mph*
Fuel capacity	150 gallons
Range	700 miles

*These speeds, although taken from Bellanca information, appear to be exaggerated for a steel tube, wood and fabric airplane.

A row of Bellanca 28-90 Flash *airplanes marked for export delivery to Air France. Photo courtesy of Frances Rees Hayford.*

1936 BELLANCA Model XSOE-1

In 1935, Bellanca bid on a U.S. Navy design to build an observation scout seaplane. Since the Navy design included scaled drawings, Bellanca was required to finish the design and enter production for one airplane. The U.S. Navy, under contract no. 38637 designated the airplane XSOE-1 (X for experimental, S for Scout, O for Observation, and E for Bellanca). This airplane was fitted with a Wright *Cyclone* R-1820-84 engine. The airplane was grossly overweight and many attempts to rig the wings proved futile due to inadequate tie-rods. The airplane, serial number 503 (Navy 9728), was completed on March 28, 1936. It was never test flown. On July 15, 1938, George Haldeman taxied the XSOE-1 up the Delaware River to the Philadelphia Navy Yard where it was scrapped. The Navy should have named it the "Penguin."

1937 BELLANCA Model 28-92

Alex Papana, a renowned Rumanian aerobatic pilot who performed at air shows with the German-built *Jungmeister*, wanted a fast, single-seat airplane for his personal use. Once again, it was Giuseppi Bellanca who was approached to build the airplane. Meetings with Papana concluded the airplane would be a single-seat, tri-motor monoplane with retractable landing gear and cantilever wing. The two outboard engines were 250-hp supercharged inverted *Menascos* and the center engine was a 420-hp supercharged V-12 *Ranger*. All three engines were equipped with controllable-pitch propellers.

The airplane was registered in Rumania as YR-AHA and test flown by Papana from the Bellanca Airfield in New Castle in 1937. Alex Papana was a superb aerobatic pilot who could put his *Jungmeister* through any aerobatic maneuver of the day with precision. However, he did not have a history of flying fast, supercharged airplanes with controllable pitch propellers. On his first test flight from Bellanca Airfield, he over-boosted the engines and overshot the runway on landing. Fortunately, the airplane was not damaged, except for burned-out engines.

Art Chester was flown in from the Menasco factory, and he overhauled the engines at the Bellanca facto-

The XSOE-1 was built for the U.S. Navy as an Observation Scout plane by Bellanca Aircraft in New Castle, Delaware. Photo from the Daniel Koble collection.

The Bellanca model 28-92 racer.
Top Photo from the National Air & Space Museum. Bottom Photo from the Daniel Koble collection.

ry. The incident caused a flap between Bellanca and Papana which caused the latter to cancel the order. Giuseppe Bellanca harbored deep feelings about his airplanes. If a pilot abused a Bellanca, and Giuseppe heard about it, it was a certainty the pilot would hear from him.

Giuseppe Bellanca then decided to prepare the airplane for the 1938 Bendix cross-country air race and reregistered the airplane in the U.S. as NX2433. Although off to an excellent start, pilot Frank Cordova was forced to land in Cleveland with a burnt-out *Menasco.* The engine was so badly damaged it had to be removed from the airplane and weights installed to balance the airplane. Cordova flew the airplane back to Delaware on the two remaining engines for repair.

By 1939, the airplane was ready for another attempt at winning the coveted Bendix cross-country event. Art Bussy piloted the Bellanca racer to a second place victory behind a new military Seversky fighter. The Seversky was the first in a series of military fighters which would evolve into the mighty P-47 *Thunderbolt* used in WWII. Bussy flew from Los Angeles to Cleveland in 8 hours and 21 minutes averaging 244.486 mph. After refueling in Cleveland, Bussy flew on to New York to finish. Second place prize money was a mere $5800, which barely paid the operating costs for the race.

After this historic flight, the Bellanca racer was stored in a hangar on Bellanca Airfield and never raced again. It was reported in a 1968 issue of *Sport Flying* that the airplane was sold to the Argentine Government. This is the only information uncovered concerning the fate of this remarkable airplane.

A NEW AIRPLANE FOR A NEW ERA

Sales of the infamous Bellanca CH, in spite of having gone through many changes since its introduction in 1928, had now dried up. Now obsolete, the CH series production was limited to special orders only. The single-engine Bellancas had stretched the limits of breaking almost every aviation record in the books. Other manufacturers, like Lockheed, Beechcraft and Spartan, offered the wealthy buyer sleeker, faster, and roomier airplanes. The airlines and air freight market was now dominated by all-metal airplanes built by Lockheed and Douglas.

Bellanca was in trouble and had to come up with a new product which he hoped would sell in large numbers.

Buyers appeared to have an insatiable quest for speed and economy. For 1937, Bellanca focused on the low cost, three-seat airplane market. The airplanes in this market, like Stinson, were of old designs. They went through annual upgrades, much like the Bellanca CH, but still did not offer the owner high-speed travel at low cost. Giuseppe Bellanca decided he would challenge this market with a completely new design that would outclass the competition by a wide margin.

1937 BELLANCA 1937 Model 14-7

Although there were several manufacturers offering many different versions of what they felt the private owner wanted, Bellanca's new airplane design would incorporate many of the ideas he had put into production on previous models. This is where he held steadfast to his steel tube, wood and fabric airplane construction with which he and his employees were familiar. However, the all-new 14-7 would incorporate some innovations that had never been offered on a Bellanca before. The cantilever 14-7 wing would be constructed without the use of wires or struts for supports. The airplane would also have a manually-operated, retractable landing gear to improve speed. On the ground, or flying, it would give the impression of being fast, and it was.

The first 14-7 prototype was produced in early 1937 and was issued serial number 1001. The prototype unfortunately fell short of meeting Bellanca's goals. The 14-7 was built with a flat wing (no dihedral) and fixed landing gear. All indications suggest that it was fitted with the five-cylinder, 70-hp, LeBlond 5E, radial air-cooled engine. The wing was constructed with wood spars and a plywood leading edge, then covered with fabric. LeBlond later sold out to Rearwin, who built the *Cloudster* and the *Sportster,* monoplanes. Bellanca continued to use the LeBlond/

Bellanca 14-7 takes off at the Chicago International Air Show, 1937. Note the tail skid, no outboard stabilizer fins, and no wing flaps. Photo courtesy of Frances Rees Hayford.

Rearwin engine (now named the Ken-Royce after Rearwin's two sons) in 14-9 production airplanes.

The first 14-7 was lost in a high-speed dive test killing test pilot Cecil Hoffman when the fabric ripped off the wings. To overcome the flight problems which were evident with the prototype, serial number 1002 was given four degrees dihedral and the entire wing was now covered with plywood before it was covered with fabric. The added dihedral caused the landing gear to appear spraddle-legged. Although it looked awkward, it did not affect ground handling and was never changed. Two years of extensive test flights by test pilot Holger Hoiriis, in serial number 1002, included spins. Entry into spins with 1002 came far too early to be acceptable by the CAB. Ironically, serial number 1001 did not have this trait and was approved in the spin category by the CAB before it crashed. Wing root and cowling fairings were changed to no avail. Finally, two kidney-shaped outboard stabilizer fins were added. This modification, along with limited elevator up-travel when the landing gear was retracted, virtually eliminated the spins. These small end-plates would become a distinctive trademark for this Bellanca model. All this testing consumed two years of time and funding that Bellanca could ill afford.

It is interesting to note that Giuseppe Bellanca approached the problem of spins in the same manner as Fred Weick, designer of the *Ercoupe.* In 1937, Weick also installed dual outboard fins on the prototype *Ercoupe* and limited elevator up-travel to eliminate spins. It is not known if Bellanca and Weick discussed the problem jointly. Although the Bellanca 14-7 would never be certified for production, it was the beginning of a long line of personal aircraft which would evolve into the finest Bellancas ever built.

BELLANCA 14-9 Junior

According to original factory literature, published February 1938, the Bellanca *Junior* would be offered with two engine options. The model 14-7 was equipped with the five-cylinder 70-hp LeBlond 5E and the model 14-9 was equipped with the seven-cylinder, 90-hp LeBlond 7F. The first 14-9 *Juniors* were not equipped with wing flaps. Wing flaps were added after serial number 1032 in 1940. Following Bellanca tradition, changes were incorporated during production as the airplane was put into service. The fail-safe chain and jack-screw retractor required 37 turns to tuck the landing gear into the wing wells and was virtually foolproof. Failure to retract the gear would reduce the airspeed by 20 mph. The fac-

This 1940 model 14-9 Bellanca Junior, *serial number 1033, was the first one to be built with wing flaps. Photo from the Daniel Koble collection.*

tory literature also states the retractable landing gear was an option. In spite of this notification, no fixed gear models were known to be produced.

The little three-seat Bellanca *Junior* was a small airplane. Much like a sports car, it was comfortable once you were planted in the seat. It was definitely not an airplane for large people. The cabin width was a mere 35.5 inches wide. Headroom space was also a problem for those over six feet tall with a cabin height of 43.5 inches. Controls were of the dual stick with rudder pedals variety. The right stick was removable for added passenger comfort. Fuel was located in one 22-gallon tank in the left wing only. The fuel gauge was of the external plunger type and viewed by the pilot through the side window.

What the *Junior* lacked in creature comfort and space, it made up in speed by cruising 120 mph with a mere 90-hp engine. However, the *Junior* was still an under-powered airplane, especially at 1650-lb gross weight, with three people and full fuel. Although the prototype flew in 1937, the CAB approved the first three-passenger Bellanca *Junior* for production on August 24, 1939.

The following specifications were copied from Original factory literature dated February 1938. In spite of the factory literature detailing the availability of the model 14-7, none were known to be built.

MODEL 14-7 SPECIFICATIONS (Feb. 1938)*

Wingspan	34'3.5"
Length	21'3"
Height	6'4"
Empty weight	912 lbs
Gross weight	1650 lbs
Wing area	140.2 sq/ft
Engine	LeBlond 5E
Number of cylinders	5
Horsepower @ 2075 rpm	70
Ceiling	12,000 feet
Maximum speed	115 mph
Cruise speed-75% power	105 mph
Rate of climb	600 fpm
Range	500 miles
Landing speed	45 mph
Fuel capacity	22 gallons
Passengers	3

*Specifications taken from factory literature.

Three of the 38 *Juniors* produced were 14-9L models (serial no. 1035, NC25313, serial no. 1036, NC25314 and no. 1037, NC25315), with the 95-hp Lenape radial air-cooled engine in early December 1940. These three airplanes were a special order for State Airlines, Inc., Charlotte, North Carolina.

Bellanca had a history of changing engines to improve performance or meet cost targets. Bellanca built one model 14-10 (serial no. 1039, NX25307) with the Lycoming 0-235A1 opposed four-cylinder, 100-hp engine. A full page advertisement was placed in the June 1941 issue of *Flying and Popular Aviation* showing the airplane in flight. Only one *Junior* was equipped with the Lycoming engine. Since the advertisement for the Lycoming-powered *Junior* appeared in June 1941, the impending war would probably be the reason no others were produced.

In late December 1940, the Franklin six-cylinder, flat-opposed, air-cooled, 130-hp engine was installed. This model was designated 14-12F3 and remained in production until the war broke out in 1941.

MODEL 14-9 SPECIFICATIONS (Feb. 1938)

Wingspan	34'3.5"
Length	21'3"
Height	6'4"
Empty weight	943 lbs
Gross weight	1650 lbs
Wing area	140.2 sq/ft
Engine	LeBlond 7F*
Number of cylinders	7
Horsepower @ 2250 rpm	90
Ceiling	14,750 feet
Maximum speed	130 mph
Cruise speed-75% power	120 mph
Rate of climb	800 fpm
Range	500 miles
Landing speed	45 mph
Fuel capacity	22 gallons
Passengers	3

*Also named the Ken-Royce.

1939 BELLANCA Model T14-14

The U.S. Army made an open request in 1938 to all airplane manufacturers that the Air Corps needed a new primary trainer to train pilots. The specifications called for a tandem-seat, open-cockpit trainer capable of withstanding the rigors of student pilotage. The fly-off would be held at Wright Airfield, Dayton, Ohio, on May 15, 1939. Bellanca joined seventeen other manufacturers in vying for the acceptance of the Army Air Corps contract for a primary trainer with the entry of the Bellanca T14-14. It is interesting to note in the construction photo that another fuselage is hanging from the wall. This suggests the possibility of a second airplane's being produced. However, no records could be located to confirm the existence of the second airplane.

Bellanca built one airplane, NX28994, which used the wood wing, including flaps, from the model 14-9 *Junior.* The six-cylinder, 0298 Franklin 130-hp air-cooled engine was used for power. In typical Bellanca tradition, the fuselage and empennage were welded steel tubing covered with fabric.

After the fly-off, the U.S. Army Air Corps accepted the Fairchild PT-19 *Cornell* as their trainer. Fairchild would build 4889 PT-19s which could be purchased after the war for as little as $500. Approximately 272 remain in service today. As for the lone Bellanca entry, it is in original condition in a hangar in Alabama as this is written.

SPECIFICATIONS T14-14

Seats (in tandem)	2
Wingspan	34' 10"
Length	23' 8"
Height	70"
Engine	Franklin
Cylinders (opposed)	6
Horsepower @ 2550 rpm	130
Rate of climb (gross weight)	650 fpm
Fuel capacity	40 gallons
Range	500 miles
Oil capacity	8 quarts
Empty weight	1223 lbs
Gross weight	1950 lbs
Maximum speed	170 mph
Cruise speed	110 mph
Stall speed (flaps down)	48 mph
Stall speed (flaps up)	50 mph

This photo shows the T14-14 under construction. Photo from the Daniel Koble collection.

The sole Bellanca T14-14 trainer as it appeared in 1939 prior to the fly-off. Photo from the Daniel Koble collection.

The Bellanca entry for a short takeoff and landing (STOL) was the YO-50. Photo from the National Air & Space Museum.

BELLANCA Model YO-50

Undaunted, Bellanca once again, in 1940, submitted an entry to the U.S. Army Air Corps for a STOL (short takeoff and landing) observation aircraft (YO-50). The Army Air Corps decided they needed a STOL airplane after seeing the German-built Fiesler-Storch perform from unimproved battlefield sites. The Bellanca-designed, high-wing monoplane was powered by an inverted 420-hp model V-770-1 engine. It was also equipped with full-span wing slots and Fowler-type trailing edge flaps. The design allowed the airplane to land at a ridiculous speed of only 27 mph. Bellanca built three YO-50s, serial no. 40-741 through 40-743. The Army Air Corps decided to purchase the Stinson L-1 entry, which was a make-over of an existing civilian airplane, which cost considerably less than the Bellanca entry. The specifications and ultimate fate of the three YO-50 aircraft are unknown.

As the war progressed, the Army purchased additional liaison aircraft from other manufacturers. Stinson, (a division of Consolidated Vultee Aircraft) was first with the L-1, followed by the Taylorcraft L-2, Aeronca L-3 and Piper L-4. Bellanca, who sorely needed this contract was once again left out in the cold. However, the war years would change Bellanca's problems from no production to excessive production.

THE BELLANCA WAR YEARS

Prior to WWII, Bellanca used a synthetic resin thermo-bonding material for gluing wood joints, called "casein." But this resin, made of animal and vegetable adhesives, was prone to mechanical failure due to repeated expansion and contraction of the joints. In addition, decomposition due to bacteria, molds, and fungi created additional shortcomings. Wood had never been a problem in aircraft construction. It was the glued joints which were questionable. Although Bellanca took extensive precautions by controlling the heat and humidity while the glue was setting, it was a labor-intensive ordeal. The glued joints would have to be cured for ten hours at 140 degrees. After curing, Bellanca painted all the wood with a moisture-proof shellac to ward off moisture. Not one Bellanca was known to have a structural failure at this time. However, the crash of a Fokker, which took the life of Knute Rockne the famous Notre Dame football coach, created a ripple effect on the safety of wood when the CAB grounded the Fokkers due to structural failure at the glued joints.

Pennsylvania Coal Products Company introduced a new product called *Penacolite* in 1938 which had all the properties of a synthetic glue and only took 70 degrees for four hours to cure. More important, the glue was also impervious to moisture, fungus, alkali,

The Bellanca-built AT-21, nicknamed The Gunner, *was used as a trainer for gunners and bombing practice during WWII. Photo from the Daniel Koble collection.*

acids, and bacteria. Bellanca reported that the glue joint was as strong as if it were cut from a single tree. Tests of glued joints were placed on the roof of the Bellanca factory by Andrew Haiduck, engineer, exposing them to all the elements of nature for as long as two years. In addition, the glued joints were placed in boiling water for several hours. The tests proved beyond any doubt the glued joint was stronger than the wood itself. With Bellanca's history of building quality aircraft with wood, Fairchild Aircraft allowed Bellanca to join them in the production of the Army Air Corps AT-21. The AT-21 was a multi-engine advanced trainer that was used for bombing and gunnery training. Pilots often referred to these airplanes as "Rhapsody in Glue."

AT-21 SPECIFICATIONS

Engines	Ranger, V-770
Number of cylinders	12 each
Horsepower	520 each
Wingspan	53 feet
Wing area	375 sq/ft
Empty weight	8,637 lb
Gross weight	11,287 lb
Maximum speed	210 mph
Cruise speed	184 mph
Rate of climb	1,000 ft/min
Service ceiling	20,000 feet
Crew	5

Pre-manufacturing of wiring harnesses was created by women employees. This elaborate system checked out all the circuits before they were installed in the airplane. Photo from the Daniel Koble collection.

A new 100,000 square-foot building was erected to produce the AT-21. The outer skins were made of preformed plywood to save aluminum for war plane production. Photo from the Daniel Koble collection.

To accommodate production of the AT-21, major changes were incorporated in the Bellanca factory. Another 100,000-square-foot plant was built on the site and named "Plant 2." This plant would be used solely for the production of the AT-21. *The Gunner,* as it was called, would be built from as many non-strategic materials as possible. The outer skins were made of preformed molded plywood to save aluminum for the airplanes that were built for combat. The AT-21 contract would account for $16,000,000 of revenues for Bellanca. In addition to the plant, 2500 feet of runway were paved to allow the finished bombers to depart from the factory site. Bellanca Airport would now become the first, privately owned airport in Delaware with an all-weather landing surface. The first AT-21, armed with four 30-caliper machine guns, departed Bellanca Airport January 2, 1944. In addition to a multi-engine trainer, the AT-21 was also adapted for target towing. Of the 175, AT-21s constructed during the war, 106 were built by Fairchild, 39 by Bellanca and 30 by McDonnell. None are known to have survived.

BELLANCA-BUILT SUB-ASSEMBLIES

Bellanca also built sub-assemblies for Republic, Glen L. Martin, Grumman, Douglas, RCA, Philco, and the military in plant number one. In addition to wood and metal, Bellanca laboratories created moldable plastic laminates. At the peak of production, Bellanca would hire 3000 employees to produce a wide variety of aviation-related materials. Gun tur-

Bellanca's postwar entry was the improved 14-13 Cruisair. *Photo by Hans Groenhoff.*

rets, wing floats for the Martin TBM and JRM (Mars), bomb-bay doors, gasoline tanks, ailerons and flaps for the C-46 and C-109, and engine cowlings are a few of the items manufactured by Bellanca. In one military contract after the war, Bellanca produced the high-speed XM-24 tow target. The XM-24 was twenty feet long and was towed at 400 mph behind an Air Force B-47. This V-tail design was the forerunner of the lifting body that was tested by NASA.

POST WAR BELLANCAS

The aviation industry as a whole felt there would be a postwar boom beyond everyone's wildest dreams. The theory was that the thousands of military flight instructors and pilots would want airplanes after returning to civilian status. The military flight instructors would be carving out new privately owned, public-use airfields all over the country to continue their profession. Manufacturers, including Bellanca, sent out thousands of marketing letters to confirm their convictions. The war was won by air power, and the romance of the aviators coming home would hopefully create a new market for buyers. Manufacturers assembly lines geared up for volume production to meet the anticipated demand. What was not taken into consideration was that all the other aircraft manufacturers geared up for this demand as well. Additionally, the War Assets Administration was selling military aircraft for virtually nothing. In a couple of years, production exceeded demand by a wide margin and airplane inventories overflowed. The predicted boom indeed happened, but it was short-lived.

After the war, Bellanca resumed production with an improved version of the 14-12F3, which was designated model 14-13 *Cruisair.* The 14-13 was produced with the 150- or 165-hp Franklin six-cylinder engine and was now a true four-place airplane. The triple tail was also enlarged, and the kidney-shaped outboard fins became oval. The Bellanca *Cruisair* base price was $6,350 with a list of many options, including an electrically-operated, retractable landing gear for $350.00. However, the fail-safe, manually-operated version was preferred.

Bellanca faced heavy competition with the $6,350 *Cruisair* when Beechcraft offered the all-new *Bonan-*

The last Delaware-built Bellanca was the 260 hp model 14-19 Cruisemaster. *Photo courtesy National Air & Space Museum.*

The AviaBellanca Sky Rocket, the all-composite airplane completed by Augie Bellanca. Photo from the Daniel Koble collection.

za for virtually the same price. The *Bonanza* offered many new features, such as, tricycle-retractable landing gear, and all-aluminum construction. The *Bonanza* was also roomier, faster, and more powerful. For the first time in Bellanca history, Giuseppe built airplanes faster than they could be sold. This created a fleet of airplanes that sat outside in the open weather while Bellanca looked desperately for sales. Many were flown across the Delaware Bay to Cape May, New Jersey, to be modified with a baggage compartment and outside access door to match the competition's standard features. To add to Bellanca's grief, the United Auto Workers, local 810, organized a union at the Bellanca plant. For the next six years Bellanca operated in the red. This, and an unforeseen economic downturn, ruined Bellanca Aircraft.

The Bellanca airplane production by 1950 was reduced to a mere trickle. Massive layoffs followed. Bellanca was not alone. Piper, Aeronca, Stinson, Ercoupe, Taylorcraft, Globe, Culver, Navion, Cessna, Funk, Porterfield, Rearwin, and Fairchild all fell on hard times. Only four would weather the storm in the small personal airplane market: Piper, Beechcraft, Mooney, and Cessna.

Modifications and changes in engine horsepower were made to the *Cruisair* over the years in hopes of increasing sales. The *Cruisemaster,* with a larger engine, followed in a vain attempt to recapture market share. A lack of capital negated any opportunity to design a new airplane and have an ATC approval. The last production airplane produced by Bellanca in Delaware was the *Cruisemaster,* 14-19-2. This model had the 230-hp, Continental 0-470 engine installed and is now a highly sought antique. Bellanca saw the handwriting on the wall and sold the Bellanca Aircraft Company in 1954. On October 23, 1956, Giuseppe Bellanca sold the physical plant, land, and remaining military contracts to Piasecki Helicopters (renamed VERTOL), Morton, Pennsylvania. Bellanca only produced 103 *Cruisemasters* during the 1950s. At the end of 1995, approximately 75 were still registered with the FAA.

Northern Aircraft, Inc. purchased Bellanca Aircraft in 1954 and moved all the tooling to Alexandria, Minnesota. Northern built 44 *Cruisemasters* there. The same year, Northern merged with American Aviation and renamed the company Downer Aircraft Industries. Downer produced the airplane as the model 14-19-3 *Viking* and installed a 260-hp Continental engine. Downer continues to use the world-famous Bellanca name in what will eventually be the last of the Bellanca; the *Viking.*

After the sale of the company, Giuseppe Bellanca retired to his Eastern Shore home on the Sassafras River near Galena, Maryland. Giuseppe and his son August (Augie), formed a new company *AviaBellanca, Inc.* and proceeded to design a new fiberglass composite, six-place airplane named the *Sky Rocket.* Unfortunately, Giuseppe passed away from leukemia in 1960, at the age of 74, before the airplane was completed. Augie completed the prototype in 1973 and started FAA certification. NASA also evaluated the *Sky Rocket* in April of 1983. The conclusion of the NASA evaluation confirmed that the laminar-flow wing on the *Sky Rocket* increased the cruise range by twenty-five percent. The sleek low-wing *Sky Rocket* broke five FIA records in 1975.

	OLD RECORD	NEW RECORD
FIA Class C-1d (3858-6614 lbs)		
500 km closed course (mph)		
	304.1 *Aerostar*	327.7 *Sky Rocket*
1000 km closed course (mph		
	305.2 *Aerostar*	313.5 *Sky Rocket*
FIA Class C-1c (2204-3857 lbs)		
100 km closed course (mph)		
	260.5 *Ambrosini*	285.3 *Sky Rocket*
500 km closed course (mph)		
	255.8 *Ambrosini*	298.4 *Sky Rocket*
1000 km closed course (mph)		
	216.1 *Ambrosini*	302.8 *Sky Rocket*

The FAA certification process (ATC) is so complex and lengthy, it is not something one person can afford financially. At this writing, the AviaBellanca *Sky Rocket* resides in a hangar at Summit Airport, Middletown, Delaware, while Augie Bellanca seeks venture capital to begin production of the airplane in kit form.

Giuseppe Bellanca was an aeronautical genius. He spent his entire life searching for the pure streamline. He was a shy person who sought no personal publicity for his efforts. His only fault, if you can call it one, was his constant search for perfection. He was an artist. He openly supported women in aviation. He met Elinor Smith in New York when she was flying her father's Waco biplane at the age of fifteen. In a flight from Roosevelt Field, Eleanor Smith, at age 17, broke the Women's Solo Endurance Record in 1929, flying a Delaware-built Bellanca CH-300. Other company pilots included Clarence Chamberlain, Mr. Shirley Short, J. Allison Buck, George Haldeman, Clyde Panghorn, and Roger Williams.

Fortunately, the Delaware investors saw this talent and allowed Bellanca to excel in what he did best; design and build airplanes. Giuseppe Bellanca was nominated, with 107 others, to the prestigious National Aviation Hall of Fame in 1993. Only four were selected to receive the honor. Giuseppe's son, Augie, accepted the award posthumously on behalf of his father.

Between 1928 and 1954, Bellanca Aircraft, Inc. produced approximately 3000 aircraft of various types in Delaware. August "Augie" Bellanca donated all the Bellanca records and artifacts to the Smithsonian Institute in July of 1994. The collection consists of approximately 423,000 individual historical drawings, photos, models, trophies, etc. The specifications and performance data for some of the airplanes built by Bellanca will be unavailable to researchers until NASM catalogues the collection and allows the public access. As this is written, NASM has not started on the Bellanca collection.

In-depth research has asserted that although many aviation writers call Giuseppe Bellanca an aeronautical genius, they mention he was a poor businessman. However, it should be pointed out that Bellanca survived the depression years; whereas Henry B. du Pont's Verville failed. Waco, Mahoney/Ryan, Fleet, Kinner, Buhl, Standard, Travel Air, Howard, Monocoupe, Spartan, Rearwin, Pitcarin, and many other great names in aviation went bankrupt, but Giuseppe Bellanca survived. Although G.M. sold the company and has passed away, the Bellanca *Viking* is still being produced as this is written. That represents 70 years of Bellanca production history!

BELLANCA RECORDS

1922-Bellanca CF won thirteen first place prizes in national and international contests.

1923-Bellanca CF won the Efficiency Contest, National Air Races, St. Louis. This airplane is restored and on display at the Paul E. Garber National Air & Space Facility, Silver Hill, Maryland.

1925-Bellanca WBI won the Efficiency Contest, National Air Races, New York.

1926-Bellanca WBII won *ALL* the Efficiency Contests, National Air Races, Philadelphia.

1927-Bellanca WBII *Columbia* established a new World Endurance Record, 51 hours 30 minutes.

1927- Bellanca WBII *Columbia,* piloted by Chamberlin & Levine, established a new World Distance Record-New York to Eisleben, Germany, 3911 miles.

1928-Bellanca CH-200 won both Efficiency Contests, National Air Races, Los Angeles.

1928-Bellanca J *Pathfinder* named *Reliance,* Piloted by Royal V. Thomas, established a new Solo World Endurance Record, 35 hours 25 minutes.

1928-Bellanca J *Pathfinder* named *Rosemarie* set new World Endurance Record, 59 hours 7 minutes. Piloted by Edward Schlee and William Brock.

1929-Bellanca CH-300. First nonstop flight New York to Cuba in 12 hours 56 minutes. George Haldeman, pilot.

1929-Bellanca J *Pathfinder* named *North Star* flew from Maine to Spain, 3400 miles in 30 hours 30 minutes. Piloted by Roger Williams and Lewis Yancy.

1929-Bellanca J *Green Flash* set new Solo Endurance Record of 35 hours 33 minutes. Piloted by Martin Jensen.

1929-Bellanca CH-300 set Solo Endurance Record for women, 26 hours 23 minutes. Piloted by Elinor Smith.

1929-Bellanca CH won five 1sts, three 2nds, two 3rds, and two 4ths at National Air Races.

1929-Bellanca CH won Cleveland to Buffalo Efficiency Race.

1929-Bellanca CH won the Aviation Town & Country Club Efficiency Contest, New York.

1929-Bellanca CH won the *Detroit News* Efficiency Contest, Cleveland.

1929-Bellanca CH won 1st and 2nd place, Ford Reliability Tour.

1930-Bellanca CH *Pacemaker* established new World Altitude Record of 30,418 feet. George Haldeman, Pilot.

1930-Bellanca CH *Pacemaker* established new Women's Altitude Record of 27,418 feet. Elinor Smith, Pilot.

1930-Bellanca WBII *Columbia.* Nonstop flight from New York to Bermuda and return. Piloted by Roger Williams, Errol Boyd and Harry Conner.

1930-Bellanca CH won both Efficiency Contests, National Air Races.

1930-Bellanca CH won for third consecutive time, *Detroit Daily News* Efficiency Award. This win allowed Bellanca to retain permanent possession of the trophy.

1930-Bellanca CH seaplane won the Imperial Oil Company Trophy for speed and efficiency.

1930-Bellanca WBII *Columbia* Third trans-Atlantic flight Harbor Grace, Newfoundland, to Scilly Isles, England.

1930-Bellanca CH won 1st and 2nd place, Ford Reliability Tour.

1931-Bellanca J2 *Diesel* established non-refueling World Endurance Record of 84 hours 33 minutes. Piloted by Walter Lees and Frederick Brossy.

1931-Bellanca J *Liberty.* NR797W. Fourth trans-Atlantic crossing, Newfoundland to Germany. Piloted by Holger Hoiriis and Otto Hillig. Airplane is reportedly in a Polish Government museum.

1931-Bellanca J *Miss Veedol.* Fifth trans-Atlantic crossing from New York to Cardigan, Wales. Piloted by Clyde Pangborn and Hugh Herndon.

1931-Bellanca J *Cape Cod.* Sixth trans-Atlantic crossing and World's Long-Distance, Nonstop Record. New York to Istanbul, Turkey. Pilots Russell Boardman and John Polando flew 5,011 miles in 49 hours.

1931-Bellanca J *Miss Veedol.* First trans-Pacific flight from Japan to Washington. Piloted by Clyde Pangborn and Hugh Herndon.

1933-Bellanca J *Litvanica.* Seventh trans-Atlantic flight New York to Pomerania, Germany. Piloted by Stephen Darius and Stanley Girenas.

1934-Bellanca J. Eighth trans-Atlantic flight. New York to Lahinch, Ireland. George Pond and Cesare Sabelli, pilots.

1934- Bellanca J *White Falcon.* Ninth trans-Atlantic flight. Newfoundland to Caen, France. Piloted by Benjamin and Joseph Adamowicz.

1934-New York to Newfoundland, Iceland, Orkney Islands, Scotland, Dutch East Indies, Manila, etc.- August-November, 1934 - pilot was Dr. Richard Light-Bellanca CH-400 *Skyrocket,* serial no. 811, NR10794, seaplane.

1936-Harbor Grace, Newfoundland, to London, England-October 30, 1936, fastest trans-Atlantic flight-13 hours 17 minutes, Capt. James Mollinson in Bellanca 28-70 *Flash,* serial no. 902, NR190M.

1939-Bellanca *Three-engine racer.* Second place Bendix Trophy Race from Los Angeles to Newark, N.J., averaging 248 mph.

1947-Bellanca *Cruisair, Sr.* First place, Beechcraft Trophy Race, Miami, Florida.

Starting with the proven design of the WB-II *Columbia,* Giuseppe Bellanca created over the years a complete line of record-breaking airplanes in Delaware. The following list contains the model, CAB issued ATC number and dates of issuance of Bellancas built for sale in the U.S.A. The CAB did not require an ATC number on experimental aircraft or aircraft sold outside the U.S.A. The following production numbers, and specifications listed, are as accurate as the information available allowed.

DELAWARE BELLANCA PRODUCTION

MODEL	NAME	REGISTRATION	MO/YR	QUANTITY	HP	PRICE
Model J	Pathfinder	EXP	9-1927	U	V	V
CH-200		ATC-47	6-1928	18	220	$14,500
TES	Blue Streak	EXP	5-1929	1	2-600s	U
CH-300	Pacemaker	ATC-129	5-1929	60	300	$14,950
CH-400	Skyrocket	ATC-319	4-1930	30	420	$17,950
CH-Sr.	Skyrocket	ATC-550	5-1937	U	550	$21,150
CH-400	RE-1,2,3	NAVY	1932	3	450	U
CH-300W	Pacemaker	ATC-328	6-1930	8	300	$18,400
CH-420E	Pacemaker	ATC-476	4-1932	7	420	V
XSE-1	Experimental	NAVY	12-1932	1	450	U
XSE-2	Experimental	NAVY	1934	1	450	U
CH-D	Skyrocket	ATC-480		U		V
P-100	Airbus	ATC-360	8-1930	1	600	U
P-200	Airbus	ATC-391	1-1931	4	575	$32,900
C27-A	Airbus	Army	1933	14	670	U
66-70	Aircruiser	ATC-563	3-1935	U	715	$32,900
66-75	Aircruiser	ATC-		1	760	V
66-76	Aircruiser	ATC-		1	750	V
66-85	Aircruiser	ATC-	1940	1	850	$35,800
Cuban	Aircruiser	ATC-		1	850	U
CH-31-50	Skyrocket	ATC-565	4-1935	6	525	$21,950
CH-31-55	Sr Skyrocket	ATC-565	4-1935	10	550	V
XSOE-1	Experimental	NAVY	3-1936	1	550	U
CH-31-55	JE-1	Navy	1937	1	570	U
CH-31-42	Pacemaker Sr.	ATC-578	7-1935	4	420/450	$16,800
28-50	Irish Swoop	EXP	-1934	1		U
28-90	Flash	MIL	-1936	42	900	U
28-92	3-engine	EXP	-1937	1	250/350	Racer
14-7	Cruisair Jr.	EXP	-1937	2	90	U
14-9	Cruisair Jr.	ATC-716	-1938	41	90	$3,150
T14-14	Air Corps	ARMY	-1939	1	130	U
14-10	Cruisair	ATC-	-1940	1	100	U
14-12F3	Cruisair	ATC-745	-1941	13	120	$4,500
YO-50	STOL	ARMY	-1942	3	400	U
AT-21	Bomber	ARMY	-1943	39	2/400s	U
14-13-2	Cruisair Sr.	ATC-773	3-1946	588	145	$5,950
14-13-3	Cruisair Sr.	ATC-	6-1947	U	150	U
14-13-4	Cruisair Sr.	ATC-	10-1948	U	150	U
14-19	Cruisemaster	ATC-	9-1949	97	190	U

U=Unknown V=Varied EXP=Experimental MIL=Military

CHAPTER 6

All American Aviation

Richard C. du Pont, 1911-1943

"Yesterday I experienced one of the greatest thrills of my life, soaring! We went down to a steep bluff overlooking the beach and flew up and down the edge of the cliff. It was the most wonderful sensation in the world, flying up and down the beach 200 feet above the cliff top without a sound. The airspeed is only 30 mph. So you don't have to wear goggles. This is the most wonderful sport in the world!"

And so with this partial letter to his mother, Richard du Pont let his family know how he felt about flying. Richard C. du Pont was born in Wilmington, Delaware, January 2, 1911. Like many young men interested in flying, he was first exposed to aviation by building and flying model airplanes. His older brother, A. Felix du Pont, Jr., who held an Air Transport rating, gave Richard his first airplane ride. Although Richard earned his pilot's license in 1930, his love for flying centered around the world of soaring in motorless aircraft. At seventeen he acquired his first glider and felt the thrill of riding thermals. From that day on, soaring would become a passion for satisfying his appetite for flight. His first glider was a frail, homebuilt contraption where the pilot sat out front on a bare frame with wings and tail. He learned to fly it when he was on vacation in Cape Cod in his early teens. Unfortunately, at that age those beautiful summer vacations come to an end all too soon. Richard's soaring would end when he had to return to Tower Hill School in Wilmington in the fall. He also attended the Salisbury Connecticut School. After he finished his senior year at Hun High School in Princeton, New Jersey, his brother Felix taught Richard powered airplane-flying before he left for college at the University of Virginia in 1930.

There is little doubt that Richard picked the University of Virginia because of the terrain. It would make a beautiful spot to go soaring among the mountaintops and play among the clouds. It was not long before Richard established a soaring club on campus. There was a definite conflict between his studies and the campus soaring club. The studies lost out to soaring, and after two years, Richard transferred to the Curtiss-Wright Technical Institute of Aeronautics in Glendale, California, to study aeronautical engineering.

Gliders and sailplanes are one and the same. The difference is in the way each is flown. The glider, or sailplane, is a heavier than air aircraft. Therefore, once airborne, the pilot seeks air currents that are rising faster than his craft is descending. In soaring,

Hawley Bowlus and Richard du Pont pose with the Bowlus "Dragonfly." *It crashed on takeoff at Elmira in 1933, with Richard and Felix du Pont aboard. Neither was injured, but Felix was heard to say, "Someone get me out of this violin case." Photo from the du Pont Collection.*

the air is called "green air." If there are no rising air currents, the craft is gliding. If the craft enters air currents from time-to-time which increase the altitude, the glider is soaring and is now considered a sailplane. Soaring is an exciting sport. Finding an invisible upward air current, and holding on to it, is sheer excitement for the pilot. The objective of the sailplane/glider designer is to achieve the lowest possible sink rate for the craft. The true challenge in soaring begins with the design.

It was at Curtiss-Wright that Richard met Hawley Bowlus and Martin Schempp, headmasters of the school. Hawley had been responsible for hands-on supervision and construction of the *Spirit of St. Louis* after Donald Hall finished the drawings. He also designed and built his own sailplane in which he set an American Endurance Record of nine hours and five minutes in 1929. Hawley also taught Charles Lindbergh and his wife Anne how to fly sailplanes in 1929. Anne Lindbergh was issued women's license number 1 for glider pilots. It was also Hawley Bowlus and Martin Schempp who taught Richard du Pont the fine art of designing, building, and flying these graceful motorless aircraft. Richard wrote home in 1932;

> Gosh it is wonderful out here. This is the kind of life I enjoy more than anything else. It seems funny to actually look forward to going to classes. A feeling I have never had all the years I have been going to school and college. The engineering course is so interesting all I do in my spare time is sit around with Hawley and talk about wing design, fuselages, sinking rates and gliding angles.

Many letters of this nature were sent home to his mother and father. The many personal family letters reviewed for research confirm the closeness of the du Pont family. A. Felix, the father, allowed his children to seek out their own profession in life. Ironically, of the four children, all chose flying except one daughter, Lydia.

The completed Albatross *fuselage and empennage show the detail and sleek lines of this graceful sailplane. Photo from the du Pont collection.*

Richard paid room and board to bunk in with headmaster Hawley Bowlus on the family farm in the California valley. Since Hawley's job at the Institute only paid $150 a month, the extra income certainly must have come in handy. Richard was responsible for washing and ironing his own laundry and keeping his room clean as part of the agreement. In one of the letters father Felix wrote to his son Richard, he elaborated on how important it was for Richard to be like the other students. He also wrote showering praise and support for his son's school grades, which now reflected his keen interest in aeronautics. Richard's letters home bore out his total commitment to school and learning. It was inevitable that Richard du Pont and Hawley Bowlus would design and build their own glider. It would not be just any glider, but a glider that would set records in the world of soaring champions.

The original design of the *Albatross* series was the brainchild of Hawley Bowlus and Martin Schempp around 1930. But lacking the proper funding, the *Albatross* lay dormant until revived by Richard du Pont's interest and was completed in 1932. Although father Felix felt gliders would not provide an income to support Richard's adult life, he did feel positive about the education it would offer his son, and supported the project both enthusiastically and mone-

The prototype (straight wing) Albatross *built by Hawley Bowlus and Richard du Pont in 1932. Photo from the du Pont collection.*

tarily. In the October 7, 1933, issue of the Wilmington publication *Views,* A. Felix du Pont wrote a two-page feature article on the merits of soaring. Although not a pilot himself, (his interests were nautical) he was well-versed on the subject.

BOWLUS ALBATROSS SPECIFICATIONS

Wingspan	62'
Wing area	204.4 sq/ft
Sinking speed	2' per second
Glide ratio	19:1
Length	24'
Height	7' 3"
Weight	347 lbs
Fuselage construction	Monocoque
Fuselage covering	plywood
Split trailing edge flaps	30% of span
Landing speed (with flaps)	28 mph

Winds rushing off the Pacific Ocean toward land slammed into the side of California cliffs and shot upward. The cliffs became favorite spots to launch sailplanes. One *Albatross* flight was scheduled to launch from the noted "Hollywood Riviera" cliff. With Martin Schempp at the controls, the *Albatross* was launched and flown out to sea beyond the rising air currents. Unable to return, Schempp ditched in the Pacific Ocean and the *Albatross* was destroyed. A second *Albatross I* was constructed with a slight wing dihedral starting half-way out to improve stability.

Richard entered *Albatross I* (G12271) in the Fourth Annual Soaring Society of America (SSA) Nationals held July 1933, in Elmira, New York. The SSA origin started in Elmira in 1929 with its first national soaring meet. By 1933, the SSA convinced the government Civil Works Administration the site needed major improvement to accommodate their National meet. The project was completed by the 1933 event and was named "Harris Hill." Elmira is now recognized as America's Soaring Capital.

Richard du Pont, in a national display of publicity, had his Bowlus-DuPont *Albatross I* towed behind his *Stearman* from Glendale, California, to Wilmington, Delaware. Don Stevens flew the *Stearman* and Richard flew the *Albatross I.* Although the event did not disclose any records being broken, more than likely it has never been repeated. The Elmira location has recorded many historic soaring events in U.S.

An aerial view of Harris Hill, Elmira, New York. This area is still considered the home for glider pilots today. Photo from the du Pont collection.

Ernest Schweizer helped design this launching mechanism attached to a 1927 Chevrolet. The handle was moved laterally to assure the line spooled evenly on the drum. Photo from the Ted Bellak collection.

history. To house the history of soaring, the SSA erected a museum there that is open to the public. Richard announced, in 1936, that he had aspirations of attaching his glider to a balloon and being released at 30,000 feet in order to soar 1500 miles. This never became a reality. However, Richard du Pont would leave his mark in the SSA Museum by setting several soaring records, serving as president of the SSA, and being nominated to the SSA Hall of Fame.

The Bowlus-DuPont *Albatross I* and *II* was unique in that it did not use wing spoilers to control descent. The *Albatross* used split wing flaps. Another first in sailplanes was the use of a stabilator in place of the normal combination stabilizer and elevator. A control wheel, instead of the stick, was another first on the *Albatross.* No one can dispute the fact the Bowlus-DuPont *Albatross* was the most impressive sailplane at the meet. In the fourteen-day event, there would only be four days of quality soaring weather. Richard's only record was for soaring to 4,334 feet. A record for the meet, but 456 feet short of Jack O'Meara's U.S. record. Richard's performance would earn him second place in the 1933, SSA Nationals.

Unhappy with his results at the Nationals, Richard assembled four crack pilots after the meet to invite them to an informal gathering at the Swannanoa Country Club near Waynesboro, Virginia. The last week of September the meet got underway with Richard du Pont, Warren Eaton, Gus Haller, Heath McDowell, and Emerson Mehlhose. On the fourth day, Mehlhose broke the U.S. Soaring Distance Record by flying his Wolf *Hawk* 71 miles to Front Royal. (The old record was 66 miles, held by O'Meara.) However, he would only enjoy the record for one day. Richard du Pont broke the record the next day, September 21, 1933, when he soared from Waynesboro, Virginia, to Frederick, Maryland, a dis-

The Bowlus Albatross I *in flight is truly a magnificent sight. Photo from the du Pont collection.*

Ted Bellak built this model of Richard's Bowlus-DuPont Albatross II. *It was later marketed as a kit model through Sunrise Model Airplane Company, New York. Photo from the Ted Bellak collection.*

tance of 122 miles. However, it fell short of the world record of 136 miles held by Guenther Groenhoff of Germany by 14 miles. Richard could have broken the German record if rain and fog had not forced him to land at Detrick Airfield in Frederick.

Apparently, father Felix felt comfortable with airplanes in terms of safety. In January 1934, he flew to Miami to bid "bon voyage" to his son Richard and daughter Alice. They departed in Alice's new Waco F2, open-cockpit, floatplane for a three month tour of South America. Alice, now twenty-one, had received her license at seventeen and was an accomplished pilot in her own right. (Alice was also a flying voice against prohibition, vocally denouncing its existence as she toured the country in her airplane.) Richard and Alice returned to Miami on March 8, 1934. Alice later wrote a detailed feature story, with photos of their trip to Brazil, for *"The Flying Sportsman"* magazine.

Twelve days after the Brazilian tour, on March 20, 1934, Richard married Allaire Crozer, of Chestnut Hill, Philadelphia, who was a close friend of Alice's in school. They were married in Florida at the Trinity Episcopal Church, Miami. They used the same Waco F2 floatplane used on the South American trip to depart Miami for a honeymoon in the Bahamas. After their honeymoon, they departed Miami for Glendale, California, to oversee the completion of the Bowlus-DuPont *Albatross* production.

Bowlus-DuPont Sailplane, Inc. was incorporated in Delaware in 1934. The company was structured as follows; Richard C. du Pont, president; A. Felix du Pont, Sr., vice president, Allaire Crozer du Pont, secretary-treasurer. The directors were Richard C. du Pont, A. Felix du Pont, Sr., Warren E. Eaton, Hawley Bowlus, and Lewis B. Barrington. Richard agreed to consign four *Albatross IIs* to be built with a "gull-wing" design. G13780 Bowlus-DuPont *Albatross II*

Richard and Allaire du Pont toured California in this sleek 1934 side-by-side, two-seat, open-cockpit, Kinner B-2 Sportwing. *Only eight of these classy monoplanes were built. Photo from the Hagley Museum & Library collection.*

was built for Richard du Pont, using mahogany plywood to skin the fuselage. Another *Albatross II* (G13763) was delivered to Warren Eaton, president of SSA in May 1934. It was built using spruce plywood on the fuselage. Warren's *Albatross II* was named *Falcon.*

Richard brought both Bowlus *Albatross I* and *II* to the 1934 SSA Nationals held at Elmira in June. The Bowlus-DuPont *Albatross* was the center of attraction at the Elmira event. There was little doubt the other pilots knew they were up against a formidable competitor. On the first day of the Nationals, Richard piloted his *Albatross* from Elmira, New York, to Somerset, New Jersey, for a total of 155 miles. Five days later, in another contest, flying the second *Albatross,* he established a new National Altitude Record of 6,224 feet. His combined total points for the Elmira meet earned Richard the 1934 SSA National Championship. The title carried with it possession of the coveted "Evans Trophy" for one year. At the same meet, Richard's wife Allaire broke the women's duration record by soaring over four and one-half hours. Unfortunately, her sailplane did not have a barograph on board to make it official, and the old record prevailed.

Richard was in his element when he flew a glider. He was one of very few who could perform aerobatics flawlessly in a glider. Hawley Bowlus signed a contract with the National Air Races (NAR) in 1934 to put on a glider exhibition flight for $250. When the glider pilot assigned could not fly, Richard took his place because he knew Hawley needed the money. Richard flew the exhibition flight at the NAR in 1934, which included rolls, flying upside-down, as well as inside and outside loops. Hawley Bowlus and Martin Schempp, both accomplished

Warren Eaton's Bowlus/DuPont Albatross II *at Harris Hill, Elmira. Photo from the Ted Bellak collection.*

Lewin Barringer's, gull-wing, Bowlus-DuPont Albatross II, *with cockpit cover removed, in flight at Elmira, New York. Photo from National Air & Space Museum.*

sailplane pilots in their own right, marveled at Richard's ballet in the sky.

Richard's Bowlus-DuPont *Albatross II* was later sold to Jack O'Meara in 1936. O'Meara flew it until 1939 when it was returned to Hawley Bowlus for restoration. After the restoration, Hawley flew it on several occasions before O'Meara passed away. It was then put in storage until 1952, when it was purchased by Stewart Baxter. The historic Bowlus-DuPont *Albatross II* had suffered major deterioration over the years from neglect, and Baxter left it in storage. It was eventually resurrected in 1981 by Raul Blacksten and Steve Lowery, of Maywood, California, who are restoring it to original condition. The restoration is a very long and costly undertaking. To further create originality, the registration number was traced to a Continental Airlines 727. The restoration staff sent a letter to Continental requesting they relinquish the number for its historical significance. Continental replied that N13780 was scheduled for heavy maintenance in May 1993, and the number could be transferred at that time. An *Albatross I* is on display at the SSA Museum in Elmira, New York. Another is on display at the NASM Paul Garber Restoration facility, Silver Hill, Maryland.

Richard du Pont devoured every bit of information available on sailplanes. The Germans were known to excel in the design and craftsmanship of building these elegant machines, and they held the world records to prove it. One such sailplane caught Richard's fancy in 1935. Richard purchased, from Gus Haller, a German-designed *Hawk,* which was demonstrated at the Elmira soaring event. Richard bought it at a depressed price when the pilot ran out of sky and landed it in a river. It was trailered back to Wilmington to the DuPont Motor Car Company plant for Ted Bellak and Don Stevens to refurbish. When finished, Richard flew it in 1936 to second place at the Nationals held in Elmira.

The Bowlus-DuPont Sailplane Company only lasted until September 1936. During that time, only four *Albatross IIs* were built. In spite of performance records, orders were not forthcoming due to the depression. On July 14, 1936, Richard and Allaire boarded the *Hindenberg* for a flight to Germany to

The gliders were placed on special trailers and towed by this 1936 Ford pickup. Inscribed above the bird on the door is, "On the Wings of an Albatross." A winch was installed in the bed of the truck to launch the gliders. Photo from the Ted Bellak collection.

study soaring and purchase a Schempp-Hirth *Minimoa.* (Remember, Schempp was the co-designer of the *Albatross.*) On the flight overseas they became friends with Captain Ernst Lehmann. Allaire remembered they ran into foul weather over London and spent the entire day flying up and down the Thames waiting until it was clear enough to land. After a short tour of Europe, they returned to the U.S.A. on the *Hindenburg* and landed at Lakehurst, New Jersey. The *Hindenburg* made its last crossing when it crashed at Lakehurst on May 6, 1937. Captain Ernst Lehmann was severely injured, and Richard and Allaire flew to New York to visit him in the hospital. Unfortunately, Captain Lehmann passed away shortly after their visit.

Prior to the crash of the *Hindenburg,* many aviation experts held the belief that the airplane could never replace the dirigible. The airplane was too noisy, too frail, too unreliable, did not have the range, and could not carry passengers in luxury and comfort. The *Hindenburg* accident was the greatest aviation tragedy of its era and the publicity spelled an end to the dirigible.

After Richard and Allaire returned to Delaware, Ted Bellak requested, and was given, permission to go to Germany as a visiting student to oversee the construction of the *Minimoa.* The *Minimoa* sailplane was only built to order, and Richard would have to wait almost a year for delivery. After the *Minimoa* was delivered to the U.S., Richard du Pont flew it in the 8th National Soaring meet in 1937. He managed to break his own altitude record by soaring to 6700 feet in the *Minimoa.* However, he could not beat his opponent, Peter Riedel, a German pilot, who captured the National Distance Record by soaring from Elmira to Long Island, New York, a distance of 182 miles in his own *Minimoa.*

Although the Bowlus-DuPont Sailplane, Inc. was dissolved, Richard decided to return to Delaware and build another sailplane in the defunct DuPont Automobile factory at the foot of Commerce Street, Wilmington. The U.S. Navy was using outdated Franklin gliders, and Richard heard they were interested in looking for something better. The initial order was to be for six gliders. The DuPont Glider & Supply Company was incorporated in Delaware on September 18, 1936. Heath McDowell, a graduate from the University of Michigan, was hired to oversee construction of the DuPont *Utility.* Naturally, Hawley Bowlus and Richard would be responsible for the design. Ted Bellak and Don Stevens were hired to work with McDowell and maintain Richard's other sailplanes. The *Utility* was test flown at Elmira by Richard du Pont and Heath McDowell in 1937. The *Utility* performed very well, but the Navy's order never materialized, and no others were produced. The fate of the DuPont *Utility* is unknown beyond this point.

Since production of the *Utility* never materialized, Richard decided to move the maintenance operation from the Wilmington location to the du Pont estate *Granogue* in the summer of 1937. The glider main-

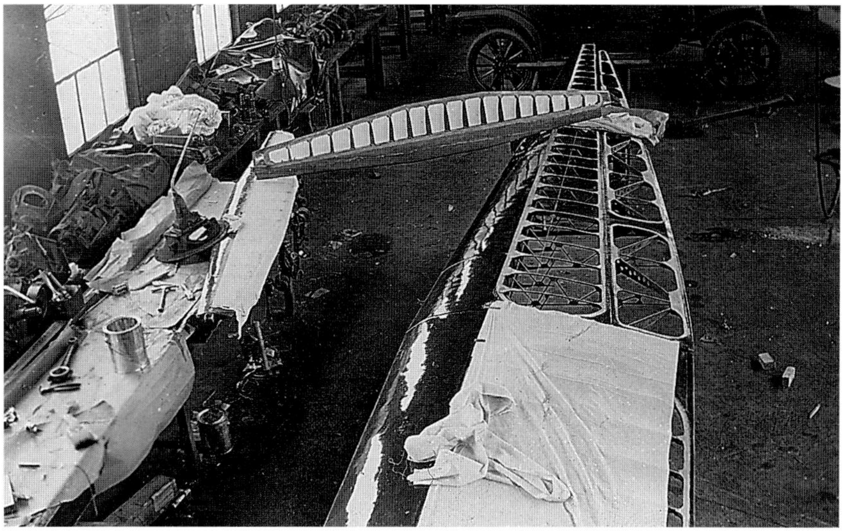

The defunct DuPont Motor Car Company building was used by DuPont Glider and Supply Company for building the DuPont Utility *and for maintenance on gliders. Photo from Ted Bellak collection.*

The DuPont Utility. *Only one was known to be built. Photo from the Ted Bellak collection.*

Richard du Pont's Bowlus-DuPont Albatross II *in the Granogue shop for a complete rebuild by Ted Bellak (far right). Photo from the Ted Bellak collection.*

tenance crew was now down to one. Don Stevens elected not to move to *Granogue* and moved back to his native state, California. Ted Bellak was given a free hand to restore an existing barn into a maintenance facility for Richard's sailplanes. Ted not only maintained and repaired Richard's sailplanes, but also custom-designed and built trailers to transport them around the country.

Richard C. du Pont served as president of the Soaring Society of America from 1937 through 1939. After his term as president was over, Richard decided to drop out of soaring and commit full-time energy to his newly acquired "All American Aviation" business. Richard's last remaining sailplane, the *Minimoa,* was sold to Ted Bellak, who moved to Michigan.

ALL AMERICAN AVIATION

Many believe the airmail pickup system developed by Richard C. du Pont and All American, was Richard's idea. History, however, records that Orville and Wilbur Wright demonstrated a similar system as early as 1908. The low power and lack of performance of the aircraft created too many obstacles for the frail machines, and development ceased.

Roger Wolf Kahn swoops down with his Bellanca CH for a successful pickup at Hoover Airport, Washington, D.C., June 1933. Note that Cabot's system is completely mobile and can be towed behind a car. Photo from National Air & Space Museum.

According to a newspaper article of June 8, 1918, Alvin I. Marvel, Wilmington, also introduced an air-mail pickup system. The account describes the system as able to pick up and drop mail in a single pass. Marvel had progressed beyond the stage of drawing plans and displayed a fully assembled model of the finished product. No other information could be located beyond this newspaper article.

Another early pioneer was Godfrey L. Cabot, a Boston scientist and engineer. Cabot felt the mail pick-up system used by the railroads could have application to aircraft as well. After successfully picking up a 155-lb package in Boston Harbor on October 2, 1917, he applied for a patent in the fall of 1918. Cabot's patent was approved by the U.S. Patent office in 1920. Cabot also enlisted the financial aid of band leader Roger Wolf Kahn to help demonstrate his system. Roger was the son of a wealthy New York banker. In 1928, Kahn purchased a Bellanca CH-300 *Pacemaker* and had Cabot's aerial pickup-system installed. At the New York Aviation Meet in 1930, Kahn flew the Bellanca and made several flawless promotional mail pickups. With this success, they went to Washington to demonstrate the system at Hoover Airport (now Washington National) before government officials. The demonstrations were continued throughout the East in 1930 through 1933 in an effort to convince the Post Office Department to sanction its use. In spite of flawless demonstrations, the Post Office Department failed to endorse the system.

After a change in the administration in 1933, Cabot and Kahn demonstrated the system once again to the new Postmaster General Farley. This demonstration was one of the feature events to be held at a much-publicized air show on Hoover Airport the last week of June 1933. The Bellanca *Pacemaker* and its pilot would once again show off their stuff and amaze the large crowd with a successful performance.

At the opposite end of the North American continent, Dr. Lytle Adams, a dental surgeon and inventor in Seattle, was working on a system quite similar to Cabot's. Adams is credited with the invention of the spiral plow and a rotary planting machine which he patented and manufactured in East St. Louis.

Adams recognized the potential of the airplane as a working piece of machinery. He felt the airplane's most important asset was its speed. He also felt two things hindered the development of the airplane; the lack of airfields, and turnaround time on the ground. Since the airmail system, as it was conceived, only serviced the major cities, the smaller towns were left out. Most small towns did not have the population or the funds to build an airfield.

Dr. Adams dealt with these concerns and felt an airplane could swoop down low over an open field and pickup and drop mail in a single pass. Adams worked in his spare time (in his dental practice office building in Seattle, Washington) building many models to prove his concept. He applied for, and received, patents on his designs in 1928 and 1933. Although he worked hard to overcome the engineering problems, which included absorption of the kinetic energy during pickup, financing was his major obstacle.

Living in Seattle gave Adams the opportunity to discuss his invention with William Boeing, who manufactured airplanes nearby. Boeing offered the talents of his engineering department to help develop a full-scale prototype of the system. The prototype was mounted on a turntable which allowed it to be turned into the wind. This novel idea allowed the pickup plane to operate at the lowest possible ground speed during the crucial pickup/drop phase. In August of 1928, a series of pickups were made, and a 98 percent success rate was achieved picking up a 35-pound mailbag.

The successful pickups encouraged Adams to go to New York City to establish contacts for financing. After Lindbergh's flight from New York to Paris in 1927, aviation became a hot item for investors. Through connections in New York, Adams met Charles Lindbergh, then a trustee for the Daniel Guggenheim Fund. Lindbergh was sufficiently impressed with Adam's idea to obtain an interview with Harry New, the Postmaster General. New was less impressed than Adams and Lindbergh. However, he agreed to fit the apparatus into a *Travel Air* monoplane not for the pickup of mail but to see if it would enhance an endurance record attempt by picking up food and fuel without landing. Tests in December 1928 and January 1929 proved successful in regard to pickups and drop offs, but the attempt to utilize it for an endurance record was scrubbed for reasons unknown.

Some have said that Adams and Cabot, although working on similar systems, were unaware of the other's involvement. This theory is difficult to accept since they were both actively seeking funds and demonstrating their systems publicly during the same period. Newspaper accounts of their activities were published often.

Adams continued to press forward. He showed the device at the New York Aviation Show in 1929. It was at this show where he met Clifford Ball, who had acquired Contract Air Mail (CAM) route 11 between Pittsburgh and Cleveland. Although this route was considered short, it was one of the profitable ones. Ball felt the Adams pickup system could be of benefit to his CAM route 11 by reducing landings and takeoffs. Ball agreed to supply partial funding to install the system and demonstrate it to the Post Office Department for acceptance.

During the month-long trials, structural problems surfaced in the airplane due to stresses and shock during pickup. Worse were the two near-fatal accidents, when the pickup was missed which flung the heavy ball, used as a cable weight, into the propeller. In another incident, the cable snapped and jammed in the elevator causing the airplane to go into a dive. On both occasions, the skill of the pilots avoided a crash. Adams made many modifications over the next year in an effort to improve the system. Nonetheless, Ball called the whole thing off after a year. There was an insufficient volume of mail to justify the cost. But there were other problems as well. First, there was a clash of personalties. And second, the Post Office was not impressed and refused to endorse its use. In fact, they stated flatly they had no intentions to inaugurate such a service.

Adams pressed forward in spite of the obstacles. Through the depression years of 1929 and 1930 he tinkered with the system in an effort to overcome some of its shortcomings. He would not find support for two agonizing years until he met Thomas Braniff, of Braniff Airways from Oklahoma City. Braniff offered financial support to conduct tests which proved reasonably successful. However, the Post Office once again refused to approve the system, and Braniff ceased further funding.

Godfrey Cabot and Adams continued to seek venture capital to promote their systems. In desperation, Adams made direct contact with Eleanor Roosevelt in hopes of convincing her the poor people living in small towns were denied access to air mail service. Eleanor turned the request over to her brother, Hall Roosevelt. Hall was not impressed with Adams, or his ideas, and agreed with the Post Office (the expense of installation was $3500 to $6000 each) that it would be too costly to operate and maintain.

Adams, still obsessed with the idea his system was a good one, pressed on. In New York he found financial backing from Arthur Davis, president of Arma Engineering Corporation, in Brooklyn. Adams and Davis created a new company they named "All American Aviation, Inc." and incorporated the company in Delaware on March 5, 1937. Davis received one-half ownership in the Adams patent for his financial assistance. Although the charter gave the Corporation broad powers, Davis and Adams decided to use All American Aviation, Inc. as a holding company for the patents and the issuance of licenses for their use. A separate corporation was formed at the same time and acted as the operating arm named "Tri-State Aviation Corporation, Inc." Although both companies were Delaware corporations, the base of operations for All American would be Irwin Airfield in Pennsylvania.

West Virginia was one of the states that wished to benefit from the aviation explosion. In an effort to boost the State's aviation progress, Congressman Jennings Randolph invited Lyle Adams' Tri-State Aviation to move to the newly-opened airport at Morgantown in the fall of 1937, which he did. Two mail-order houses, Montgomery Ward and Kauffman's department stores, began shipments with Tri-State. By 1938, Tri-State Aviation routes grew to include Pittsburgh, Charleston, and Baltimore. Three Delaware-built Bellancas and one Stinson were operating the routes with quick stops in-between the major cities when required. The airmail pickup system was not put into service at this time due to lack of federal approvals and the need for more testing.

Tri-State Aviation carried almost anything that would fit into the airplanes, including passengers. Although passengers fares contributed the least of the total income, they were more than welcome aboard. Tri-State operated without government subsidies although the carrying of airmail and cargo contributed to their survival. To say it was a profitable venture would be a bit optimistic.

An unplanned opportunity surfaced for Adams when Eleanor Roosevelt accepted an invitation to host a luncheon for the National Editorial Association in West Virginia. Tri-State offered to fly the First Lady for the round trip and the offer was accepted. Adams and Norman Rintoul, chief pilot, flew to Hyde Park and spent the evening with the Roosevelts before leaving the next morning for Morgantown. That evening Adams was introduced to Franklin D. Roosevelt, Jr., who had recently married Ethel du Pont of Greenville, Delaware. Ethel was the cousin to Richard C. du Pont, nationally known glider pilot and aviation enthusiast. The connection was made, and Richard du Pont was given an invitation to see the airmail pickup system in action.

Du Pont and a group of advisers, which included Charles Wendt, du Pont's chief financial advisor, went to Morgantown to witness the system in operation firsthand. The demonstration included three pickup attempts. The first two failed, and the last, although successful, was anything but ideal. Charles Wendt and the group were not impressed and made their feelings known to Richard on the return trip to Delaware. In spite of their objections, and objections from family members, Richard felt the technical problems could be overcome and the project was worth pursuing.

Adams, chronically short on funds, had sought venture capital for fifteen years and was running out of options. When Richard du Pont offered to buy fifty-one percent of the venture, Adams was all too eager to accept. Du Pont's first objective was to rid Adams from the outstanding debts he incurred trying to promote the system. To accomplish this, Richard du Pont, on August 16, 1938, advanced Adams $45,000, repayable in two years at 1.75 percent interest. Adams signed over to du Pont three of his patents, which were the most critical to the pickup system, as collateral. Also included in the agreement were the sole rights to any new patents which might be assigned in the future.

In a separate agreement, du Pont would also have the option, in the two-year-period of the note, of acquiring forty percent of Tri-State Aviation class A non-voting stock and fifty-one percent of the class B voting stock. This option was available to du Pont in exchange for cancelling the August 16, 1938, note. In addition to the above, du Pont agreed to pay Adams a further $45,000 for the tangible assets of Tri-State Aviation.

To seal the above agreements the du Ponts required a restructuring of the capitalization of All American Aviation stock. The stock would now consist of 250,000 shares of class A non-voting stock with a $1.00 par value. There would also be 100 shares of class B voting stock, with no par value. Adams would receive 120,000 shares of class A stock and 49 shares of class B stock. This stock agreement put Richard du Pont firmly in the position of being the majority stockholder. In addition to Adams stock, it was agreed Adams would be employed as "technical advisor."

The first board of directors meeting of All American Aviation, Inc. was held in Wilmington, Delaware. Not surprisingly, Richard du Pont was elected president; Lytle Adams, vice president; and Charles Wendt, secretary-treasurer. Arthur Davis, the New York financial investor, sold his shares to Adams, who in turn sold them to du Pont. Davis was no longer associated with the company. In his place, Charles Wendt was given the position of director. This put Adams in the uncomfortable position of having a worthless vote.

By the next board of directors meeting, Adams had countered by talking Davis into buying back twenty-four shares of his voting stock which would give him a place on the board. This move would allow Adams to check the du Pont vote in the event he wished to do so. The du Ponts countered the move in the same manner and elected A. Felix du Pont to the board of directors. The du Ponts now controlled three out of five votes. Checkmate!

During a February 1939 board meeting, it was decided to offer Godfrey Cabot 750 shares of All American stock and a director's position on the board for the patent rights to his system. It was felt some of Cabot's catapult-system patents and other parts could be adapted into the overall design. This move also angered Adams who felt the system could be designed without Cabot's patents. This caused more dissension between du Pont and Adams.

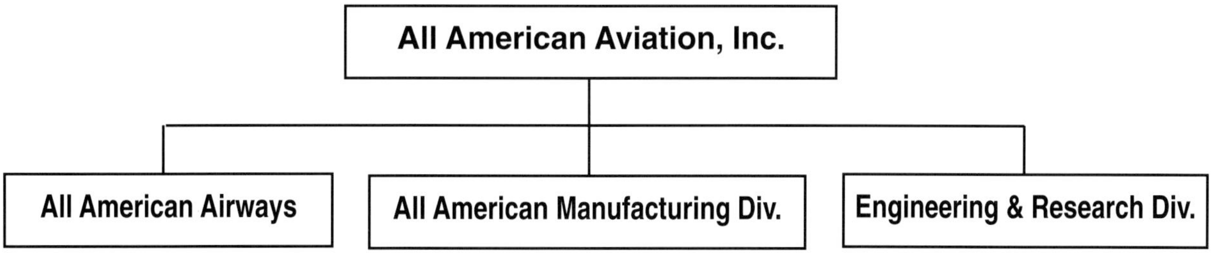

Much to Adams' displeasure, Richard du Pont also moved immediately to engage the services of Kirkham Engineering and Manufacturing Company, Farmingdale, New York, to redesign the system. He also assigned several All American employees to work with Kirkland engineers to speed up the process. The first stages of redesign took place on the du Pont estate, *Granogue,* in the building formerly used to service and store Richard's gliders. By the late fall of 1938, Richard established the Engineering Division and a newly formed Manufacturing Division, which were located on the DuPont Airfield.

Needless to say, Adams was furious about anyone's attempting to make changes in his design without his permission. Adams wrote Richard du Pont a fiery letter in which he demanded no changes be made to the system without his personal approval. He also wanted every All American employee to work directly under his supervision. Adams' attitude placed a great strain on Richard du Pont's relationship with Adams, which could only become worse.

Adams felt he had nurtured the pickup design for fifteen years. He had refused to give up in face of rejection from the Post Office Department and financial supporters. He was the vice president and technical advisor of All American Aviation, and he felt he was being left out. Although changes in the system had to be made, it hurt Adams deeply to see others take the helm when he felt light could finally be seen at the end of the tunnel.

Richard du Pont felt differently. He knew that Adams had spent fifteen years working on the prob-

The first version of the aerial pickup system designed by All American's Engineering Department. Note the hand-crank, which would give the crewman a real workout bringing the mail aboard. Photo from Hagley Museum & Library.

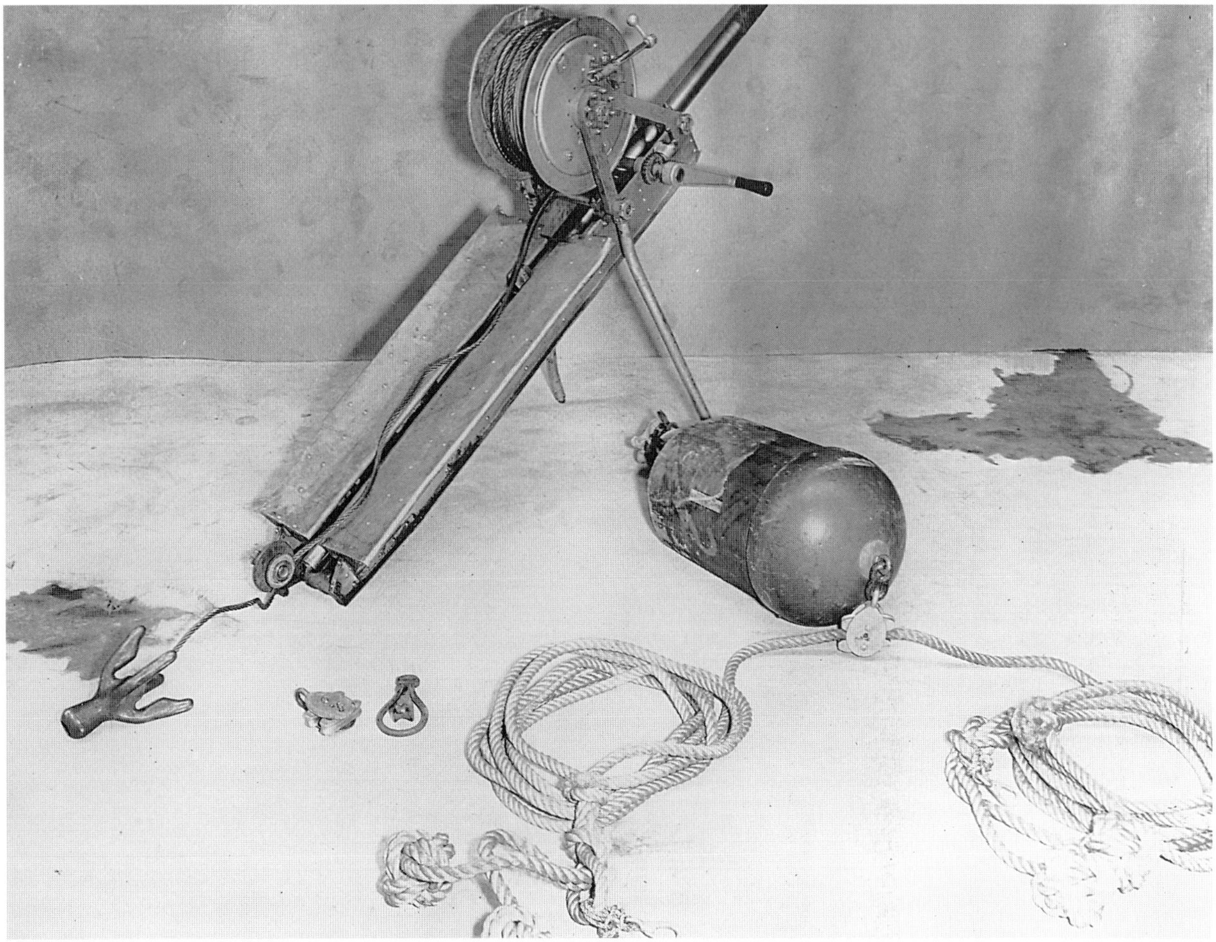

lems but had yet to solve them. Richard also knew that he was supplying the funds and wanted the system in operation as quickly as possible. After all, the ultimate goal was to service the entire country. The sooner it became operational, the sooner funds would start offsetting expenses. Richard assigned Adams to lobbying Congressman Jennings Randolph (WV), for support of their idea, in part to keep him away from Wilmington.

With Lytle Adams lobbying, Congressman Jennings Randolph had legislation passed called the "Experimental Air Mail Act." Under the terms of the legislation, funding would be allowed for one year to prove the airmail pickup concept. The Act provided the Post Office Department funding for autogiros and other apparatus to enhance the carrying of airmail to/from rural areas. The Post Office Department also was to approve, and/or deny, route requests. Much lobbying was done by Adams since legislation was sorely needed by All American to prove the worth of their planned operation.

To the Post Office Department's displeasure, pressure was applied to have two contracts assigned for nonstop aerial pickup and delivery of mail by December 1938. The first two routes were numbered 1001 and 1002 which would total 878 miles of rugged Appalachian Mountain Terrain. Route 1001 would pay thirty-two cents per air mile and route 1002, forty-three cents. The harshness of the terrain on route 1002 allowed the higher compensation. The terrain alone would tax the capabilities of men and machines. Since All American was the only company able to provide the service, it was simply a matter of paperwork to approve the agreement. All American Aviation posted the appropriate bonds and signed the Post Office Department contracts December 27, 1938.

The first informal demonstration of the aerial pickup system was made November 5, 1938, on Mr. Charles C. Kurtz's property located two miles outside the Wilmington City limits on Concord Pike. Two demonstrations were announced, the first at 10 a.m. and the second at 2:30 p.m. (The Stinson pickup airplane departed from the DuPont Airfield for the two demonstrations.)

The reasons the Stinson SR-10C *Reliant* monoplane was selected by All American Aviation to perform the task were many. With its 260-hp Lycoming,

This All American Stinson Reliant *has just dropped the incoming mail (arrow) and has snagged the line for picking up the outgoing mail. Photo from the State of Delaware Public Archives.*

Cachets, such as this, would serve two purposes; increase volume and become philatelic collector memorabilia. From the State of Delaware Public Archives.

radial, air-cooled engine the mighty Stinson could cruise at 155 mph and still land at a docile 55 mph. Special modifications were made to attach the winch inside the airplane for the pickup system. The landing gear struts were fitted with knife-edges to sever the transfer line in the event the airplane descended too low during the pickup run. Pilots enjoyed the smooth controls of the mighty Stinson, which allowed the airplane to respond to any maneuver with a light but positive touch. The huge gull-wing was fitted with speed arresters, which permitted the airplane to nose down at a high angle without gaining speed. These marvelous airplanes proved time and time again they could fly in the most adverse climate conditions and terrain imaginable.

Du Pont would hire only pilots with at least ten years flying experience and four thousand hours of flying time. Nationally acclaimed pilots like Holger Hoiriis, Tom Kincheloe, Jimmy Piersol, Cammy Vinet, Norm Rintoul and Gerald McGovern flew the *Reliants.* The aircraft and ground crews were all selected from the very best obtainable.

The first formal demonstration of the new pickup system was held at Coatsville, Pennsylvania, on March 6, 1939. Two practice runs were made early in the morning and they were flawless. The formal runs would be made at four o'clock in the afternoon before a public audience. More than twenty-five hundred people assembled at the airport to observe the combination pickup and delivery of the mail. The first attempt was marred because of tangled lines, but the second and third attempts were flawless. Other locations were selected later along the routes for demonstrations to the public before the service would officially begin in May. Newspapers headlined these events positively, thus drawing larger crowds at each demonstration.

All American would start operations with 28 full time employees. The average distance between pickup points would be a mere seventeen miles. In some cases the mileage would be as little as five. Route 1002 from Pittsburgh to West Virginia was inaugurated May 12, 1939. Any politician or dignitary, who had anything remotely to do with making it happen, showed up at the various pickup points for the momentous occasion. Newspapers announced the event in advance, which drew thousands of spectators. The loads were extra large for the opening event since special cancellations provided by the Post Office for first-day cachets for philatelic collectors were included.

After six weeks of operation, All American lost over six thousand dollars. By the end of the first year, the losses would reach over fifty-three thousand dollars. These losses were in stark contrast to the 91.6 percent completion rate and 75,000 lbs of mail carried. The Stinsons, manned by their professional pilots, logged 438,145 miles of flying without a single accident. This must be considered nothing short of spectacular considering the weather and the terrain.

However, in spite of the financial losses, the Post Office Department refused to increase the rates. In desperation, Richard du Pont announced to the board that the company must expand the operations to create enough volume to operate profitably. To achieve this goal they would offer license agreements to carriers who had operational routes. They would also make a formal request to the Post Office Department for additional routes.

To emphasize the viability of the system, All American held a demonstration on the Mall *at night* in front of the U.S. Capitol during the National Postmasters Convention. Although everything worked like a well-oiled Swiss watch, the Post Office Department felt the expenditures did not warrant the investment and failed to approve expansion of the system. To make matters worse, they did not guarantee they would continue the funding of the existing routes once the year contract was up. The Post Office Department felt they were in no position to fund an operation, no matter how successful, which incurred such huge deficits. It now became apparent the Post Office Department's position of a resentful attitude in supporting the aerial pickup-system prevailed.

As feared, when the year of experimental airmail pickup service was completed, the Congressional funding ceased. It was now up to the Post Office Department to continue the service and provide the funding. Since All American felt the Post Office Department was anything but thrilled with the idea of continuing the airmail pickup system, it sought new markets.

All American decided to apply for feeder lines carrying passengers *with* the airmail to major terminals to offset expenses. Test flights were made *with* passengers to address any concerns while the airplane zoomed down to pickup mail en route. To suggest that a non-flyer would feel comfortable while the aircraft made low-level passes picking up mail surely needed to be addressed thoroughly. Passenger statements would be vital in an application for approval to the Post Office Department.

In any event, du Pont kept his focus on the same dream which Adams had for years; to have air mail service available to everyone. In fact, du Pont was so committed to the system he already had visions of international applications through leases of the system outside the United States. An attempt was made to establish such a system in Brazil. However, it failed since the Brazilian government was unable to organize the operation.

While the Post Office Department was considering discontinuing the mail pickup operation, All American filed an application with the Civil Aeronautics Administration (CAA) for the passenger routes. Along with the existing routes, Richard du Pont requested the CAA expand All American routes to an additional six. These flights would include carrying passengers, airmail pickup, and cargo. He would, in his proposal, convert All American aircraft to carry passengers as well as swoop down to pick up airmail en route to hub terminals. Hub terminals were constructed throughout the country to accommodate larger airliners that could not operate from smaller airfields. This operation of small feeder airlines, he suggested, would bolster the major hub terminals with more passengers and increase their business in the process.

Not everyone was pleased with the idea. One in particular was Lytle Adams, still owner of Tri-State Aviation, a feeder airline service operating from Morgantown. Adams assembled other feeder airline carriers who shared his feelings and made valid complaints at the hearings. The group took the position that, since the Congress funded the Post Office Department and gave them control of the airmail pickup system for one year, it was now the Post Office Department's responsibility under the Act to continue or discontinue the service. In addition, they asserted that the CAA was not authorized to approve *any* airmail routes. It was simply not within its jurisdiction. They also argued that the expansion would create financial drains on the existing small feeder lines not subsidized by the Post Office Department.

The whole mess ended up on Capitol Hill where several bills were introduced to resolve the situation. Heavy lobbying on each side caused considerable grief for elected officials. To add to the confusion, the CAA and the Post Office Department were designating routes in the same airspace. The CAA felt the Post Office Department did not have the authority to designate air space for any reason. This question had to be resolved by Congress. And there were many other differences to be solved between the parties involved. Finally, in June 1939, a compromise bill passed the House and the Senate and was sent to the President.

July 2, 1939, Franklin D. Roosevelt signed the new Act into law. The Act consisted of two sections. Much to All American's pleasure, the first section broadened the definition of air transportation to include mail, passengers, and freight. It also created a Civil Aeronautics Board (CAB) to regulate services. The CAA would no longer have control over air routes and their functions. Their duties would be limited to licensing pilots and aircraft, issuing Approved Type Certificates (ATC) to airplane manufacturers, and rulemaking. The second part of the legislation repealed the legislation which had given the Postmaster General authority to establish and supervise the airmail pickup system. This would now be regulated by the CAB.

The CAB was given broad powers which extended to routes, tariffs, mergers, and virtually every other aspect of airline operations. Because the legislation was rammed through Congress to resolve the immediate problem, it was not perfect. The legislation, as interpreted, allowed the CAB total control over air carriers. The Act virtually wiped out the free market system for independent air carrier operations. It also created an additional burden on aviation by dealing with two federal agencies; the CAA and the CAB.

This was good news and bad news for All American Aviation. Du Pont would now be required to apply to the CAB for the permanent routes he requested and he would have to learn to deal with the CAB bureaucracy in every facet of its aviation operation. The CAA would have final authority over the airplanes and the pilots who flew them. The CAB assigned Robert Bartoo the All American requests, and he took his job very seriously. In spite of All American's connections in Washington, it would not be a cakewalk for Richard du Pont and All American Aviation to obtain their requests.

The first blow came when Bartoo dismissed All American's request to include the carrying of passengers during the mail pickup operations on the grounds of safety. This setback came as a surprise since All American had demonstrated the operation successfully. All American had also applied for several new routes. Once again Bartoo denied the proposed Pittsburgh-Cleveland airmail pickup route since he felt they were already well-served by existing carriers. He also denied the Williamsport-Philadelphia-New York run on the same grounds. The proposed Jamestown-Buffalo run also received a thumbs down because Bartoo felt it was too close to existing carrier lines. However, Bartoo did authorize All American five airmail pickup and express freight routes which would take them to points in Pennsylvania, Virginia, West Virginia, Ohio, Delaware, and New York. The five routes would serve approximately two million people and total 1365 air miles, serving 86 cities and towns with freight and airmail. Passenger-carrying and night operations were disallowed.

All American used the generic title "Airway to Everywhere" in their promotions. This map depicts the routes All American flew as of 1940. Map from the du Pont collection.

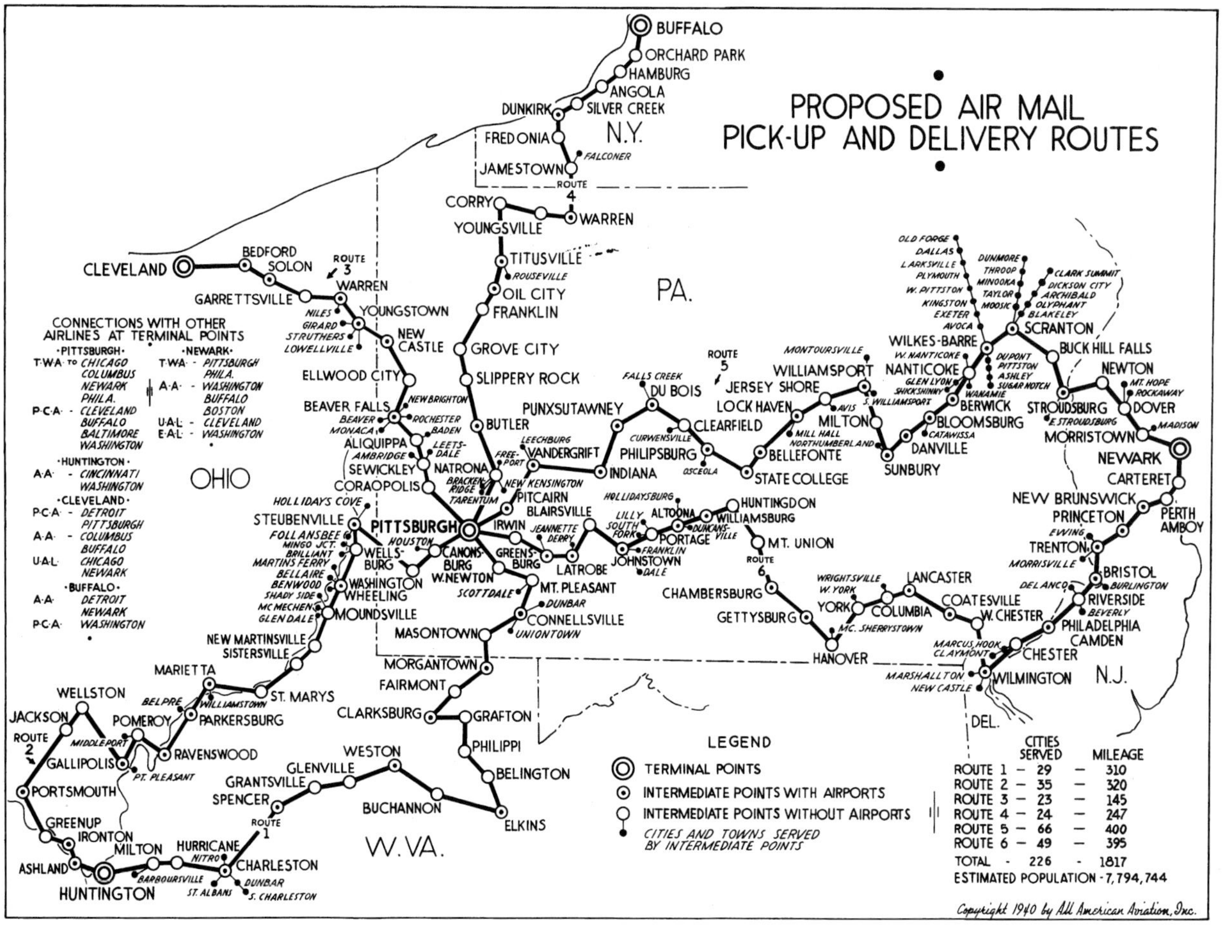

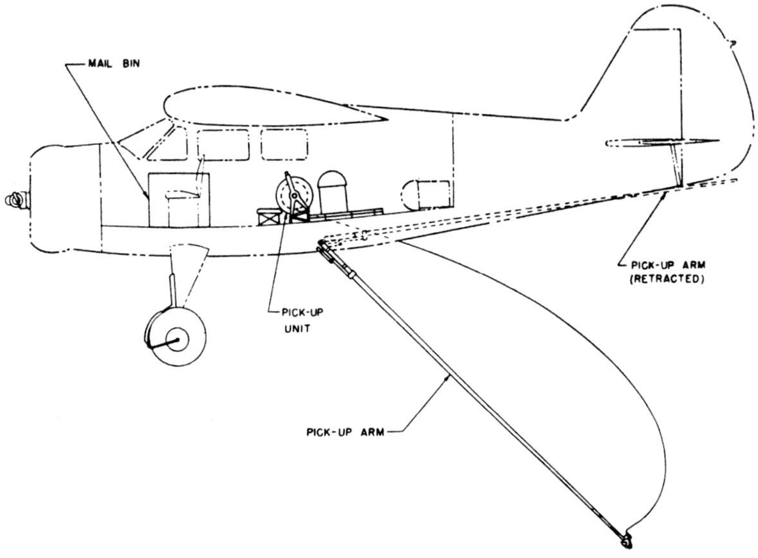

The model 4 version of the pickup system used a boom to drop the lines away from the propwash to eliminate lines being tangled. Drawing from the Hagley Museum & Library.

All American was not totally unhappy with the five routes the CAB granted. The transition, however, took three months of government paperwork before All American was fully operational. During this time, du Pont made the decision to keep the three All American divisions intact and retain all employees in spite of the great expense involved. He also took advantage of restructuring the company. The executive offices would be located in Wilmington, Delaware. All flights would originate from the Allegheny County Airport in Pennsylvania. All maintenance of aircraft would be performed by the Manufacturing Division located on the DuPont Airfield in Wilmington. Engineering and Research would also occupy offices on the DuPont Airfield.

The Engineering Division of All American replaced the hand-operated winch with an electric-powered system. The oil-air shock absorber was also replaced with a clutch-brake system which automatically took up the shock during pickup. A fifteen-foot retractable boom was also attached to the airplane forward of the center of gravity to eliminate the problem of propwash tangling the lines. Changes to the ground support system were also addressed. The ground system was simplified to the extent that it required only two, twenty-foot bamboo poles which were attached to aluminum bases. This was in sharp contrast to the elaborate and expensive system patented by Adams. Virtually all known problems were resolved by the end of 1940. It was now a rare occasion when a pass for pickup was aborted or had to be repeated.

The one problem, which appeared to be insurmountable for the Post Office Department and All American, was expenditures exceeding income by a wide margin. The Post Office Department would have liked nothing better if the air pickup system simply died. The people who were being served, however, had positive feelings about the service. It gave them a feeling of being part of the modern era although they had no airfield of their own. On most occasions the townspeople would come to the pickup site to watch the Stinson *Reliant* swoop down and snatch the mail and fly off. In many cases these small towns did not have a movie house. Therefore, the mail pickup was the most exciting thing to happen in a day.

All American, specifically Richard du Pont, refused to accept the fact the return on investment could not be overcome. Du Pont's position could have stemmed from the fact that if the airmail system failed, he would have to admit his advisors were right at the onset. Richard du Pont felt the solution to the problem was simply volume or, better said, correcting the lack of it. He planned two approaches to solve the problem. Winter months reduced daylight flying hours which eliminated longer routes.

Du Pont proposed night pickup to improve the service which would in turn attract more customers. Included in the petition to the CAB, All American requested permission to carry Railway Express Agency (REA) packages on the mail pickup routes.

A night pickup demonstration was planned at the Bellanca Airfield, New Castle, Delaware. Night pickup would allow the mail, which was sorted during the day, to depart the same day, eliminating the need for it to stay in the Post Office overnight. Lighted poles were used to give the pilot better reference for pickup than daylight operations. Although the demonstration was flawless, the CAB refused to allow All American to include night pickup in their operations. Their decision was based on safety, especially in the areas of the Allegheny Mountains. The CAB decision also put additional pressure on All American to complete the route pickups during the short days of winter. However, all was not lost when the CAB approved the Railway Express Agency package-carrying request.

In spite of the setbacks, All American showed its first profit of $3,824.17 for the year ending June 30, 1941. Although profits are a good sign, they are only one part of the financial statement. The current debt-asset ratio was an appalling .81, indicating a desperate need for operating capital. To obtain the needed capital, All American offered 20,000 shares of class A stock to be sold at $4.00 a share.

THE WAR YEARS

In the meantime the war was escalating in Europe. Hitler's military accomplished a spectacular victory in early WWII by invading the island of Crete with troop-carrying gliders at night. This did not go unnoticed by the allies. Britain and the U.S.A. immediately embarked upon a military glider program.

Richard du Pont and his All American engineers embarked on an ambitious program to adapt the airmail pickup system to pick up gliders. The first attempt was made May 30, 1942, on the DuPont Airfield. The airplane was a standard Piper *Cub* with the propeller removed. An adapter was made to

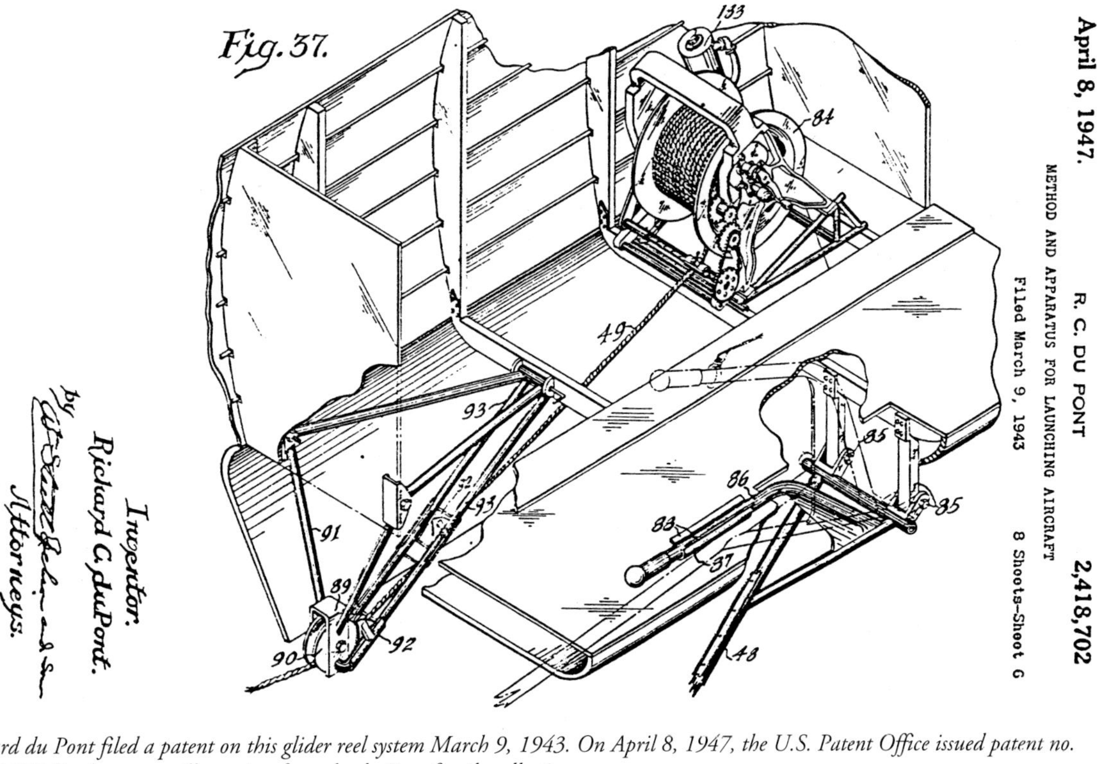

Richard du Pont filed a patent on this glider reel system March 9, 1943. On April 8, 1947, the U.S. Patent Office issued patent no. 2,418,702 for the system. Illustration from the du Pont family collection.

allow the pilot to release the tow cable once the Piper obtained the designated altitude. The major problem to overcome was the absorption of the kinetic energy when the pickup plane snatched the line as it flew by at 95-120 mph. This was accomplished in several ways. The tow line was specially made by the DuPont Company from unspun Nylon. This gave the line considerable stretch to take up shock. The unit mounted in the tow plane was equipped with a large reel to accommodate 500 feet of line. A friction clutch allowed the line to unspool as the pickup line snagged the ground line. Finally, there was a brake that brought the entire system to a halt before all the line ran out. The big Stinson swooped down for the pickup, hooked the line strung between two poles, and, as if by sheer magic, started to move the Piper very slowly until airborne. Richard was the pilot in the Piper. He was elated at the complete lack of shock at the point of pickup.

On June 4 and 5, Colonel Dent, Chief of the Experimental Glider Unit at Wright Field, Dayton, and Lt. Charles Decker were invited to participate in the pickup event on the DuPont Airfield. After the two days of testing without a single mishap, the men returned to Dayton very satisfied with what they came to preview.

By late July 1942, Richard invited over twenty-five Army officers to the DuPont Airfield to observe a dual-pickup. Two Pipers were stationed at the launch point. The tow plane swooped down and picked up the first plane. The tow plane made a single circle around the Airfield and swooped down once again to pick up the second Piper. Needless to say, it was a spectacular event, and the military was well-pleased with the results.

All American continued to improve the system to accommodate heavier pickup loads by using larger military airplanes. Propellers were removed from fighter aircraft, towed aloft, and released. Visions of the military were to build troop-carrying gliders and release them to land fighting-troops. The glider could be landed in a small field and picked up to be used again.

Richard du Pont was the first to try the towing of the Piper Cub, *sans the propeller, to the military. Photo from the du Pont collection.*

The Army Air Force experimental glider in which Richard du Pont lost his life. Note the retractable tricycle landing gear. Photo from the du Pont family collection.

Lewin Barringer, a long time personal friend and soaring buddy of Richard du Pont, was promoted from the National Guard to major in the Army Air Corps glider program at the onset of the war. Major Barringer's mission was to coordinate glider activities for the invasion of Sicily and he worked closely with Richard. Unfortunately, Barringer, on January 29, 1943, on a flight from Puerto Rico to North Africa, in which the Douglas C-47 encountered a tropical storm, was lost. On July 9, 1943, the military utilized gliders in the invasion of Sicily. The use of the gliders was not as effective as planned, and the military moved to correct the problems.

Richard C. du Pont felt honor-bound when summoned by General "Hap" Arnold to accept a position as a civilian advisor to their glider program. Richard resigned his presidency with All American and was immediately sent to Africa and Sicily to review the glider invasion program.

The facts which unfolded were as follows. On the night of the invasion, 144 Wacos and 8 British Horsa gliders were towed aloft by Douglas C-47s for the invasion. All the gliders were flown by British pilots who had less than five hours flying time in the Wacos. The mission was given the green light in spite of the stormy weather and high winds. The tow plane pilots were also unprepared for these conditions. Dust, smoke, anti-aircraft fire, and poor weather made navigation extremely difficult. The glider pilots, unaccustomed to the wind, released from the tow plane too soon, and 76 gliders with over 200 men were lost when they came down in the rough ocean. Many of the gliders were shot down by our own anti-aircraft fire from ships because they were mistaken for enemy aircraft. Richard made notes on every crashed glider that made it to land. Before leaving Sicily on September 6, he said openly, "I have seen enough of the gliders to know that something drastic has to be done in training and technique if they are ever going to be a very big thing in warfare."

Upon his return to the U.S., Richard gave his report to General Arnold. The next day he left for March Airfield in California to review the progress of the Bowlus/Criz-built XCG-16 experimental military glider.

On September 11, 1943, Col. Ernest Gabel, pilot, Howard Morrison, copilot (and a test pilot), C. Chandler, Paul Wells, Harry Perl and Richard du Pont took off for a test flight in the XCG-16. Loaded along with the six men was 850 lbs of buckshot ballast. At 1500 feet the glider started to porpoise, which caused the ballast load to shift. The tow line was released and the glider entered into a flat spin. Although Richard du Pont and two others managed to bail out, Richard's parachute failed to deploy fully due to the low altitude, and he perished. The other two parachuted to safety. Three other men aboard the glider could not deplane in time and died on impact. A military investigation followed the tragedy. A full account of the inquiry can be reviewed from documents available at the

Eleutherian Mills Historical Library, Greenville, Delaware. Richard's death was a major loss for All American and the du Pont family.

Richard C. du Pont, posthumously received several honors. On November 15, 1944, Richard was awarded the Evans Glider Trophy for the greatest individual contribution to glider development. In addition to the Evans Glider Trophy, the Guggenheim Aeronautical Laboratory dedicated the "Richard C. du Pont Memorial Room" for his contribution to aeronautical science. Present for the formal presentation were Allaire and A. Felix du Pont. The room would be used for aeronautical engineering students and for symposia in aeronautical science. His greatest award was presented to the family by General "Hap" Arnold. It was the Distinguished Service Medal, signed by President Franklin D. Roosevelt. The inscription reads as follows:

> For exceptional meritorious service to the Government in a position of great responsibility from April 20, 1943, to September 11, 1943. In the capacity of special assistant on the Army Air Forces program, Mr. du Pont displayed exceptional professional ability and inspiring leadership in planning and supervising the effective utilization of military gliders and related equipment. Under his skillful direction, full exploitation of the large glider for airborne operations was rapidly attained. He effected important changes in training technique which constituted practical application of current combat requirements personally observed in the Sicilian campaign. Charged with overall supervision of research in the field of unpowered flight, he maintained close coordination with civilian contractors in their efforts to improve glider equipment. While participating in an experimental flight of a new type of glider, Mr. du Pont was killed at March Field, California, on September 11, 1943. His extensive background of glider flight experimentation, coupled with his intelligent appraisal of the immediate requirements of the glider program of the Army Air Forces, contributed materially to the successful accomplishment of a mission of the greatest importance to the war effort of the United States Government.

World War II affected almost every company in the country. All American was no exception. Airmail pickups increased dramatically due to families writing servicemen. In spite of the increased volume, the

General "Hap" Arnold presents Allaire du Pont, A. Felix du Pont, and eight-year-old "Kip" du Pont (Richard's son) with the Distinguished Service Medal. Photo from the du Pont family collection.

UNIT	MAX. GLIDER WEIGHT	SPEED	WT. OF UNIT	CABLE SIZE	CABLE LENGTH	CABLE BREAKING STRENGTH	DIMENSIONS: WIDTH	DIMENSIONS: HEIGHT	DIMENSIONS: FORE & AFT	K.E. CAPACITY
Model 4	550 lbs. 250 kg	70 m.p.h 110 km/h	95 lbs. 43 kg	1/8 in. 3.18 mm	700 ft. 213 m	2000 lbs. 910 kg	15 in.* 394 mm*	22 1/8 in.* 570 mm*	19 1/2 in.* 495 mm*	100,000 ft-lb. 13,800 m-kg
Model 15	1500 lbs. 680 kg	120 m.p.h. 195 km/h	144 lbs. 65 kg	1/8 in. 3.18 mm	1000 ft. 305 m	2000 lbs. 910 kg	23 in. 585 mm	24 in. 610 mm	24 in. 610 mm	720,000 ft-lb. 100,000 m-kg
Model 40	4000 lbs. 1800 kg	120 m.p.h. 195 km/h	275 lbs. 125 kg	1/4 in. 6.35 mm	1000 ft. 305 m	7000 lbs. 3200 kg	27 1/2 in. 700 mm	26 in. 660 mm	26 in. 660 mm	1,930,000 ft-lb. 265,000 m-kg
Model 80	8000 lbs. 3600 kg	120 m.p.h. 195 km/h	620 lbs. 280 kg	3/8 in. 9.5 mm	1080 ft. 330 m	14,400 lbs. 6500 kg	40 in. 1020 mm	31 1/2 in. 800 mm	30 in. 760 mm	3,610,000 ft-lb. 500,000 m-kg
Model 81	8000 lbs. 3600 kg	135 m.p.h. 215 km/h	778 lbs. 355 kg	3/8 in. 9.5 mm	1080 ft. 330 m	14,400 lbs. 6500 kg	33 in. 840 mm	29 1/2 in. 750 mm	30 in. 760 mm	5,000,000 ft-lb. 690,000 m-kg
Model 120	12,000 lbs. 5450 kg	140 m.p.h. 225 km/h	1250 lbs. 565 kg	1/2 in. 12.7 mm	1175 ft. 360 m	22,800 lbs. 10,300 kg	44 in. 1120 mm	40 in. 1020 mm	39 in. 990 mm	7,710,000 ft-lb. 1,065,000 m-kg
Model 160	16,000 lbs. 7250 kg	140 m.p.h. 225 km/h	1900 lbs. 860 kg	9/16 in 14.3 mm	1350 ft. 410 m	35,000 lbs. 16,000 kg	45 in. 1140 mm	36 in. 915 mm	52 in. 1320 mm	10,500,000 ft-lb. 1,450,000 m-kg
Model 200	20,000 lbs. 9000 kg	150 m.p.h. 240 km/h	3100 lbs. 1400 kg	9/16 in. 14.3 mm	1350 ft. 410 m	35,000 lbs. 16,000 kg	49 in. 1245 mm	62 1/2 in. 1590 mm	74 1/2 in. 1890 mm	15,000,000 ft-lb. 2,070,000 m-kg

*Excluding Electric Motor.

A Model 40 with drum guard removed. Emergency rewind crank in foreground.

A Model 80, complete, ready for shipment.

Cable payout side of a Model 160. Operator's side to the right.

Airmail Pickup Division lost $74,000 in the most prosperous year ending June 30, 1945. For reasons only known to management, All American continued to pursue this single purpose product line which was a constant drain on company funds. The number of employees had also grown from the original 28 to 306. Another drain were the legal expenses incurred whenever the company wished to make a change in operations, which were under government control.

However, the overall situation was to improve during the war years. The Manufacturing & Development Division secured government contracts for aircraft sub-assemblies, and a separate contract for training pilots for the Army Air Corps which operated from the DuPont Airfield. Graduates from the flight training program were transferred to Air Transport Command (ATC). The Manufacturing & Development Division for the period 1944-1945 posted earnings of $850,000.

The Engineering Department, also under government contracts, developed and tested rescue equipment, arresting systems, and seat ejectors for the Army Air Corps. One of their outstanding accomplishments during this time was the human pickup. Initial pickups, using containers with instruments enclosed, indicated a seventeen-G positive force at pickup. This was well beyond human endurance. However, changes in the harness and transfer line brought the Gs down to seven. Tests continued with dummies and a live sheep, selected because of its fragile bone structure. The next test was scheduled with a human-size chimpanzee named "Barbette." However, when test pilot Norman Rintoul learned of the planned event, he replied, "That chimp better know how to fly, cause when she comes aboard, I'm out of here!" Much to the delight of Rintoul and Barbette, the mission was scrubbed.

It would take over 100 successful pickups before a human being was allowed to make the attempt. A paratrooper, Lt. Alexis Doster, volunteered, and the stage was set at Wright Field, Dayton, Ohio, on September 5, 1943. Norman Rintoul streaked across the field in the Stinson *Reliant* and made a perfect snatch. Doster was aboard the airplane in less than

By 1961 the entire DuPont Airfield complex was occupied by all American Engineering. Photo from the Hagley Museum & Library.

three minutes. In spite of this engineering accomplishment, the system was not put into service. The introduction of helicopters made the method obsolete before it could be used.

All this growth during the war years for All American was not easily accomplished without other serious hardships. Many experienced personnel were obligated to serve in the armed forces and therefore left the company for the duration. There was also an immediate need for more space for manufacturing and classrooms for pilot training. Atlantic Aviation Service (not to be confused with Atlantic Aviation) owned by H. B. du Pont, who also owned the DuPont Airfield, built a 30,000-square-foot building there and leased it to All American to fulfill their needs.

All American's Engineering Division also played a major role during WWII in the rescue of military personnel. One particular incident stands out when a Douglas C-47 crashed in a remote valley in New Guinea. The three injured survivors were stranded, and paratroopers were dropped to assist in the rescue attempt. Helicopters had not been developed at this point to accomplish the mission. The paratroopers took forty-nine days to manually cut a swath through the jungle long enough for a glider to land. The successful rescue was made with a Boeing B-17 *Flying Fortress* equipped with an All American-designed pickup system.

POST WAR ADJUSTMENTS

The end of the war created major readjustment problems for businesses as well as returning servicemen. Most government contracts were for the duration of the war only, and they abruptly ended, creating an idle work force. Management scrambled to find ways of creating markets in a peacetime economy. Many businesses, which operated solely on wartime contracts, simply shut down completely.

All American's three divisions each had specific problems for management to address. They did not have prewar customers who would renew business. All American Airways initial focus was on the airmail pickup system which had been a real financial loser. To make matters worse, airmail pickup dropped off dramatically without military pickups and families writing servicemen.

The Manufacturing Division lost their war contracts and started to look into fabrication of domestic products for survival. This Division was fortunate to find a few domestic accounts producing non-aviation items such as trusses, jackets for water heaters, and castings.

The Engineering and Research Division took the smallest cut. The government needed to address many problems associated with the new age of supersonic aviation. A whole new world of aeronautical engineering was about to unfold. The All American Engineering Division utilized the now-vacated Sussex County Airport, testing rapid acceleration/deceleration equipment utilizing rocket sleds. Catapult systems were engineered for aircraft carriers. Rapid deployment of materials from low-flying aircraft was tested and proved. In-flight retrieval of satellite capsules was another interesting project.

AN AMBITIOUS VIEW OF THE FUTURE

All American's president, Hal Bazley, decided to invest some of their wartime earnings in new quarters. Corporate offices would be established in a building, in need of renovation, located at 210 Greenhill Avenue, Wilmington. One additional hangar would be built on the DuPont Airfield in anticipation of the postwar boom. The board also approved the sale of one million shares of common stock for working capital.

All American, in an effort to bolster revenues for the airmail pickup system, petitioned the CAB to approve eight new routes in New England and New York and an additional twenty-three in the Southwest. Additional applications were filed to approve thirty-five new routes in the Great Lakes and Middle Atlantic States. In addition, citing their wartime record of performance and safety, they reapplied for permission to carry passengers en route. This was a very ambitious undertaking and, if approved, would require a huge expenditure for 128 new aircraft needed to fulfill the obligation. The Canadian-built Noordun *Norseman V* was selected as the replacement for the ageing *Reliants.* Fortunately, All American only purchased one, as it proved to be too heavy on the controls to accomplish the low-level maneuvers safely. Once this was confirmed, the *Norseman* was placed on the market and sold. There appeared to be nothing available that could replace the ageing Stinson *Reliants.*

Hal Bazley decided upon a new, but expensive, approach to convince the CAB to approve their applications. All American's board approved his idea to purchase two new Beechcraft twin-engine model D18C aircraft and have them fitted with the pickup system. Bazley felt the multi-engine airplane would add a high degree of safety to hopefully sway the CAB to approve their request to include passengers during pickup operations. The order was placed November 1945, and delivery of the first airplane was made in May 1946. Unfortunately, problems surfaced with the hydromatic propellers and the installation of the pickup system. These problems caused testing to be delayed, and the demonstrations for the Post Office Department were shelved temporarily.

By June 1946, All American was once again experiencing liquid cash problems. The board approved the sale of 100,000 shares of common stock at $9.50 per share to pay for new equipment and needed operating capital. A portion of these funds was earmarked for a newly proposed Lockheed *Saturn,* which was being developed specifically for feeder airline service. Unfortunately, Lockheed only produced one airplane and decided there were insufficient orders to go into production. Hal Bazley was unshakable in his conviction that the CAB would eventually allow All American to include passengers on their pickup runs with the Lockheed. Charles

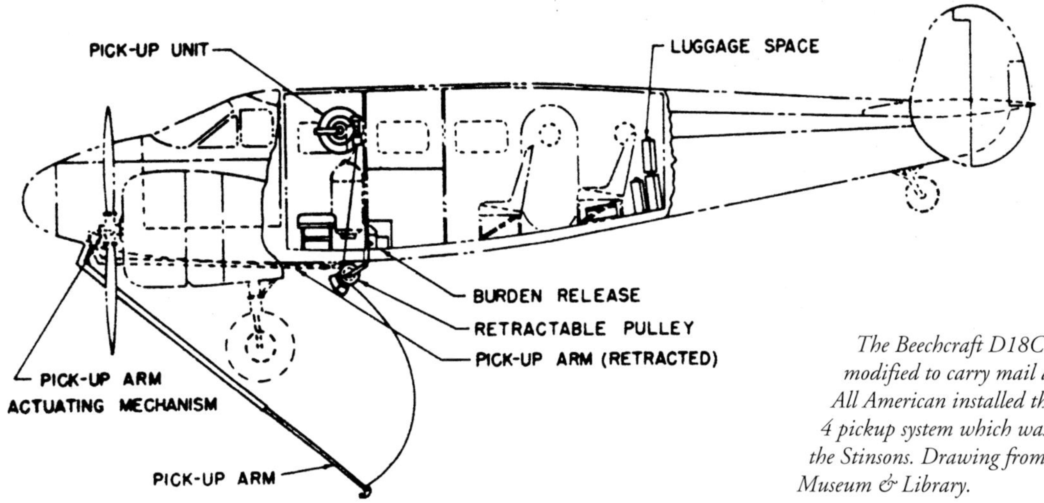

The Beechcraft D18C was extensively modified to carry mail and passengers. All American installed the same model 4 pickup system which was installed in the Stinsons. Drawing from the Hagley Museum & Library.

Wendt felt otherwise and had felt for some time that All American's continued focus on the airmail pick-up system would eventually bankrupt the company.

Privately, Charles Wendt searched for a new leader to replace Hal Bazley. This action was thoroughly thought out before any effort was made. Bazley's focus was on continuing Richard du Pont's dream. Charles Wendt felt if All American were going to survive, new leadership, with new ideas and vision of the future, would be necessary.

Robert MacLure Love, a graduate from Princeton and MIT, had learned to fly in 1929 and served as a colonel in the Army Air Corps during the war. More important was his background in commercial air-carrier operations. He had, by 1932, organized and operated a successful airline between Boston, Springfield, and Albany. When the war broke out in 1941, Robert Love played an important role in establishing the Air Transport Command (ATC). During this period he also worked closely with American Airlines president C. R. Smith. Although Wendt considered others, Robert Love was his first choice.

When Charles Wendt met Robert Love in New England for the first time, he found Love was making plans to start an air carrier operation of his own. When Love was promised a free hand in turning All American Aviation around, and was given the option of buying 15,000 shares of company stock, Love accepted the challenge. In the change of command, Bazley returned to the position of vice president in charge of operations and moved to Pittsburgh. Bazley was not unhappy with this arrangement since he felt from the very beginning his position as president was temporary. Wendt took on an additional role as vice president of the Manufacturing Division, as well as treasurer of All American Aviation, to reduce costs. In addition, Wendt wooed A. Felix du Pont back on the board to serve as a director. Harry Lunger, esq., a du Pont relative, was named corporate attorney. Allaire du Pont, who had inherited Richard's stock, also served as a director.

With the new board in place, a thorough evaluation of the three operations of All American Aviation ensued. Although All American was the big loser, the Manufacturing Division was also losing money. The Engineering Division was the only profitmaker. The board agreed with Love, that an honest attempt should be made to save the three divisions in light of the vast investment over the years in equipment and personnel. The board approved the sale of 100,000 shares of common stock, which sold for $9.75 per share. This would be used for operating capital and the purchase of the two new Beechcraft D-18s previously ordered by Bazley.

Although the next fiscal year recorded almost 1.8 million revenue miles flown, and almost 94 percent reliability, losses totaled over $94,000. A study of the problems, and how they could be corrected, was undertaken. The number of miles flown and reliability meant little if the mailbags were not full. The men in uniform were home and the airmail volume had dropped off significantly. The Post Office Department was again reluctant to support the CAB in allocating new routes and increase prices in the face of lower volume.

All American decided to appeal to the Post Office Department for a retroactive increase in their services to offset losses. These appeals took many months of legal testimony to grind through the bureaucratic process before decisions were handed down.

The singular, and most important, restriction which All American wanted lifted by the CAB was the inability to carry passengers during mail pickup operations. Once again, in September 1946, the All American board of directors agreed to hold demonstrations to prove their concept was safe. The proper dignitaries from the CAB, Congress, the Post Office Department, and the news media were invited to fly in the new twin-engine Beechcrafts while mail was deposited and picked up. Eighteen separate flights departed from Washington National Airport for Hybla Valley Airport in Virginia, and Clinton Airport in Maryland for mail pickup demonstrations. Although the pickups were flawless, and there were no complaints of passenger discomfort, the approval All American wanted so desperately was not granted.

At the suggestion of A. Felix du Pont, Sr., who was also a director on the board of Piasecki Helicopters, All American proposed the same request using Helicopters instead of airplanes. The CAB also denied approval of this application. Another setback came when a request for airmail pickup was denied All American in the Southeastern United States. The CAB designated feeder airline routes to Piedmont and Southern Airways, citing their experience in their native territory. The CAB also felt the feeder lines could accommodate the Post Office Department's mail pickup service during their normal stops. When this request was also denied All American, it finally became evident the joint operations were not negotiable in the eyes of the CAB and the Post Office.

However, the board recognized a pattern emerging in the CAB allocation of routes. The CAB was showing favoritism toward those with a history of aviation in their local area. These regional air carriers, more often than not, also received approval for additional routes in their area when applications were submitted.

The mid-Atlantic feeder line routes had not been allocated by the CAB at this time. All American Aviation finally agreed the airmail pickup system, which was Richard du Pont's dream of "airmail service to everywhere," was all but dead. Applications were submitted to the CAB early in 1947 for mid-Atlantic routes for passenger service. With the applications, All American agreed to phase out the airmail pickup system if the passenger routes were granted. More than likely, there were shouts of joy from the CAB and the Post Office Department when this news was heard. Allocating the passenger routes to All American would free the federal government from the huge financial burden of financing a system which they felt had no chance of becoming self-supporting.

A. Felix du Pont, Sr. and his son Felix Jr. both served as directors on the board of Piasecki Helicopters and All American Aviation. This, the CAB felt, could be construed as conflict of interest in spite of the fact Piasecki only built helicopters for the military. To appease the CAB, the du Ponts sold their stock to Robert Love and were no longer affiliated with All American Aviation. This acquisition of stock by Love placed him firmly in control of All American Aviation and its three divisions. Allaire du Pont, Richard's widow, also sold out and no longer served on the board.

Robert Love had a definite plan to change the company. The new board would be made up of people who had no previous ties with Lytle Adams, Richard du Pont, or the pickup system. Love needed a supportive board which would look to the future and would not dwell upon the past. The only exception was Charles Wendt, treasurer, who had sought out

Love for the task at hand. It would not be an easy task to rebuild, and there were no guarantees.

One major obstacle remained for Robert Love. The third quarter financial statement, ending March 1947, disclosed All American Airways and the Manufacturing Division posting losses. All American Aviation, in spite of the Engineering Division being profitable, posted a staggering $156,660 in losses. Once again All American was cash poor. The financial bleeding had to be stopped immediately! The Manufacturing Division, which was losing money, was shut down and the equipment sold at auction, which raised $141,000. The Engineering Division, which was profitable, was retained as an operating division. A welcome additional lump sum income of $461,000 came from retroactive back-payment increases previously applied for, from the Post Office Department, by All American for its airmail pickup services.

With the company in a new, positive cash position, and the CAB happy with the board personnel, the door seemed open for a change. It was only a matter of the CAB approving the routes. This would take longer than expected, and the ageing Stinsons and the two Beechcrafts continued the airmail pickup until the CAB and All American's lawyers finally completed acceptable language which would dissolve the airmail routes and relieve the government of any future litigation.

Not surprisingly, on February 19, 1948, the CAB announced that All American was selected over other applicants as the carrier for the Mid-Atlantic routes. On October 4, 1948, Love changed the name to All American Airways, Inc. Six separate routes were approved. These were: (1) Washington, Frederick, Hagerstown, Martinsburg, Cumberland, Connellsville, Uniontown, Pittsburgh. (2) Washington, Baltimore, Dover, Bridgeton, Millville, Vineland, Atlantic City. (3) Pittsburgh, Johnstown, Altoona, Harrisburgh, Lancaster, Wilmington, Philadelphia, Camden. (4) Pittsburgh, Butler, Oil City, Warren, Bradford, Jamestown, Dunkirk, Buffalo. (5) Pittsburgh, Wheeling, Parkersburgh, Athens, Chillicothe, Cincinnati. (6) Washington, Baltimore, Easton, Cambridge, Salisbury, Ocean City, Rehoboth, Georgetown, Cape May, Dover, Wilmington. Rehoboth, Ocean City, and Cape May were serviced only during the summer months. The total distance for the six routes was calculated to be approximately 1500 miles.

Although things were looking up, the company was not yet solvent. An additional $250,000 was needed to purchase the Douglas C-47 (military version of the civilian DC-3) military surplus airplanes. Among other economic decisions, the All American Aviation corporate office building in Wilmington was put up for sale. The profitable Engineering Division was retained. Plans were to sell the ageing Stinsons and Beechcrafts during a gradual shutdown of the airmail pickup system. The last run would be on the original route between Pittsburgh and Jamestown, New York, on June 30, 1949. Norman Rintoul, who piloted these sturdy airplanes since the beginning, purchased the last Stinson and donated it to the Smithsonian. It is on display at the NASM Paul Garber Facility, Silver Hill, Maryland.

ALL AMERICAN AIRWAYS

A leased hangar at Washington National Airport would become the corporate office and base of operations for All American Airways, Inc. This location, Robert Love suggested, would put the company closer to the CAA and CAB, which it must continue dealing with over the years.

In spite of selling off all the tangible assets to raise cash, All American still needed $800,000. This was made possible by securing a loan from the Reconstruction Finance Corporation, which would allow funding for the purchase of eight, military-surplus Douglas C-47s for $33,000 each. These aircraft would be returned to Douglas to be converted for civilian use. The cost for the total rebuild of each airplane by the Douglas factory was $78,000. The estimate of eight airplanes was found to be too conservative, and three more had to be ordered before operations started. Although the airmail pickup system was now history, All American gave preferred employment options to existing employees of All American Airways. Although most were heartbroken to see the pickup system die, the closeness of the years of working together allowed the unit to work together and accept the transition. Retraining of personnel was implemented immediately to assure a smooth transition. Stinson pilots were now trained in the DC-3s. Ground crews were trained to service aircraft, passengers, and handle baggage. It would indeed be a totally different world for everyone.

By March 7, 1949, an All American Airways, Inc. Douglas DC-3, departed Washington for Cumberland-Hagerstown-Martinsburg-Pittsburgh. Brief celebrations took place at each stop along the route. The Douglas DC-3s were so thoroughly redone it was not possible to distinguish the airplanes from new ones. They presented a good image for the company.

Flying an airline schedule at low altitudes is not a small task to accomplish. In these early days of airline operations, airliners had to contend with the low altitude weather. More often than not the flight line between two points was anything but straight. The eleven airplanes were utilized to the maximum to complete the seven routes. (An additional route was approved later between Pittsburgh and Buffalo.) Ground crew training paid off with two minute turn around times to load passengers and baggage. All American Airways was committed to making an impression on the flying public, and they worked hard to please. During the first year All American carried an average of 12,000 passengers each month, and the fleet of airplanes amassed 9,000 hours in the same period. However, in the first nine months the company lost $587,000. The problem was immediately identified when the seating capacity was calculated at a dismal 22.5 percent. Love identified the routes of low volume and immediately petitioned the CAB to allow All American to suspend service to those areas. This action corresponded with a layoff of personnel and reduced the losses to $250,000 for the first year of operations ending June 30, 1950.

Robert Love initiated a massive advertising campaign which included flight-reservation connections with TWA flights. This action virtually doubled the number of seat reservations. The 1951 annual report confirmed the results of Love's efforts by posting a net profit of $4,660. Although this may appear paltry now, it was a quantum leap for All American Airways management.

The board of directors had discussed on many occasions the merits of separating All American Airways and the Engineering Division into two separate companies. The Engineering Division, although aviation related, had its major contracts with the military. None of its services were related to aerial transportation of passengers. Now that the company offices were located in Washington, the subject was once again addressed about whether the two locations (Wilmington and Georgetown, Delaware) of the Engineering Division should be retained. The proposal was submitted to the stockholders for approval. For legal and tax reasons, the Engineering Division was sold to a wholly owned subsidiary on October 31, 1952. The name was also changed to "All American Engineering Company," and Charles Wendt was named as president.

This board action virtually separated the two companies into two distinct entities. Since the airline operated throughout the Allegheny Mountain area, the idea for a regional name for identity was selected. Effective January 1, 1953, All American Airways, Inc.'s name changed to Allegheny Airlines, Inc. Under Love's leadership, (now chairman of the board) and newly elected president Leslie Barnes, the company would focus its energy toward a singular niche; the carrying of passengers by air. The aggressive board appeared to do everything right, and the company prospered and expanded over the years by taking over Lake Central Airlines in 1968. This was followed by the acquisition of Mowhawk Airlines in 1972. By 1983, the company grew to such proportions that the Allegheny Airlines, Inc. name was considered too regional and the name USAir, Inc. was selected. What was once conceived by Richard C. du Pont as a national airmail pickup operation now became a jet-propelled national airline. The year 1984, USAir flew seventeen million passengers and boasted an incredible 121.6-million-dollar profit.

ALL AMERICAN ENGINEERING COMPANY

The three test tracks

Test track number one was located in front of the control tower and ran parallel to the 5000-foot run-

The main building and control tower at Georgetown. The Piper Super Cub *was used on many occasions to fly to/from the DuPont Airfield in Wilmington. Photo donated by Jean Cochran.*

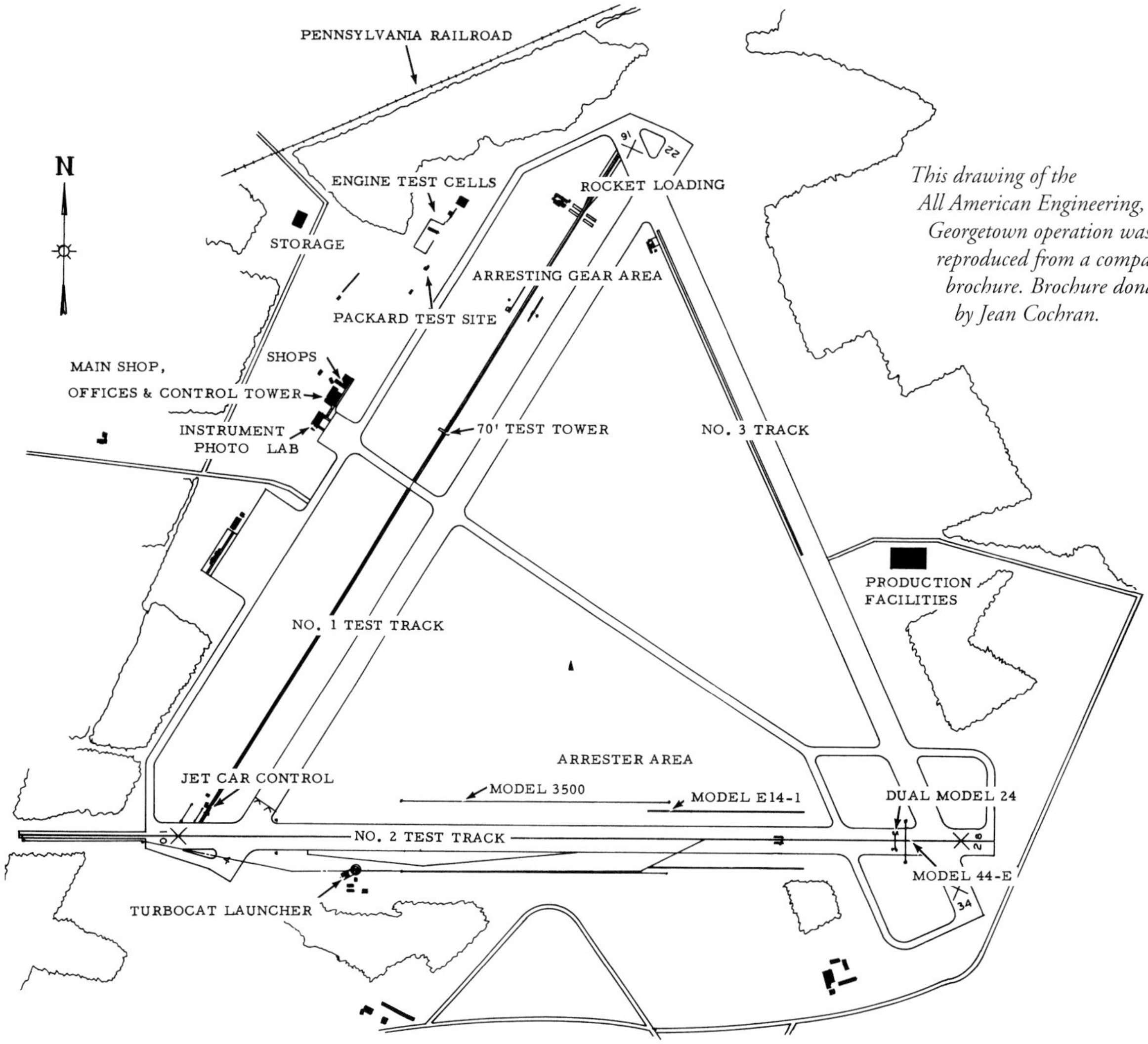

This drawing of the All American Engineering, Georgetown operation was reproduced from a company brochure. Brochure donated by Jean Cochran.

A simulated jet fighter launched down the test rail is captured in the All American designed runway overrun restraint system. Photo from Jean Cochran.

way 4-22. It used a single surface-mounted steel rail for guidance and braking tests. This track was 4500 feet long and was also used to test energy absorbers. A 70-foot-high test tower over the track was used to calibrate airflow through parachutes.

Test track number two was 6,000 feet long and was installed flush in the centerline of runway 10-28. A number of All American's arresters were located on this track including the 200,000,000 plus foot-pound Model 3500, High G Decelerator for testing aircraft structures. It had the capability of inflicting 1,250,000 pounds of retardation force with controllable deceleration forces to 25 Gs. It was also able to reproduce kinetic energy up to 8,000,000 foot-pounds.

Test track number three was located on runway 34-16 and was used for catapult testing. All American was heavily involved in structural testing of airframes involved in catapult launches of aircraft. The All American catapult design was accepted and put to use by the U.S. Marines for their SATS (Short Airfields for Tactical Support) program. Power sources used for testing on the three tracks were rockets, jet-powered cars, and a Turbocat, 50,000-hp gas-turbine accelerator.

In addition to the three test tracks, Georgetown housed the Packard Test Site, so named because of its power plant. Powered by a Packard marine engine, the test device would permit close observation and measurement of wire ropes under various speeds and loads. This device was also used to develop a capstan cable-drive system which operated at 1500 rpm, and to conduct feasibility tests on the tailhook-arresting-gear engagement at speeds of 230 mph. On the same site there were two jet or rocket engine test cells, one of which was used in the development of the world's first vectored thrust VTOL (Vertical Takeoff and Land) aircraft, the Bell X-14.

Charles Wendt, president of All American Engineering, guided the Wilmington-based company through years of prosperity. All American-designed recovery equipment successfully snatched the first *Discoverer*

Perfecting the seat ejector at the Georgetown Airport was a major step in saving pilots' lives in the event a low altitude bailout was necessary. Photo from the Hagley Museum & Library.

XIV space capsule in mid-air with a Fairchild C-119. Countless lives were saved due to the company's development of retrieval systems whose embryo was the airmail pickup system. The Mid-Air Retrieval System (MARS), developed in the 1960s, snagged many parachutes deployed from satellites. A fifty-pound winch named "Mighty Mite," mounted in a helicopter, had the capacity to pick up a staggering 6,000 pounds of dead weight. The aircraft tail-hook, which snags the line on an aircraft carrier, is another All American contribution to aeronautics.

Although All America Engineering perfected the seat ejector, they were not the first to attempt saving people from falling aircraft. As early as 1933, the *Washington Daily News* reported the testing of a device installed in an eight-passenger Buhl aircraft. The system, activated by a lever by the pilot, would open a side door and fling the entire seat (with the passenger attached thereto) out into the void. Sleeping passengers would not be required to be awakened. A static line would automatically pull the ripcord to activate the parachute, allowing the passenger to float gently to earth. A passenger who might become unruly on his way to Miami, might find himself dangling from a pine tree in North Carolina. It is doubtful passengers thought this addition to be an enticing reason to purchase an airline ticket. Although the system worked, the apparatus failed to be accepted by the traveling public.

A much later attempt, by another company, was a pod design. In this configuration the entire cockpit section was deployed at the command of the pilot. Although the system proved successful at high altitudes, it failed to pass the test when activated at low altitudes. Another consideration was the weight of the pod also required a huge parachute to allow a safe rate of descent.

To support and collect data, All American established two separate groups. The Instrumentation Group was mobile as they were housed in air-conditioned trailers. These were equipped with the latest in technology, including racks of signal conditioners, oscillographs, motor generators, power supplies, and data storage recorders which captured vital information. These trailers could be moved to any location in the country to support off-site testing programs. Typical of such equipment use was the special console for the arresting gear/tailhook programs tested on the Boeing 720 and the FAA's C-131.

The Instrumentation Group, which was a highly sophisticated photographic group, was formed to provide useful optical data. Their inventory included: a Hulcher, 70mm, electric sequence motion-picture camera capable of 1-20fps; a Kodak hi-speed, 3000fps camera; and a Photosonic, 16mm IB camera (6-1000fps) capable of surviving high G-loads. These cameras were supported by a wide selection of still cameras, still and motion film development systems, movie projectors, etc. This group was also mobile as it was housed in a specially engineered trailer capable of full operation without an outside power source.

In addition to surface testing, All American was heavily involved with testing crosswind landing gear, in-flight refueling and rearming, airborne cargo delivery systems, rocket assisted takeoff, and vectored thrusting. Aircraft used for testing purposes included everything from Piper *Cubs* to multi-engine jets, supersonic fighters and helicopters.

The 22,000-square-foot production building, located off runways 16-34, was used to manufacture a variety of low-volume components. Guide rails, drive cable assemblies, rigging and packing of aerial balloon stations were but a few. High-tolerance machine work, welding, mixing plastics, fabricating, castings, sewing nylon cargo-harnesses were also products produced by this facility.

All American became a subsidiary of International Controls, Inc. in early 1970, and a name-change was recorded as All American Industries, Inc. Charles Wendt, after serving the company since 1939, retired as president shortly thereafter. The new company then consolidated all operations in New Jersey. Although this is not a complete accounting of all of All American Engineering's accomplishments, it does reflect some of the important goals they achieved. What started as a small division of All American Aviation by Richard C. du Pont, to create an operational airmail pickup system, the Engineering Division grew to leave an indelible mark in world aviation history.

This modified Fairchild *C119* Flying Boxcar *deposits a Ford pickup truck on the runway of Georgetown Airport in Sussex County in one of many tests performed by All American's Engineering Division. Note the nonstandard jet engine mounted to the fuselage for additional power. Photo from Hagley Museum & Library.*

This Bellanca Cruisemaster *is equipped with retractable hydro-skis. At least 15 mph would have to be maintained to keep the airplane afloat. The system, installed on Stinson STOLs, saved many downed pilots' lives during the Korean conflict when they landed and took off on mud flats. Photo from the Hagley Museum & Library.*

The early days of testing hydro-skis were not always successful as this modified Piper Cub *decided to flip on its back. Note the tricycle landing gear. Photo from the Hagley Museum & Library.*

Scale models were used for testing prior to moving to full scale. The U.S.N. was looking for better ways to stop aircraft from crashing off the end of an aircraft carrier in the event of an overrun. Photo from the Hagley Museum & Library.

A record that will probably never be broken. All America's Engineering Division received this plaque from General Ben Funk (right) and Captain Vomadeau (left) on October 30, 1961, after 100 successful aerial pickups had been accomplished. Over 12,000 aerial recoveries were completed before the program ended. Photo from the Hagley Museum & Library.

CHAPTER 7

Three For Delaware

In spite of the Neutrality Act passed by Congress in 1937, which forbade American businesses to supply war materials to any nation engaged in war, Washington, by 1940, accepted the fact that neutrality was no longer an option. The Atlantic Seacoast would be vulnerable if the United States were forced into a military conflict. The 76th Congress passed Public Law No. 812 to remedy the situation which called for federal funding of such airports. The decision was made to allow the Civil Aeronautics Authority (CAA) to offer federal funding to states bordering the East Coast in order to build the civil air terminals.

The federal aid was very enticing since the only requirement was that a local government (state, county, or city) own the land. The CAA engineers would design the entire infrastructure, and construction would be bid out to contractors. Once the facility was complete, maintenance of the buildings and grounds would become the responsibility of the municipality. In the event of war or national emergency, the Government would lease the airports for the duration. In Delaware, the CAA approved New Castle County, Kent County, and Sussex County as sites for municipal airports.

Governor Walter W. Bacon and the legislature declined the CAA offer. The state of Delaware did not want to accept the permanent financial responsibility of three airports. This decision should not be blamed on Governor Bacon. Historically the state of Delaware has not looked favorably toward financially supporting aviation. The state, however, suggested to the counties that they could pass legislation to allow the counties to enter into the CAA agreement. (Counties in Delaware did not have "home rule," therefore they were not authorized to enter into such an agreement without the approval from the legislature.) Both the New Castle County and the Sussex County Levy Courts agreed to the state proposal. The Kent County Levy Court declined. Although no reasons are documented, it is possible Kent County also felt the long-term financial commitment was

Original concept drawing of Dover Airport dated 1940. From the Delaware Public Archives.

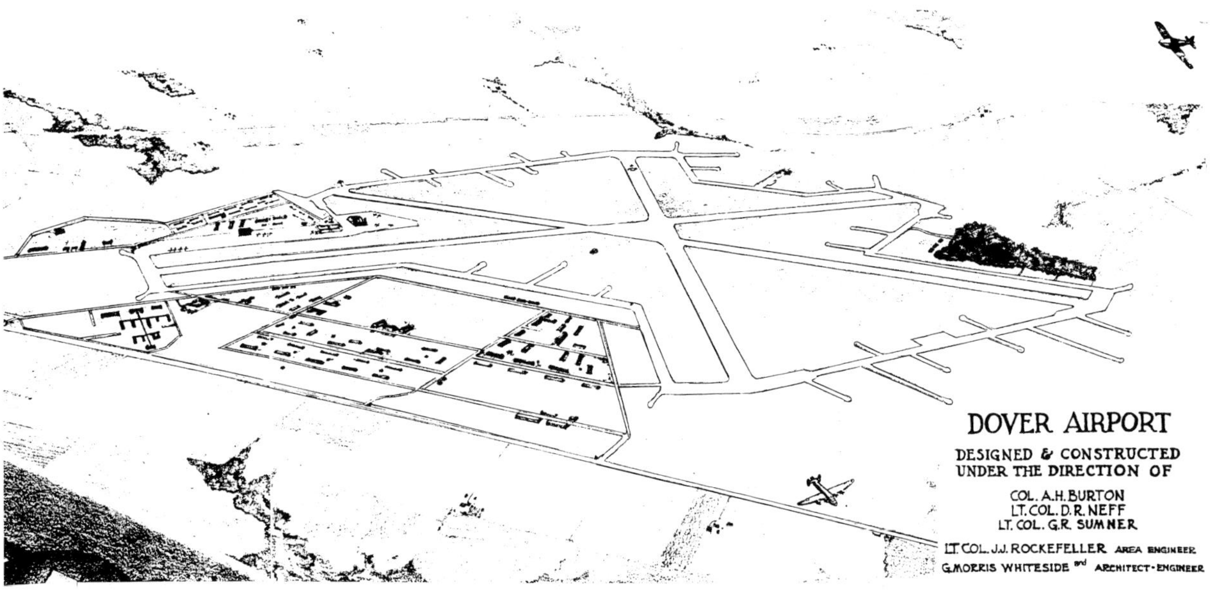

beyond their means. In addition, the local Dover Airfield appeared to fulfill the needs of the community. After Kent County rejected the offer, it was passed on down to the city of Dover which accepted it. Since the Enabling Act of 1931 (see page 56), which allowed cities to purchase land for the construction of airports, special legislation by the state was not required for Dover to proceed. The city of Dover would become the first of the three to break ground for construction.

Site locations and zoning were reviewed for the three locations. Since each location consumed a large amount of territory, several parcels of land were joined together to create each airport. This in itself was no small task, since agreements needed to be reached between the buyer and the seller. In some cases, agreements could not be reached, and the process went to the courts under the condemnation law.

➲ Dover Municipal Airport

The city of Dover accepted the CAA offer in 1940. After approval of the site, the city purchased approximately 587 acres from Charles Paradee, Arthur Davis, and Michael Scanlon, southeast of Dover, for approximately $35,000 in September 1940. The first stage of construction called for three hard-surfaced runways and a foundation for a hangar. The three hard-surfaced runways were laid out in such a manner to allow pilots to take advantage of optimum wind conditions. Bids for construction were requested on February 5, 1941, and construction was underway in March. The projected cost of the project was $563,000, which was borne by the CAA. Bureaucratic red tape created sporadic releases of funds, and many delays were incurred during initial building.

On December 7, 1941, the Japanese bombed Pearl Harbor, and Congress declared war on Japan on the 8th of December. In rapid succession, on December 11, Germany and Italy declared war on the United States of America. In three days, the United States of America entered into global warfare. The Department of War immediately leased the three unfinished Delaware airports from the municipalities. All three airports were renamed as follows; New Castle Army Air Base (NCAAB), Dover Army Air Base (DAAB) and Georgetown Navy Air Base (GNAB). The Army Corps of Engineers worked around the clock to finish the airports and make them operational. Meanwhile, German submarines played havoc on our merchant marine ships, sinking them

The first aircraft to be assigned to DAAB was the 0-47, Owl. *Its mission was submarine patrol. Photo from NASM.*

without any confrontation whatsoever. Invasion of our shores on the Atlantic by the enemy was considered a real possibility.

The Army Corps of Engineers came to Dover to complete the infrastructure in record time by working around the clock, seven days a week. During the war, the U.S. Government purchased additional land bordering Dover Army Air Base (DAAB) from adjacent owners for expanding operations. This is interesting since federal land would now border city-owned airport property. This action may have contributed to the Air Force decision to retain Dover as a military facility after the war.

A few days before Christmas 1941, the first personnel from the 112th Observation Squadron (Ohio National Guard), a federalized National Guard unit, arrived at DAAB. Their mission was to fly 0-47 Owl observation aircraft on anti-submarine patrol. The 103rd Regiment, Corps of Engineers, an engineering group of the Pennsylvania National Guard, arrived a few days after the 112th Ohio Guard. Their mission was to build an infrastructure on the base to include roads, barracks, mess hall, hangars and commissary.

The first building to be constructed was a single-story hangar 150 feet wide, 160 feet deep and 42 feet high, with a sliding door opening of 28 feet. The completed structure had no internal columns and was designated "Hangar 295." The hangar is now listed in the National Register of Historic Places and is occupied by the Air Mobility Command (AMC) Museum. Attached to the north side of the hangar was the power plant. Designed to be an integral part of the hangar complex, it was a two-story, gabled-roof building with cinder block walls. On the north side of the power plant was a workshop building. It was connected to the hangar by a long narrow corridor that formed part of the east wall of the power plant. The shop building was a simple steel truss, single-story, gable-roof building, framed with studs and plasterboard covering.

DAAB commissary (circa 1942) allowed military personnel the convenience of shopping on base and also enjoying substantial savings. Photo from the Delaware Public Archives.

A row of Republic P-47B Thunderbolts *of the 125th on the line at DAAB. This is one of the first variants of the P-47. This model was know as the* Razorback. *Photo from the Delaware Public Archives.*

Aerial view of DAAB circa 1942. The original rocket test center buildings (upper left) are now occupied by the Air Mobility Command (AMC) Museum. Photo from the Delaware Public Archives.

A P-47D Thunderbolt *from Dover AAB unleashes a salvo of air-to-surface rockets. Photo from the Delaware Public Archives*

On August 1, 1943, the 312th Army Air Corps BU (Base Unit) operations started combat training of Republic Aviation P-47 *Thunderbolt* pilots at Dover. The massive seven-ton P-47 *Thunderbolt* (nicknamed the *Jug*) was the heaviest fighter produced during WWII. In spite of its weight, the airplane was a very stable platform on the ground as well as in the air. By April 10, 1944, the 312th Base Unit (BU) became the 125th BU and continued training P-47 pilots in wartime exercises. DAAB flight instructors would train over one thousand P-47 pilots before the end of WWII. The P-47 *Thunderbolt* today is one of the rarest of warbirds in spite of the fact that more were built than any other U.S. fighter. The heavy fighter, with its 2300-hp Pratt & Whitney radial engine, could pass through 400 mph in a dive quicker than any other fighter of the time. In spite of its bulk and weight, the airplane was a pilot's dream to fly. It was light on the controls, had ample power, and had a faster roll rate than the famous, British-built, Supermarine *Spitfire*.

The 39th, 78th, 79th, and 80th Bombardment Groups flew North American B-25 *Mitchell* antisubmarine patrols from DAAB in response to calls from the CAP. Later, some aircraft from the 4146th Base Unit were fitted with air-to-surface rockets for experimental testing. However, the B-25's most historical achievement during WWII was the raid on Tokyo, Japan, by the "Doolittle Raiders." Pilots flew their twin-engine bombers off the aircraft carrier *Hornet* a mere three months after the Japanese attack on Pearl Harbor. Admiral Halsey, who headed the strike force, though born in New Jersey, claimed Wilmington, Delaware, as his home. Each of the B-25's Wright engines could produce 1700 hp which gave the airplane a flying speed of 272 mph. Although designated a bomber, the B-25 was also used extensively in low-level strafing and skip-bombing operations.

The Army Air Force entered the field of rocket testing when, in June of 1942, the Ordnance Department brought to Wright Field, Dayton, Ohio, pictures of a 4.5-inch diameter rocket they thought could be made adaptable for launching from an aircraft. A single tube rocket was mounted on a P-40 *Warhawk* at Aberdeen Proving Grounds in Maryland for testing July 6, 1942. The initial firing was not successful. However, changes were incorporated and a second installation was made and successfully air-fired.

All machine work for fabricating the rocket mounts was produced on the base. Photo from the AMC Museum.

Efforts were then made to secure an appropriate site where rocket development could meet the criteria for future development. After reviewing several locations, Headquarters, USAAF, recommended DAAB, much to the dismay of General Barcus, who was in control of three fighter squadrons of P-47 *Thunderbolts* at Dover. He felt the DAAB air space and gunnery range would be saturated beyond its capacity. He also considered the strong possibility of First Fighter Command's losing its foothold at Dover once Material Command was entrenched. He and his staff were very pleased with Dover, and their record of performance proved it.

A study was made of other possible sites, but to no avail. So Dover it was! Dover had the low population needed to maintain secrecy. It also had the marshlands that were already earmarked for gunnery practice. In addition, the Delaware Bay offered air-to-sea rocket testing which was a high priority on the agenda. The decision was made that the rocket test center, under Material Command, would have to provide their own housing and security guards. The base commander would provide the other necessary "housekeeping" facilities.

The highly-classified secret operation at DAAB, known to very few, was the 4146th Army Air Force BU (formerly the 732nd Army Air Force BU). This unit was responsible for the research, engineering and development of the first air-to-surface rocket-propelled projectiles to be used in offensive and defensive military air operations. The original building complex, hangar 295 (later changed to 1301), became the home for special personnel capable of designing, manufacturing and installing rocket mounts on aircraft. The Unit installed mounts on the Curtiss P-40 *Warhawk,* Republic P-47 *Thunderbolt*, North American P-51 *Mustang*, Lockheed P-38 *Lightning*, North American B-25 *Mitchell*, and the Martin B-26 *Marauder* aircraft, to name a few. A surface testing site was located about ten miles from the base on the marshlands, along with designated restricted areas in the marsh and Delaware Bay for aerial gunnery and rocket firing.

The 4146th responsibilities included manufacturing mounts for various aircraft that would be uniform in design and could be field-installed. In the end, they succeeded in providing kits that could be installed on the exterior wings of aircraft in twenty-three man-hours. This included the electric-controlled firing mechanism. By October 1944, large quantities of rockets were available for use in combat.

In researching the material covering these early rockets, the following comparison chart, dated October 4, 1944, was discovered:

	Rocket	75mm gun
Rounds in feed	6	1
Rate of fire (min)	120	15-20
Weight of gun (lbs)	300	800
Installed weight	350	1270
Recoil (lbs)	50	30,000
Recoil movement (ins)	0	23
Weight of round (lbs)	36	20
Charge (lbs)	4.65	2
Weight of shell (lbs)	33	14.5
Explosive charge (lbs)	5.3	1.5
Velocity (ft/sec)	850-950	1900
Effective range (yds)	300	2-3000

This chart is interesting in that it shows the rocket had virtually no recoil. This must have been well received by pilots who complained bitterly about the vibration, recoil, and noise incurred when firing conventional weapons. This was especially true of the cannons whose recoil would shake the aircraft violently. On the other hand, the chart reveals an effective range of only 300 yards for the rocket. This appears to be a dangerously close range to a target that would in all probability explode in the pilot's face, showering his aircraft with debris.

Twenty-nine hundred rounds of 4.5-inch rockets were shipped to the China theater of operations for use in the China-Burma-India theater. This shipment was escorted by 1st Lieutenant John M. Colmant, Ordnance Department, who would supervise the installations and train personnel. P-40 *Warhawks,* P-51 *Mustangs,* and P-38 *Lightnings* were eventually fitted with multi-tube, wing-mounted rockets.

After the first mission, General Chennault, 14th Air Force Commander, sent the following radio message to General Arnold confirming the excellent results of an air-ground attack on an airfield and adjacent

The first installation of rockets on the Stinson *L-1 required relocation when the rocket exhaust burned the fabric on the wings. Photo from the AMC Museum.*

The B-25 Mitchell *was also fitted with rocket mounts at DAFB. Photo from the AMC Museum.*

The Douglas C-47 was used by CONAC and MATS at the Dover Air Base. It was assigned to the 1607 ABG between 1955-57. Photo from the AMC Museum.

The Lockheed F-80 Shooting Star *jet fighter was deployed at DAFB. The 4146 BU used the F-80 primarily for rocket testing. Note the rocket mounts under the wings. Photo from the AMC Museum.*

installations. The message dated 24 March, 1944, read in part;

> Five P-40 combat rocket missions flown which comprised of thirty-nine sorties. First mission target large undispersed aircraft and airfield installation. Sixteen P-40s half with rockets were top cover for six B-25s loaded with frag clusters. Rocket planes were last to attack from three thousand feet. Planes attacked in-line abreast, each instructed to select individual targets. Rockets fused delay fired intervolmeter automatic. Result three planes on ground destroyed, one plane damaged, one hangar left burning, with hits in barracks area with observed results. Pilots had very little practice. No difficulty found in installations or ammo. Only one dud reported and this in practice. Only difficulty is selecting proper fuse setting at start of mission.

The realization of the vast potential of airborne rockets finally came into focus, and rockets were supplied to other units including the European theater. In one account of a P-47 mission in Europe, a group of P-47s were on a low-level strafing mission when several German ME-109 fighters dived down on them from above. One P-47 was shot down as the ME-109s flew through their formation. After they passed, they did a 180-degree turn to come at the formation head on. The P-47 pilots fired their rockets at the ME-109s, which immediately scattered and did not return to engage in combat again. This is the first report of air-to-air use of rockets, although this was not their intended use. However, the action did frighten off the aggressors, which was the intent.

Today, rockets play a major role in the armament of aircraft. Surface-to-air, air-to-surface, and air-to-air rockets are now guided by electronics. Once locked on the target, avoidance is almost impossible to achieve.

Victory at Last

The German High Command surrendered unconditionally to the Allied Forces on May 7, 1945. Japan surrendered three months later on August 15, 1945. It was late in the war when the War Department released the information to the public which disclosed that Dover was used as a rocket test-center.

The DAAB, 125th Base Unit, which previously trained P-47 pilots, now assumed a new role as a pre-separation center for eligible personnel. One year later the Dover Army Air Force Base was inactivated and a token task force was regulated to housekeeping duties. A joint-use agreement between All American Airlines was signed, but short-lived. Several other commuter airlines would attempt service between 1947 and the 1980s, but all of them failed.

It was not until December of 1948 that the base was transferred to Continental Air Command (CONAC). This command utilized the twin-engine Douglas C-47 (civilian version DC-3) for awhile until CONAC again transferred on February 18, 1949, to the Ninth Air Force. Various operations were conducted by the Ninth during this relative period of inactivity including being hosts for National Guard units in summer training.

Only four years after WWII was over, the North Koreans invaded South Korea. President Truman authorized General Douglas MacArthur to use supporting ground units in Korea and the Air Force to conduct missions against specific targets in North Korea. These included the use of US warships south of the 38th parallel.

The jet age brought the familiar sounds of distant thunder to the people of Dover. The now renamed "Dover Air Force Base" became a beehive of activity with the arrival of the 336th Fighter Interceptor Squadron flying the F model *Mustang* and later transitioning to the F-86A jet fighters. After their arrival in July 1950, the 336th went through combat train-

This tandem-seat Lockheed T-33 was assigned to MATS at DAFB as the training version of the F-80. Several squadrons used the F-80, including the 46th FIS, 148th FIS, and the 95th FIS between 1952-1966 Photo from the AMC Museum.

This enterprising Air Force mechanic found a way to deal with the vast ramp area to be serviced. The McDonald F-101B Voodoo *was assigned to the 98th FIS between 1959-63. Photo from the AMC Museum.*

The F-106 Delta Dart *was assigned to the 95th FIS between 1963-71. Photographed by the author from a T-33.*

On Armed Forces Day, May 1961, Governor Elbert Carvel, assisted by SSgt Goodin, christens the KC-97 Diamond State *for the 11th Air Refueling Squadron stationed at DAFB. Photo from the AMC Museum.*

Photo courtesy of the Delaware River & Bay Authority.

The Douglas C-54G was the replacement for the aging C-47. Assigned to the 1607th ABG, it was used extensively in the Berlin Airlift of 1948-49. Photo from the AMC Museum.

ing before being sent to Korea in November. DAFB was now reactivated and assigned to the Air Defense Command. At the same time the 148th Fighter Interceptor Squadron, a Pennsylvania National Guard unit, was activated and assigned to Dover. The Guard unit arrived flying P-51 *Mustang*s until F-84s, F-86s and F-94Bs became available for transition. They would remain active until October 31, 1952, when they returned to Spaatz Field in Reading, Pennsylvania. The 46th FIS also utilized the F-94s during the same period.

The 46th FIS was deactivated October 10, 1946. When the conflict in Korea required activation of our military, the 46th FIS was assigned to the Air Defense Command (ADC). On November 1, 1952, the unit was assigned to Dover AFB flying the new F-94B jet fighters. The 46th Fighter Interceptor Squadron (46th FIS), flying Curtiss P-36s, was activated at Wheeler AAFB, Hawaii, December 1940. They were the first unit to absorb the devastating impact of the Japanese attack on Pearl Harbor. After the attack, the 46th FIS received the nickname *Skyraiders*. They went on to establish many victories in the Pacific theater during the war. Their missions were concluded with fighter support for the B-29s that bombed the homelands of Japan and brought an end to the war.

Military bases must maintain operational flexibility. Conflicts may require a base to change its entire mission and deployment of responsibilities in very short order. DAFB at this time was used for fighter interceptor squadrons (FIS) and Military Air Transport Service (MATS). April 1, 1952, once again saw a change of command and operations at DAFB when MATS assumed command. The DAFB MATS flew Douglas C-54G *Skymaster* aircraft. MATS would be responsible for flights to the Azores, Africa, Bermuda, Canada, Greenland, Iceland, British Isles, Nova Scotia, Newfoundland, the Caribbean, and areas in the Middle East.

The Douglas C-54G *Skymaster* was the military version of the Douglas DC-4 transport, the airplane of choice for the airlines. The U.S. Navy also used the Douglas, designated the R5D. The main variation in design was the dual cargo doors available on military versions to accommodate bulky cargo. In an alliance with Great Britain, France, and West Germany, 330 *Skymaster*s helped defeat the Soviet blockade from June 1948 through September 1949. Those big double doors were of major importance during rapid loading and unloading of vital cargo.

The Berlin Airlift was created when Joseph Stalin ordered his troops to close all surface transportation to Berlin from the west in June 1948. This action by the Russians made Berlin an island of two million people which was only accessible by air for over a year. Four-star General Lucius D. Clay, Commander-in Chief, European Theater, stated, "We, together with our allies, must establish an airlift to supply our occupying forces, and at the same time provide the basic living essentials to the two and half a million residents of the Western sectors of Berlin."

General Curtis E. LeMay, USAF Commander, in Wiesbaden, Germany, assigned Brig. General Joseph Smith command of the Berlin Airlift. Smith appropriately named the challenge "Operation Vittles." Under the command of General Smith and Major General William H. Tunner, Deputy Commander, aircraft serviced Berlin twenty-four hours every day to break the Russian blockade by averaging 5620 tons of supplies delivered daily. To accomplish this miraculous feat, every military installation throughout the entire globe felt the impact.

The Soviets felt the blockade would bring the allies to their knees to beg for compassion. The complete opposite happened. The allies gained tremendous prestige throughout the world while Stalin felt the pain of defeat. The Soviets were amazed at the magnitude and accomplishment of the operation. It proved, beyond any reasonable doubt, how rapidly the allies could accomplish a mission the Soviets deemed impossible. The Berlin Airlift ended officially on July 30, 1949.

On April 1, 1952, Congress appropriated twenty-five million dollars to expand and transform DAFB into an east coast port of embarkation for MATS. The DAFB mission was now redefined to airlift critical cargo and personnel to the Far East in support of the United Nations' police action in Korea, as well as to support our overseas bases in the North Atlantic, United Kingdom, Europe, and Africa. By the end of October 1952, the Azores, Bermuda,

The first airplane designed exclusively for military cargo operations was the Douglas C-124 Globemaster. *Assigned to the 1607 ATW and the 436th MAW (1954-1969), it would be the last of the radial, air-cooled engine military transports. Photo from the AMC Museum.*

Nova Scotia, Canada, Newfoundland, Caribbean, and the Middle East were included in the list of areas MATS would serve.

Also, in 1952, four additional units were moved to Dover. These included the 1737th Ferrying Squadron, the 148th Fighter Interceptor Squadron, the Air Communications Service, and the 9th Weather Group. August of 1953 saw a major consolidation of four units of the Atlantic Division, MATS, to create the 1607th Air Transport Wing (ATW), which became operational the first of January, 1954. On February 16, 1954, the 39th Air Transport Service (ATS) and the 45th ATS were redesignated and assigned to the 1607 ATS under MATS. Scheduled flights to Rhein-Main, Germany, became operational in March using Douglas C-54G *Skymaster* aircraft.

By the end of 1952, a multi-million-dollar construction program was awarded to nine individual contractors to expand the MATS airlift capability and other base-operating units facilities. When completed in December of 1953, DAFB was a self-supporting city within itself with complete aircraft services, hospital, commissary, bachelor officer quarters, officers and enlisted men's open mess, gym, dental clinic, nondenominational church, post office, schools, barracks, movie theater, fire department, NCO and Officers clubs.

NEW OWNERS MAY 27, 1953

On August 14, 1952, Brig. General Robert Eaton announced that the Air Force was taking steps to purchase the base owned by the City of Dover. After nine months of negotiations, apparently the Air Force and the city of Dover could not agree on a purchase price, and condemnation proceedings followed. The U.S. District Judge in Wilmington, on May 27, 1953, set a price of $47,000 for the 634 acres of property owned by the city and Mr. Conrad Nichols, who had sublet a portion of his farm earlier. According to James H. Hughes, City Solicitor, the figure represented an estimate of the city of Dover investment made (less CAA funding) in the facility prior to WW II.

The next MATS airplane to replace the aging C-54 was the Douglas C-74 *Globemaster I*, the first, of 14 produced, flew on September 5, 1945. Although the *Globemaster I* suffered typical contract cancellations after VJ-Day, it did leave a mark in history, on November 18, 1949, by being the first to fly nonstop across the Atlantic with more that 100 persons on board. An upgraded version of the C-74 *Globemaster I* was successfully test-flown on November 27, 1949, and called the C-124A *Globemaster II*. A production run of 204 of the "A" models were built before another upgrade was approved. The final version was a C-124C *Globemaster III*, of which the last of the 243 built was delivered in May 1955.

C-124C SPECIFICATIONS

Engines	Pratt & Whitney
Engine model	R-4360-63A
Horsepower	3800 each
Wingspan	174 ft
Wing area	2,506 sq/ft
Length	130.5 ft
Height	48.3 ft
Cruising speed	230 mph
Range	4,030 miles
Ceiling	18,400 ft
Empty weight	101,165 lbs
Gross weight	194,500 lbs

Unlike the C-54, which was a variant of the Civilian DC-4, the C-124 *Globemaster* was designed specifically for military cargo and personnel transportation. The first of these cargo aircraft to be assigned to DAFB arrived May 1, 1954. Assigned to the MATS 45th, 1st and 21st Air Transport Squadrons, Heavy, the airplane had a crew of six men. One lone C-124 on December 6, 1954, was flown by the 54th ATS on a nonstop mission from Dover to Burtonwood, England, carrying 24,000 pounds of cargo in thirteen and one-half hours. Another C-124 record was a nonstop flight of 18 hours and 40 minutes from Hickam AFB, Hawaii, to Dover, on February 7, 1960, by the 15th ATS. C-124s also responded when a request was made from Laos for assistance, in August of 1959, because of an invasion by North Vietnam. Sixteen Dover-based aircraft flew 1045

This Douglas C-133A Cargomaster, *was delivered August 28, 1957. It was christened by Mrs. J. Caleb Boggs,* State of Delaware. *The C-133 was designed to replace the C-124. Photo from the Delaware Public Archives.*

hours supporting this request for aid. On April 18, 1962, a C-124 *Globemaster* departed Dover for Central and South America carrying Colonel John Glenn's famed space capsule, *Friendship II,* on the first leg of an around-the-world tour.

It was announced on April 15, 1964, that the C-124 *Globemasters*, affectionately nicknamed "Old Shaky" had served its mission above and beyond the call of duty. After ten years of faithful service, this airplane, which was the last of the radial air-cooled engine aircraft to serve MATS, was replaced by the new turboprop Douglas C-133 *Cargomaster*. However, it was not until June 5, 1969, that the last C-124, no. 52-0950 would depart Dover to be transferred to the Air National Guard, where it joined others already in service. "Old Shaky" retired from active duty with an impressive record. Unfortunately, historic preservation of military aircraft, at this point in time, was not undertaken. The AMC Museum does not have a C-133 or C-124 for display. AMC Museum curator, Michael Leister knows where each and every survivor is located. Like a hungry hawk, Leister keeps circling his prey waiting patiently for the right moment to strike. Some day ——.

C-133 CARGOMASTER

The Korean war and the Berlin Airlift created needs for Air Force cargo-carrying aircraft. The decision was made, in the early 1950s, to seek a larger and more modern aircraft to replace the radial-engine *Globemasters*. The new Douglas C-133 differed considerably from its predecessors. A high-mounted wing supported four Pratt & Whitney T34-P-9W turboprop engines, and the main landing gear retracted into an external side-mounted pod on each side of the fuselage. The fuselage was designed in a circular section compared to the oval section of the C-124s. A two-section rear door served as a loading ramp as well as a cargo door on the port side of the forward fuselage. The 12-foot high cargo-area design

A Lockheed C-141 Starlifter *from the 436th Military Airlift Command (MAC) Dover. Photo by Harry Heist, AMC Museum.*

allowed for most military vehicles to be transported. The last three aircraft were equipped with clamshell rear doors, which allowed for the loading of a completely assembled *Titan* missile.

The first deliveries of thirty-five, C-133A *Cargomasters* began in August 1957. These aircraft were powered by Pratt & Whitney T34-P3, 6,000 shaft horsepower (shp) engines. The final run of 50 (C-133B models) were fitted with T34-P7W engines. The "W" denoted water injection which increased engine output to 7,100 shp.

The Douglas C-133A *Cargomaster*, flown by the 39th Air Transport Squadron, joined the C-124 *Globemasters* beginning on August 28, 1957. The first aircraft no. 54-0143 was officially accepted by MATS commander, Lt. General Joseph Smith. The C-133A was assigned to the 1607th ATW and the 436th MAW until 1971.

C-133 Flights to Remember ...

A C-133A flew nonstop from Dover on September 22, 1959, with a payload of 87,600 lbs, to Mildenhall, England. This was the heaviest payload ever flown overseas to date.

C-133B CARGOMASTER SPECIFICATIONS

Engines	Pratt & Whitney
Type	Turboprop
Shaft horsepower	7,100 each
Wingspan	179.7 ft
Wing area	2,673 sq/ft
Length	157.7 ft
Height	12 ft
Cruising speed	323 mph
Range	4,000 miles
Empty weight	120,263 lbs
Gross weight	286,000 lbs

In another record-breaking flight, a C-133A *Cargomaster* left Dover with a payload of 117,900 lbs of cargo, on December 16, 1958, and climbed to 10,000 feet to break the Soviet TU-104 record established in 1957.

Another impressive flight was a C-133A *Cargomaster* flight by the 39th Air Transport Squadron from

Hawaii to Dover nonstop in 15 hours and 25 minutes on June 26, 1960. It is believed that this is the longest nonstop flight ever recorded by a C-133A.

The last Lockheed C-133A departed Dover on June 30, 1971. MAC, at this point in time, no longer operated propeller-driven aircraft. The Lockheed C-141, C-5, and C-17 jet-powered aircraft would fill the MAC requirements through 1997.

C-141 STARLIFTER

The C-133 *Cargomasters* were made obsolete by the rapid improvement in the turbofan jet-powered aircraft of the mid-1960s. The Air force requested proposals from the four major U.S. manufacturers of large aircraft, Boeing, Lockheed, Douglas and Martin. Lockheed's submitted proposal was the winner in the competition with the C-141A *Starlifter*. The Lockheed was a quantum leap in technology with its high-mounted, swept-wing, and four pod-mounted Pratt & Whitney, turbofan jet engines. A total of 284 *Starlifters* were built, and delivered to the Air Force, when it was recognized that the airplane's true potential was not being utilized. In mid-1976, the Air Force issued a contract to Lockheed for a conversion to allow for in-flight refueling and a lengthening of the fuselage by 23.4 feet. The prototype, flown on March 24, 1977, proved so successful the Air Force approved the conversion of 270 *Starlifters* to C-141B models. This change increased the capacity of the airplane by as much as that of another 87 aircraft.

The first of sixteen Lockheed C-141 *Starlifters*, serial no. 64-0625, was delivered to Dover August 18, 1965, and named *First State Starlifter*. This particular aircraft completed the longest nonstop flight ever recorded at the base at this time. The *First State Starlifter* flew from Yokoto, Japan, to Dover, Delaware, in 13 hours and 10 minutes. This was the longest nonstop flight ever recorded by a C-141 *Starlifter*.

All three MATS aircraft, the C-124, the C-133A and the C-141 were operating from Dover AFB during 1965 and through 1969. This was because the MATS mission was growing faster than replacement aircraft could be built.

Effective January 1966, the Military Air Transport Service (MATS) was changed to Military Airlift Command (MAC), and the 1607 Air Transport Wing became the 436th Military Airlift Wing (MAW). Over time, other squadrons would be redesignated and accepted into the 436th MAW. The DAFB 436th Wing, with all of its support squadrons, would require over four thousand personnel when fully operational.

The responsibilities of the MAC 436th Wing grew to include deployment or evacuation of all military personnel. The return of the remains of military or civilian personnel to the DAFB Mortuary for further processing is a depressing mission. The 436th MAW would expedite the largest volume of human remains in history when they removed 913 victims of the Jonestown, Guyana, mass-suicide in 1978 and processed them through the DAFB Mortuary. The Dover mortuary was also responsible for processing the remains of the 326 people killed in the world's worst aviation accident when two jumbo jets collided on a fog-bound runway in the Canary Islands in April of 1977.

C-141B STARLIFTER SPECIFICATIONS

Engines	Pratt & Whitney
Type	TF33-P-7 turbofans
Wingspan	159.9 ft
Wing area	3,228 sq/ft
Length	168.3 ft
Height	39.3 ft
Empty weight	148,120 lbs
Gross weight	343,000 lbs
Cruise speed	566 mph
Range	2,935 miles*

**Unrefueled*

The MAC 436th MAW achieved a milestone in safety on December 31, 1966, by flying the entire year accident-free and at the same time passing 100,000 hours of accident-free flying. The 436th received the MAC Outstanding Safety Award for the year for superior performance. Safety is not some-

The Lockheed C-5 is the largest aircraft in the Air Force inventory.

thing that can be attributed to luck in flying. Safety comes from hard work and dedicated people who love their work, and it is rewarded by results. The 436th Wing would not see another accident until June 12, 1973, when the nose gear of a C-141 collapsed due to material failure. Although the aircraft sustained severe damage, there were no injuries. By 1968 Dover aircraft in one year totaled 1.69 billion ton-miles of movement. At the same time, the DAFB registered a movement of a staggering 41,824 personnel and 120,061 tons of cargo. Before the year ended, over one million tons of cargo passed through the DAFB Aerial Port. The 1966 figures represent a forty-four percent increase over the previous year. By August 31, 1977, the 436th MAW surpassed one million accident-free flying hours. Dover AFB was, without a doubt, the busiest airport in the state of Delaware.

The next generation aircraft to enter the MAC inventory was announced by Secretary of Defense Robert McNamara. The first C-5 *Galaxy* rolled off the assembly line in Marietta, Georgia, on March 2, 1968. Six squadrons of Lockheed C-5 *Galaxy* aircraft, each squadron containing sixteen aircraft, would be dispersed throughout the MAC squadrons. One squadron was earmarked for the 436th MAW, Dover.

C-5B SPECIFICATIONS

Normal cruise speed	490 mph
High speed cruise speed	540 mph
Maximum takeoff weight	769,000 lbs
Engines	four GE-TF-39 Turbofans
Fuel capacity	51,000 gallons
Thrust, each engine	41,000 lbs
Wingspan	222.8 ft
Wing area	6,200 sq/ft
Fuselage length	247.8 ft
Height	65.1 ft
Range with 261,000 lbs cargo	2958 nm
Crew	7
Upper level troop capacity	75
Lower level troop capacity	270

To accommodate the huge C-5s at Dover, the U.S. Congress appropriated $7,671,000 to DAFB for reconstruction of facilities. The first Lockheed C-5 *Galaxy* no. 90006 was delivered to the 436th Wing, DAFB, on April 16, 1971.

The people of Delaware have always been supportive of Dover Air Force Base and its personnel. The base has always contributed more to the economic stability of Kent County than the County can possibly return. From DAFB Archives, dated December 1986, the 436th MAW released its annual fiscal year Economic Resource Impact Statement. It revealed the monetary impact of the 436th on the state of Delaware exceeded 600 million dollars.

It is ironic, considering the monetary impact on the state, that General Fred W. Vetter, 436th Wing Commander, would designate October 9, 1969, as "Salute to Delaware Day." The event was designated to show the Wing's appreciation to the citizens of Delaware for the seventeen years the wing had been operational at the base. (One would think it would be the other way around.) Although the base held "Open House" events over the years, this was the first occasion that the Governor and the citizens of Delaware were formally honored by the base. The day was filled with action, including a military parade, aircraft flybys, Golden Knights parachute jumps, and the acrobatic team of the USAF *Thunderbirds*. The following statement is reprinted from the Dover AMC museum archives. "This day was by far the most successful public event in the history of Dover Air Force Base."

The Lockheed C-5 *Galaxy* (nicknamed "Fat Albert") is the fourth generation cargo aircraft designed strictly for military use. It has the distinction of being the largest aircraft in the USAF inventory. The airplane fuselage is so huge, in proportion to the landing gear, it gives the impression of a group of ants trying to run off with a loaf of bread.

On April 28, 1973, it was announced that the Lockheed C-141 *Starlifters* at Dover would be transferred to Charleston, South Carolina, and the thirteen Lockheed C-5s in Charleston would be transferred to Dover. The projected date of the exchange was completed August 29, 1973, when the last C-5 no. 680212 departed Charleston for Dover. The Dover 436th MAW would now only support one aircraft, the C-5, and would become the only C-5 base in the USAF. The construction of the all-new, state-of-the-art, air-freight terminal to support the C-5's missions was completed, November 30, 1973, at the cost of $5,246,700. This cost was in addition to the previous work required to upgrade the base infrastructure to accommodate the C-5. The base now included sixty-four hundred personnel and occupied a total of 2,247 acres.

From 1973 on, the military mission at Dover Air Force Base was singular in its objective. The Military Air Command outlined a mission statement to all of the Wings. Condensed, it asserted that MAC was ready to move virtually anything, at a moments notice, to anywhere on the globe.

C-5 Historical Missions

A Dover C-5 departed Calverton, Long island, N.Y., loaded with a replica of the *Spirit of St. Louis* for a nonstop flight to Paris for a display at the Paris air show May 27, 1977. It would be exactly fifty years since Charles Lindbergh made the first nonstop crossing of the Atlantic. This would be the first time the French would have the opportunity to see this historic airplane since it landed in Paris on May 27, 1927.

A Dover C-5 departed Chicago O'Hare International Airport, June 19, 1977, with 16 passengers, a 40-ton magnet, its special transporter, and support equipment for delivery to Moscow. The payload represented 86 tons. The C-5 was the only aircraft capable of accomplishing this nonstop mission. Two aerial refueling operations were required en route for the 5900-mile flight. Total flight time was 11.9 hours. The 17-member crew was awarded the USAF McKay Trophy for accomplishing this meritorious flight.

Six American diplomats, Joseph Stafford, Robert Anders, Cora Lijek, Mark Lijek, Harry Schatz, and Kathleen Stafford, escaped Iran with the help of the Canadians in January 1980. They were airlifted to Dover via West Germany aboard a C-5.

Five UH-1N Huey helicopters and 59 marines were loaded in one C-5 for transport to Haiti for relief after hurricane Allen devastated the island August 10, 1980.

The entire U.S. Acrobatic team and its nine aircraft were airlifted to Cherney, England, without dismantling a single aircraft, to compete in the World Acrobatic championships on July 20, 1986.

The general public has little knowledge of the tremendous commitment of personnel and equipment necessary for the Air Force to accomplish their goals. The military does not seek publicity due to the nature of many of their missions. Their cargo and destinations are mostly sensitive for reasons of safety to the crews. Operational status can change in a moment's notice when a situation so warrants. Operational readiness is a key factor in requiring all aircraft to become airborne in the least amount of time possible. This is no small feat. Flight crews are on duty-call twenty-four hours every day. The missions are global in nature. Missions include responses to natural disasters, military supply support operations, rapid deployment of personnel, mercy flights, as well as search and rescue. Cargo may include anything that can be loaded aboard the aircraft including nuclear weapons, ordinance, food, heavy equipment, aircraft, helicopters, artillery, tanks, electronics and troops.

Dover Air Force Base is the most important business asset in Kent County. As this is written, the gross income to Kent County generated by DAFB exceeds $500 million dollars annually.

DOVER AMC MUSEUM

The Air Mobility Command (AMC) is responsible for the release of many of the aircraft to the DAFB-AMC Museum. Additional funding is obtained by donations, sale of memorabilia, and a civilian organization, the AMC Museum Foundation. However, the museum is manned by a group of volunteer military and civilian personnel. Museum personnel have

The original hangar, 1301, built in 1942, is the display hangar for the Museum's collection of rare aircraft and memorabilia. Photo from AMC Museum.

restored the first hangar (1301) built on the base and have on static display the following aircraft.

Cessna 0-2 *Skymaster,* twin-engine support aircraft

Beechcraft C-45G *Expeditor,* twin-engine light transport

Douglas F-101B *Voodoo,* jet fighter

Lockheed C-141B *Starlifter,* jet-engine cargo aircraft

Fairchild C-123K *Provider,* twin-engine cargo aircraft

de Havilland C-7B *Caribou,* twin-engine cargo aircraft

Convair C-131D *Samaritan,* twin-engine cargo aircraft

Fairchild C-119G *Boxcar,* twin-engine cargo aircraft

Boeing/Stearman PT-17 *Kaydet,* open cockpit primary trainer

Vultee BT-13 *Valiant,* single-engine basic trainer

Boeing B-17G *Flying Fortress,* four-engine bomber

Consolidated AT-6 *Texan,* single-engine advanced trainer

Douglas C-47A *Skytrain,* twin-engine cargo aircraft

Douglas C-54M *Skymaster,* four-engine cargo aircraft

McDonald F101B *Voodoo,* jet fighter

North American P-51D *Mustang,* WWI fighter

Lockheed T-33A *Shooting Star,* jet fighter/trainer

Sopwith *Pup,* WWI open-cockpit, bi-wing fighter

The museum also displays over 500 scale-model aircraft and military historical items too numerous to mention. It is open to the public at no charge. Over 43,000 people visited the museum in 1997. ■

The Museum, in 1997, obtained a rare Lockheed C-121 Constellation *similar to this one that is undergoing restoration for static display. Photo by Bill McDaniel.*

➲ New Castle County Airport

The New Castle County Levy Court minutes of April 15, 1941, confirm it would accept federal funding to build a municipal airport in New Castle County. April 17, 1941, Governor Bacon signed the legislation for New Castle County to sell bonds, not to exceed $750,000, and apply to the Civil Aeronautics Administration (CAA) for funding of a municipal airport. The Levy Court issued $500,000 in County Bonds, which would mature June 1, 1961, to move the project forward, and the CAA authorized $300,000 to begin construction.

Like Dover and Georgetown, the New Castle County Airport was initially conceived as a municipal airport to serve Wilmington and New Castle County. The CAA had offered to contribute federal funds to build the airport in 1940, provided the county purchased the land. County deeds record three separate parcels totaling 380 acres of farmland, July 29, 1941. The farmland, 73 feet above sea level, was purchased for $35,000 after condemnation hearings.

Construction called for the removal of the sandy clay-loam topsoil to be replaced with two to seven feet of sand loam and gravel. This was underlaid with sand and

NCCAAB, summer 1943 , shows the completed Army Air Corps facility on the northwest side of the airport. A second transient hangar was built in 1944.

coarse gravel to create a solid base and allow for water drainage. Later tests confirmed the worst conditions, borings located water twelve feet below the surface. All four runways were constructed with a minimum of four inches of stone macadam base topped with two inches of asphaltic concrete surface. All taxiways were of similar construction. All runways were built 150-feet wide with the exception of runway 1-19, which was 200-feet wide. Runways varied in length from 5,000 to 7,000 feet. All runways and taxiways were constructed to accept loads of 100,000 lbs. In addition, the engineers installed low-intensity runway lighting, red obstruction lighting on approaches (controlled by the tower), and a rotating beacon on the airfield. Flying minimums were 500-foot ceiling and 1-mile visibility.

Construction was barely underway when the war broke out in December 1941. The unfinished airport was taken over by the Department of War and renamed New Castle Army Air Base (NCAAB). The Army Corps of Engineers moved in with construction crews to make the airport operational as soon as possible. Construction crews operated three shifts, seven days a week, due to the national emergency. Fourteen additional parcels were added by the Federal Government during their occupation to allow the airport adequate

space. After the war, these acres were deeded to the county in 1949, when they were declared surplus by the Federal Government. Further acquisitions allowed the county to control the encroachment on the property. The airport now contains approximately 1450 acres.

The New Castle Army Air Base would assume an entirely different role in the war from that of Dover or Georgetown. Aircraft manufacturing plants in the northeast included Fairchild, Piper, Curtiss, Martin, and Republic. The aircraft they produced needed to be flown from the factories to their operational destinations. This required a highly sophisticated military operation called "Air Force Ferrying Command" (AFFC). Colonel Robert Baker was named Commanding Officer of the Northeast Sector, Domestic Wing, AFFC, on January 5, 1942. New Castle Army Air Base, Logan Airport in Baltimore, and other bases in the northeast would fall under his jurisdiction. The NCAAB operation would be named "The 2nd Ferrying Group." Its mission was defined as a "ferrying operation of aircraft, personnel and cargo for the armed forces."

The 344th Base Headquarters and Air Base Squadron was also moved to NCAAB to oversee administrative, supply, food, hospital, janitorial services, and other operating essentials. This was a considerable undertaking which would require 85 officers, 580 enlisted men, and 65 civilians.

While the base was under construction, housing facilities were bare bones. Initially, most of the living quarters were in tents. The control tower was completed in July 1942. However, the tower controllers used handheld gun-lights to control aircraft since the radio transmitter/receiver had not been delivered. The radio was finally installed in the tower in September 1942. Twelve miles of interconnecting roads, to 34 barracks and other buildings, were dirt. All buildings were laid out in a hodgepodge manner as a safety precaution against enemy attack. The roads and walkways were muddy in wet weather and dusty in dry weather. In support of U.S. military leaders, the roads were named MacArthur Avenue, Mitchell Avenue, Marshall Avenue, Stimson Avenue, and Patterson Avenue. These roads, like our great leaders, are gone and are now history. Three large hangars were constructed to service the transient aircraft. To operate the base without a completed infrastructure was a challenge for the engineers, a challenge that would not be tolerated in today's world. Sewage disposal was initially dumped into Nonesuch Creek until September 1942, when the disposal plant became operational. Water was also a problem. Three wells, each up to 220-feet deep, were driven to service the complex. By the time the war ended, the Federal Government had spent $10,350,000 constructing the runways and related infrastructure on the airport.

While the Corps of Engineers was building the runways, most of the ferrying operations were carried out from Logan Airfield, Baltimore. However, the facilities at Logan were inadequate, and Colonel Baker had the group moved to NCAAB in Delaware as soon as the first runway was operational. The Delaware site also had its problems. The NCAAB was still under construction, and qualified military pilots were not available in sufficient quantity to accomplish the mission. Colonel Baker then turned to civilian pilots for assistance. He set down the following requirements for civil pilot applications into the 2nd Ferrying Command.

[1] Be between 21 and 45 years of age.
[2] Have a high school diploma.
[3] Have a commercial pilot's license.
[4] Have 300 hours flying time.
[5] Have 100 hours in aircraft over 200 hp.
[6] Pass a physical examination.
[7] Supply two letters of recommendation.

If the applicant was accepted, the pilot would receive nine days of military officer training, a uniform, and pilot wings with a large "S" in the center denoting a "Service" pilot. Rank was determined by a board of military officer pilots. Pay was $300 each month. These civilian pilots were considered "contract pilots" and were required to fly anywhere, anytime. Two of the many applicants were Randolph Holladay and Johnny "Mac" Morgan from Air Service, Inc. on Bellanca Airfield. Their civil flying business was shut down because of the war. To continue flying, they joined the 2nd Ferrying Group. Another Wilmington pilot, William Spruance, flew 367 ATC missions,

Building no. 44 is the administration building on the left and the control tower on the right. Photo courtesy of the Delaware River and Bay Authority.

many of which were over the Himalayas, called the Hump. These flights originated in Wilmington with the *Crescent Caravan,* which flew supplies to India. Once in India or Burma, supplies were flown to China when the Japanese cut off the Burma Road after the capture of Rangoon. William Spruance would make the military his career and earned the rank of Brigadier General. His last command was adjutant general for the Delaware National Guard.

One of the first ferrying orders issued to NCCAAB was for the transfer of nine Curtiss P-40 *Warhawk*s from New York to Texas. Unfortunately, there was no way to give the pilots instruction flying time in a single-seat pursuit airplane. The pilots would simply have to check themselves out by reading the flight manual while sitting in the cockpit on the ground. The pilots were soon to learn that they would be flying many different airplanes with which they were totally unfamiliar. Randolph Holladay and Johnny Mac were two Delaware pilots earmarked for this first mission.

Flying a new airplane (new in construction and new to the pilot) would become one of the most challenging demands placed upon pilots of Ferrying Command. With the exception of multi-engine aircraft, pilots flew solo. The Fairchild PT-19 *Cornell* and the Boeing/Stearman PT-17 were open-cockpit trainers, and they were generally flown to Texas where basic-training airfields were located. Open-cockpit flying in the winter is not considered pleasurable flying. With very little body movement from the pilot, the wind-chill factor penetrates the best flying suit to the bone. A flight from New Castle to Texas was also a long one in airplanes that appeared to hover at 120 mph. It was not all the fun many made it out to be. However, the men (and women) involved knew we were at war, and the job of ferrying the airplanes was something that just had to be done. When pilots were issued orders to pick up airplanes, they were on their own to find the most rapid form of transportation to arrive at the site. On rare occasions, pilots might find an airplane ride. However, the most reliable form of transportation was rail. Since train passengers normally do not carry parachutes, it was commonplace for riders to give

This bronze statue of Nancy Love, placed near the main terminal entrance, was dedicated to the memory of all the WAFS who served their country during WWII. Photo by the author.

Fifty-three years after Jacqueline Cochran created the WASPS, the Post Office Department issued this 50-cent commemorative stamp. From the author's collection.

the pilots a questionable look when they dragged their parachutes along as baggage. To add to the mystery, pilots rarely engaged in conversation with strangers because of military security.

As the war gained momentum, the aircraft became more sophisticated. The multi-engine C-46 *Commando,* A-20 *Havoc,* B-26 *Marauder*, B-25 *Mitchell,* B-17 *Flying Fortress*, B-24 *Liberator*, Martin A-30 *Baltimore,* and P-38 *Lightning* were considered huge airplanes. The P-38 was the largest of the single-seat aircraft used by the military. The Ferrying Command also flew AT-6 *Texans*, P-51 *Mustangs,* P-47 *Thunderbolts,* and P-40 *Warhawks.* It is doubtful that there was an aircraft in the Army Air Corps inventory that the 2nd Ferrying Command did not fly.

All this increased volume required additional pilots. This necessitated moving additional squadrons to NCAAB. The 4th and the 12th Ferrying Squadrons had already moved in from Baltimore. By July 1942, the 27th Ferrying Squadron also moved to NCAAB, the 63rd Ferrying Squadron on November 1, 1942, and in April 1943, the 6th Ferrying Squadron arrived. The 89th Ferrying Squadron was activated April 15, 1943. The base became a city in itself.

Women's Auxiliary Ferrying Squadron (WAFS)

In spite of the joint civil and military pilots' ferrying aircraft, they still could not keep up with the demand. A search of the CAA files disclosed there were 83 women who held commercial pilot ratings in the United States. It was a source of pilots that remained untapped. A telegram was sent, September 6, 1942, to each one of the women requesting her help in ferrying military aircraft. Requirements were almost identical to the men's civil application with two exceptions. Age was between 21 and 35 and 500 hours flying time was required to apply. Many who met the requirements left home and came to Wilmington to enlist. Nancy Love was named Commander of the first WAFS unit.

Although the Government refused to allow women to ferry aircraft overseas, they got some of the toughest domestic duty. Their first flight consisted of ferrying six Piper L4B *Grasshoppers* from Lock Haven, Pennsylvania, to Mitchell Airfield, New York. It was October 22, 1942, and the Pipers had to be started by hand since they had no electric system, heaters, or radios. Navigation was by magnetic compass and visual reference using a map. The icy winds and snow-covered terrain were a challenge for the light fabric-covered planes and their pilots. The WAFS also ferried Fairchild PT-19s and PT-17s which were open-cockpit airplanes and miserably uncomfortable in cold weather. In spite of the hazards, they made it without a hitch.

The WAFS did such an outstanding job, Nancy Love was transferred to Division Headquarters in January 1943, where she organized three more WAFS squadrons. These squadrons were located in Dallas, Texas; Long Beach, California; and Romulus, Michigan. To establish the new squadrons, Nancy Love transferred five WAFS from the Wilmington group to support her new squadrons. Betty Gilles assumed command of the Wilmington WAFS Squadron in January 1943. By August 1943, Jacqueline Cochran was named National Director and she changed the name to Womens Airforce Service Pilots (WASP). Under her command she gained acceptance to the idea of having women trained to ferry single-seat fighters and multi-engine bombers. Unfortunately, the WASP would only last until December 20, 1944, when they were deactivated and returned to civilian life. In those 16 months the WAFS flew every airplane in the Air Force inventory. The 25 WASPS stationed at New Castle were credited with delivering 3,000 aircraft during their tenure. The 1,074 WASPS carved their place in American aviation history with a beginning in New Castle, Delaware, as WAFS.

The war created some difficult logistic problems for the United States. We were an island continent at war separated by a vast ocean on both fronts. It was a formidable task to bridge this ocean of water, and ocean of air, and satisfy the appetite of war with machinery and personnel. Weather reporting could be considered unreliable at best. Navigation, whether by sea or air, was relied on by a simple magnetic compass in inclement weather. The handheld sextant was used for triangulation shots of the stars to fix position - if there were stars to see. Radios were limited to Low Frequency (LF) and had limited range. To make flights more interesting, the enemy was always a threat.

In spite of all these hazards, the bridge had to be crossed. Vital supplies had to reach the front regardless of the cost. The German submarines were sinking merchant ships faster than we could build them. And we were building one every day! Early in the war, the airplane had not been developed sufficiently so that heavy loads could be carried over long distances without refueling. By 1944, that changed when the four-engine, Douglas C-54 *Skymaster* became available.

February 6, 1944, the newly formed *Crescent Caravan*, operating C-54s from Wilmington, Delaware, would attempt to accomplish what was thought to be impossible. Colonel James Chapman, Jr. revived a plan from the old west when pony express riders were used to carry the mail. The only change in the plan would be the horse (C-54) was retained, and the rider (the crew) would change at specified points. This was a very ambitious undertaking since the plan was to fly round trips to India and Burma. One way would cover 22,600 miles and take over 53 hours flying time. Cargo would be full loads of vital supplies going over and wounded personnel upon return.

To set the wheels in motion, Colonel Chapman would require 1500 officers and 3000 enlisted men. Spare crews and mechanics would be sent to the various legs to relieve crews and service the airplanes. The C-54s were timed so the crews would have an overnight stay before another airplane and crew would arrive. Turnaround time to service the airplane and board the rested crew was two hours. Timing was very important to make the plan work, so a single trial flight was made to determine the time and distance between legs. The *Crescent Caravan* would log as high as 10,000 hours flying in a single month. Its job was completed in June 1945, after 16 months of continuous flying day and night.

The New Castle Army Air Base came to an official end on the weekend of August 3-4, 1946, when the War Assets Administration returned it to New Castle

County, with the stipulation it could return in the event of a national emergency. (The actual papers were not signed until October 27, 1947.) NCAAB military personnel were transferred to Westover Army Air Force Base in Massachusetts. The turnover was highlighted by a public two-day air show of military and civil aircraft. The NCAAB, and its many people, not only made aviation history for Delaware, but for the United States of America as well.

The airport was now run by the New Castle County Airport Commission (NCCAC) on a fiscal year basis from July 1st through June 30th. The NCCAC was responsible to the New Castle County Levy Court. Appointees were G. Roy Eaton, chairman, Robert J. Quillen, and H.B. du Pont. The airport manager was Don W. Martin, who had previously been the manager of the Detroit Municipal Airport. Martin was hired for four years with a starting salary of $8,500 annually. The airport included four runways; 1-19, 7000 feet long; 9-27, 7200 feet long; 14-32, 5000 feet long and 5-32, 5000 feet long. In the transfer of ownership there were approximately 200 buildings, which the War Assets Administration sold to the Levy Court for $7,200. The task of converting the airport from military to civilian status was enormous. Mess halls, barracks, base chapel, day rooms, and numerous other military infrastructure were not necessary for its civil use. Searching for new tenants was a priority. This was necessary to create income to make the airport self-supporting.

A concerted effort was made to deal with all the issues. The County Commissioners hired M.W. Cochran, airport consulting-engineer, to create a master plan for the airport's postwar use. The finished proposal called for a $765,000 expenditure of which almost one-half included a new combination terminal building and hotel, pilots' lounge, shop-

This 1949 photo clearly shows the U-shaped terminal building and circular driveway. The original military control tower is still operational on the far right. Photo courtesy of the Delaware River & Bay Authority.

An All American DC-3 passes over the renovated terminal building. This is the airplane loading ramp side of the building. Photo courtesy of the Hagley Museum & Library.

ping center and movie theater. All of these new structures would be located on the east side of the airport facing Route 13. None of the proposed expansion would be shared financially by the Federal Government. This left the Levy Court's assuming the entire financial burden.

The Levy Court Commissioners were less than willing to spend that kind of money on the airport in spite of the fact the Department of War spent more than ten million dollars in development. The Commissioners scaled the terminal project back to an extremely low $60,000 to renovate the existing, single floor, wooden, military terminal building. Bonds were sold to raise the money.

The original administration building was enlarged by utilizing two more existing buildings to create a "U" structure. A flagpole was erected in the center of the "U" and a circular driveway for ease of entry. The completed structure included a restaurant, waiting area, airport office and weather station. The new terminal building can be seen, left of the control tower in the 1949 aerial on the preceding page.

Interested flying businesses immediately began operations on NCCA. Flight training was offered by Superior Airways, Inc. TWA signed a contract for $80,000 annually for use of the airport facilities for scheduled airline service and their maintenance facility. TWA also installed, at its own expense, an Instrument Landing System (ILS) on runway 1 to lower the flying minimums to 200 feet and one-half-mile visibility. This was necessary to improve landing operations and reduce cancellations of flights due to inclement weather.

H.B. du Pont resigned from the NCCA Commission to negate any conflict of interest when he decided to move Atlantic Aviation to NCCA in 1948.

This aerial photo shows the two hangars built during the war that were leased to TWA. Major work on the ramp can be seen in the vicinity of the hangars. Photo courtesy of the Delaware River & Bay Authority.

Du Pont was replaced by Hugh R. Sharp, former commander of the Delaware CAP unit stationed at Rehoboth during the war. Atlantic Aviation would become one of NCCAs most valued tenants.

Many nonconforming buildings were sold and removed from the site. The Veterans Administration took over the hospital. The CAA took over the control tower, and the weather-reporting station established its District Office on the airport. The District Office would be responsible for Delaware, Maryland, and New Jersey's weather-related needs. The CAP and Air National Guard continued to operate from NCCA. Of more importance was the establishment of two scheduled airline services at the airport. All American and TWA applied for, and were granted, routes that connected NCCA with Hub terminals. The total employment at NCCA was more than 1600 persons in 1947. This was roughly one-half of the military occupation during the war.

After only two years, TWA decided to relocate its maintenance operation to Kansas City. A cash settlement of $100,000 was paid to the Levy Court for the cancellation of its lease. However, TWA could not discontinue its passenger flights due to CAA regulations requiring airlines to service small and intermediate airports.

H. B. du Pont incorporated another company called New Castle Hangar Co. Inc. to build a $170,000 facility to lease to Atlantic Aviation in 1948. Over the years, Atlantic Aviation grew like clover in a farm field. Its growth required more buildings and ground to be leased on NCCA. Atlantic Aviation obtained five separate leases from the county. All leases were identical in nature and were for a 25-year period with three 10-year options for renewal. A clause in the lease turned possession of the building(s) over to the county after 25 years.

This 1949 photo is interesting in that it shows the Route 13 overpass and ramps have not been built to access Route 141. It also shows Atlantic Aviation's first hangar and the military-style parking circles. Photo courtesy of the Delaware River & Bay Authority.

PARCEL A, including building 5028, on 4.63 acres, would become the property of the county on August 1, 1976.

PARCEL B, including building 5029 on 7.163 acres, was built in 1955 and would become county property August 30, 1980.

PARCEL C, including building 5030, on 9.585 acres, was built in 1965 and would become county property on June 1, 1989.

PARCEL D, including building 5031, on 9.823 acres, was built in 1965 and would become county property in December 31, 1992.

PARCEL E, consisting of 0.049 acres did not include buildings and was rented as parking space.

After the 25-year period, the county determined a new lease fee-based on fair market value. If the parties could not agree on the terms, either party could demand arbitration. In addition to the lease, a fuel flow fee was added amounting to $0.03 per gallon. In 1976, this amounted to $71,010.73 income to the county.

When Atlantic occupied its first hangar, it was the only building facing Route 13. All other operations, including the control tower, were located in the original Army Air Corps wooden buildings on the northwest corner of the airport. Apparently, the terms of the leases on the 31.25 acres were agreeable to du Pont. Atlantic Aviation has prospered over the years to become the largest general aviation sales and service organization in the United States and Canada.

To spruce up the place and to encourage new tenants, all thirty-two occupied buildings on the airport were painted in 1949. In addition, a special asphalt sealant was applied to protect all the runways. The

Levy Court continued to focus on aviation-related tenants. A great deal of time and money needed to be spent on the airport to convert it from military to civilian status. The many staggered, remote airplane parking pads alone took years to remove.

Then the 2237th Air Force Training Center moved to NCCA from Reading, Pennsylvania, in 1949. During July and August the NCCA became the base for Air Guard Squadron training from seven states. With the CAP also on the airport, the military once again became the primary tenant.

In spite of TWA maintenances leaving the airport, income for the 1949 fiscal year was $251,966.41. An astonishing $60,802.00 profit was realized from gross income. This included all expenses except the interest paid on the original bond issue to purchase the land back in 1941. Total military and civil flight operations at NCCA totaled 58,430 for the year. The U.S. Weather Bureau responded to 77,770 requests for weather information. The passenger terminal reported 14,041 passengers passing through and 91,127 lbs of mail were handled. In addition to mail, over 1,000,000 pounds of air express and air freight were accounted for. It was reported that the economic impact the airport had on the community exceeded $3,000,000 annually.

To service all transients, all aviation fuels were available on the airport, including jet fuel and 80/87-, 100-, and 145-octane avgas. Hangars were large enough to accommodate the four-engine Douglas DC-6 and the Lockheed *Constellation*. New military

This 1956 aerial photo shows the construction of the new control tower and terminal building nearing completion. Photo courtesy of the Delaware River & Bay Authority.

tenants in 1950 included the 142nd Fighter Squadron flying F-84s, and the 334th Fighter Wing flying F-86 *Sabres*. The Levy Court estimated the economic impact, of the 500 military personnel, on the community would exceed $2,000,000. All American Airways, American Airlines, Eastern Airlines, and TWA operated thirteen flights in and out daily. In addition to these commercial flights, Slick Airways operated an air-freight service from the airport.

On April 4, 1950, the U.S. Government entered into a lease agreement with the Levy Court to lease an area for Air Force operations. The year-to-year lease stipulated that any buildings erected by the government would remain their property. Initially, the county was responsible for the maintenance of the grounds, but this was changed in 1952, so that the government assumed this responsibility. They also extended the lease. The $47,000 annual lease agreement included unrestricted joint-use of the airport runways, taxiways and instrument approaches. Over the years, the government and the county made adjustments to the lease to meet the changing times. The current lease expires June 30, 2005.

Air shows are important to draw public attention to airports, airplanes, and the services they offer. The shows are valuable, not only to educate the public of the needs of the facility, but also of the wide range of applications in which an airplane can be utilized. The CAP, on July 25-26, 1950, sponsored an air show on NCCA that included the Air Force, Air National Guard and civil general aviation aircraft. The show was attended by thousands.

In 1952, the Airport Commission returned to the Levy Court $96,419.49 to cover appropriations, bond interest, and bond redemption to the airport from 1943-53. The airport had grown to such a point it no longer required taxpayer-funding. At the same time, the Levy court approved a request from the Commission to investigate the construction of a new terminal building and tower to replace the WWII wooden structures. Funding was also approved by the CAA to paint the runway and taxiway markings, which had become faded from use and weather.

The construction for a new $600,000 control tower, terminal building, parking lot, ramp, and access road was approved in 1955 by the Levy Court. The new terminal would be facing the southbound lane of Route 13. Bids were let and construction began early in 1956. The modern building was designed by W. Ellis Preston, R.A. and constructed by W.P.

The lobby of NCCA was befitting the modern age, with soft, recessed lighting and an abundance of glass. Photos courtesy of the Delaware River & Bay Authority.

← *Large windows in the lounge allowed family and friends the opportunity to observe the loading and unloading of passengers. Photo courtesy of the Delaware River & Bay Authority.*

Altair served New Castle, Washington, Richmond, Allentown, Wilkes Barre, and Philadelphia. It was possibly the last airline to leave NCCA. Photo courtesy of the Delaware River & Bay Authority. ↙

Haddock Construction Company. The completed project was dedicated on August 4, 1956. The modern terminal included a restaurant capable of seating 125 persons for lunch or dinner, a snack bar, lounge, observation deck, and service counters for airline operations. The control tower was manned by CAA air-traffic controllers 24 hours daily. The United States Weather Bureau also occupied space and offered 24-hour daily weather service to pilots and the public. Over the years, Runway 1 was modernized with an Instrument Landing System (ILS) with manned radar controllers, Approach Lighting System, (ALS) and Runway Visual Range (RVR). The minimum approach is listed as 200-foot ceiling and one-half-mile visibility. NCCA was considered to have one of the finest instrument approach systems in the United States.

No professional annual reports could be located to determine the financial well-being of the airport. The airport Annual Reports were combined with other reports submitted to Council by the County Executive. However a document was located which included the following fiscal year-end history.

YEAR	PROFIT/LOSS
1949	$40,179.56
1950	60,802.01
1951	12,058.07
1952	33,444.13
1953	44,984.50
1954	53,473.91
1955	29,785.58
1956	24,224.99
1957	4,356,41
1958	(35,907.03)

These fiscal year-end numbers cannot tell the entire story of what actions, if any, should have been taken to return to profitability. However, they do indicate that a thorough in-depth analysis was needed of the detailed financial statements. The public should realize that airports are built to serve the public needs. Airports are not built as profit-making facilities any more than are harbors, bus depots, or railway stations. If state government were required to justify every upgrade or repair to these facilities to ascertain profitability, we would not have the level of infrastructure we have today. The history of the privately owned, public-use airports throughout the state confirms that airports must be municipal to guarantee success.

The name of the airport was officially changed from New Castle County Airport to "Greater Wilmington Airport" (GWA) on June 1, 1961. Reasons for name changes are political and the airport would see another change to "New Castle County-Greater Wilmington Airport" (NCCGWA) in the 1970s. However, it did not last. The Delaware River & Bay Authority changed it back to "New Castle County Airport" (NCCA) when it took over. Also in 1961, a 25-year lease was issued to Skyway Motor Lodge to build a motel adjacent to the terminal, with three 5-year options for renewal.

In 1967, the Levy Court form of government for New Castle County was dissolved and replaced with a "County Council" form of government. The Council members and a County Executive were elected to their positions by the residents of New Castle County. The airport manager reported directly to the County Executive. In terms of government-run operations, the system was as good as it could get. Airport managers dread being under the thumb of several elected officials.

After the war, CAA regulations forced the airline industry to service small and intermediate airports in spite of their not being profitable. President Reagan felt the "free enterprise system" would benefit both the airline industry and the flying public and abolished the CAA control. The larger airlines moved from the small airports to the more profitable "Hub Airports." The void was almost immediately filled by commuter airlines which carried passengers to the major "Hub" terminals. NCCA is located too close to both Philadelphia and Baltimore to allow for a commuter service to operate. Therefore, TWA, All American, Altair Eastern, and American have left NCCA. The airport recovered from the loss by catering to corporate, general aviation, air charter, flight schools, aircraft sales, and maintenance services.

The new terminal location and other economic considerations made it necessary to close runway 5-23 effective December 19, 1972. The former runway was now designated as taxiway "S." The closing

The arrow points to runway 5-23 which was closed in 1972. The County Council purchased all private property between Churchmans Road and the airport. Photo courtesy of the Delaware River & Bay Authority.

of the runway also allowed for future development of airport commercial property bordering Churchmans Road.

By 1974, the *News Journal* headlined a feature article indicating the airport was a sorry mess. Potholes were in the parking lots and roadways, roofs were leaking, buildings were falling down, leases varied among tenants, and entire buildings were unaccounted for. Shocked over the news, two Councilmen, one Republican and one Democrat, prepared a thorough audit of the operation. Their conclusion was;

1 The rent structure is inconsistent to the point that "equity cannot be determined."

2 The physical condition of the buildings is "generally poor." New roofs are needed and the ceilings are falling in on three-fourths of them. Some walls are cracked, much of the wiring is hazardous and some buildings are firetraps with their exits blocked off.

3 Records are old, inadequate and contradictory to the point that Squatters could be occupying buildings without paying rent.

4 Overall management "appears unprofessional and inadequate."

The report went on to disclose that county records still include 16 airport buildings that have disappeared without a trace. There was no denying the airport's infrastructure had been neglected. It had a shabby appearance and showed signs of neglect. Considering the scope of the negative newspaper articles, a political firestorm erupted. Firings and resignations followed in rapid order. That, however, did not resolve the problem.

Representative Matushefske introduced legislation (HB-257) to amend the Delaware Code to address the situation. The bill allowed the creation of a Department of Transportation within the county. The county would have a Department of Transportation which was overseen by a Transportation Board, consisting of seven members, serving without compensation. The Board would be created to act as an advisory board. Each appointment would be for six years, staggered in such a manner that one term would expire each year. The County Executive was to appoint directors of Finance, Planning, Development, Licensing, Public Works, Police, Parks & Recreation, Transportation, and the six Advisory Board members. All appointees would serve at the pleasure of the County Executive.

This 1980's photo shows a bulldozer ready to demolish some derelict WWII buildings to make room for the new Boeing-Vertol building. If these were the type of buildings missing, it would have been a godsend. Photo Courtesy of the Delaware River & Bay Authority.

The government required Bell and Boeing to join forces to build the V-22 Osprey. *This photo is one of many test flights made at NCCA. Photo courtesy of the Delaware River & Bay Authority.*

The legislation was passed into law and signed by the Governor on May 27, 1974. This legislation required the County Executive to pick people with extraordinary skills in their fields for such departments. Locating six qualified people at an executive level who would serve on an advisory board without compensation was another challenge. The airport manager now had seven people to work with to resolve any issues. The legislation was reactionary at its worst. Sending more cooks into the kitchen will not necessarily guarantee a better meal. Two years later, the County Department of Transportation was abolished and replaced by an Airport Department on December 21, 1976. The airport manager reported only to the head of the County Executive. The County Executive reported to the County Council.

The NCCA infrastructure was not kept in prime condition over the years and fell to decay. Airport managers served at the pleasure of the County Executive. Elections play havoc with an airport manager's ability to change bosses overnight. Replacement of managers by the County Executive would surely entail other personnel changes. In a prepared speech given at an Airport Manager's Conference, at the University of Oklahoma, the speaker remarked; "More than 100 airport managers responded that their most difficult day-to-day problem was political interference. It is absolutely impossible for a business manager to run an efficient, businesslike operation if he is continually forced to grant political favors or to fight unqualified pressure group interference."

NCCA Airport managers

1947-1952	Don W. Martin
1952-1958	G. Edwin Petro
1958-1964	George J. Bean
1964-1967	Daniel F. Ginty
1967-1980	William H. Comer, Sr.
1980-1981	Robert J. Wrede (acting)
1981-1981	Michael Ratchford (acting)
1981-1984	Charles C. Darby
1985-1988	D. Drew Angeline
1988-1989	Joseph H. Toner (acting)
1989-1998	Frank J. Shahan (acting)
1990-1998	Carl L. Stacey
1998-	Rocco Tomanelli

For the next twenty years the county and airport manager tried to keep up with the changing technologies related to aviation. New installations included a helicopter pad, high intensity runway lighting, surfacing of runways, taxiways and ramps, a 36-inch rotating beacon, and much more. It appears the more they did, the more that needed to be done.

The airport continued to flounder financially and the Council looked for ways to regain the favor of the press and the people. Whatever ways were explored, it failed to happen. Bad press is one of the most difficult situations to rectify. There are many who believe business can fulfill any request and still survive. The truth is, business is very much like a human being. If a person is deprived of the proper nourishment to sustain life, they will get sick and die. If a business is deprived of enough income to meet expenses, it will also get sick and die. The decline of the airport eventually reached a point of crisis.

To remove the burden of the airport, the New Castle County Council transferred ownership of the Greater Wilmington Airport to the Delaware River and Bay Authority in 1996. The River and Bay Authority had what the County Council did not have and what the airport needed, money. The airport name was changed back to "New Castle County Airport." Mike Harkins, Director of the Delaware River & Bay Authority, immediately made funds available to improve the appearance of the airport. Buildings were painted, flower gardens were planted, signs were replaced, and potholes repaired. Harkins and his team are determined to bring the airport back from its years of neglect. ■

➲ Sussex County Airport

On May 5, 1941, the Legislature passed legislation allowing the Sussex County Levy Court to borrow $25,000 to build an airport southeast of Georgetown. Eight separate parcels of land, totaling 581.05 acres, were approved for purchase by the county. Although offers were made to the owners for the purchase of their land, the owners felt their land was worth more. To resolve the dispute, the land purchases fell into the hands of the courts when condemnation papers were filed by the county. The final results are as follows:

Halsted & Jennie Layton	110.9 acres	$ 5,000
Charles & Annie Joseph	4.2 acres	180
John & Alberta Elliott	29.0 acres	1,260
George & Elva Wilson	30.0 acres	1,080
Levi & Mary Bahn	15.75 acres	3,600
Lewis W. Briggs	87.6 acres	2,250
Harvey & Emma Wilson	88.0 acres	3,600
Farmers Bank/State of Delaware	56.0 acres	2,925
Benjamin & Bertha Ableman	135.0 acres	4,950
George & Eva Wilson	24.6 acres	1,500
TOTAL	581.05 acres	$26,345

Legal fees and advertising added an additional $1,189.92 to the cost of the purchase.

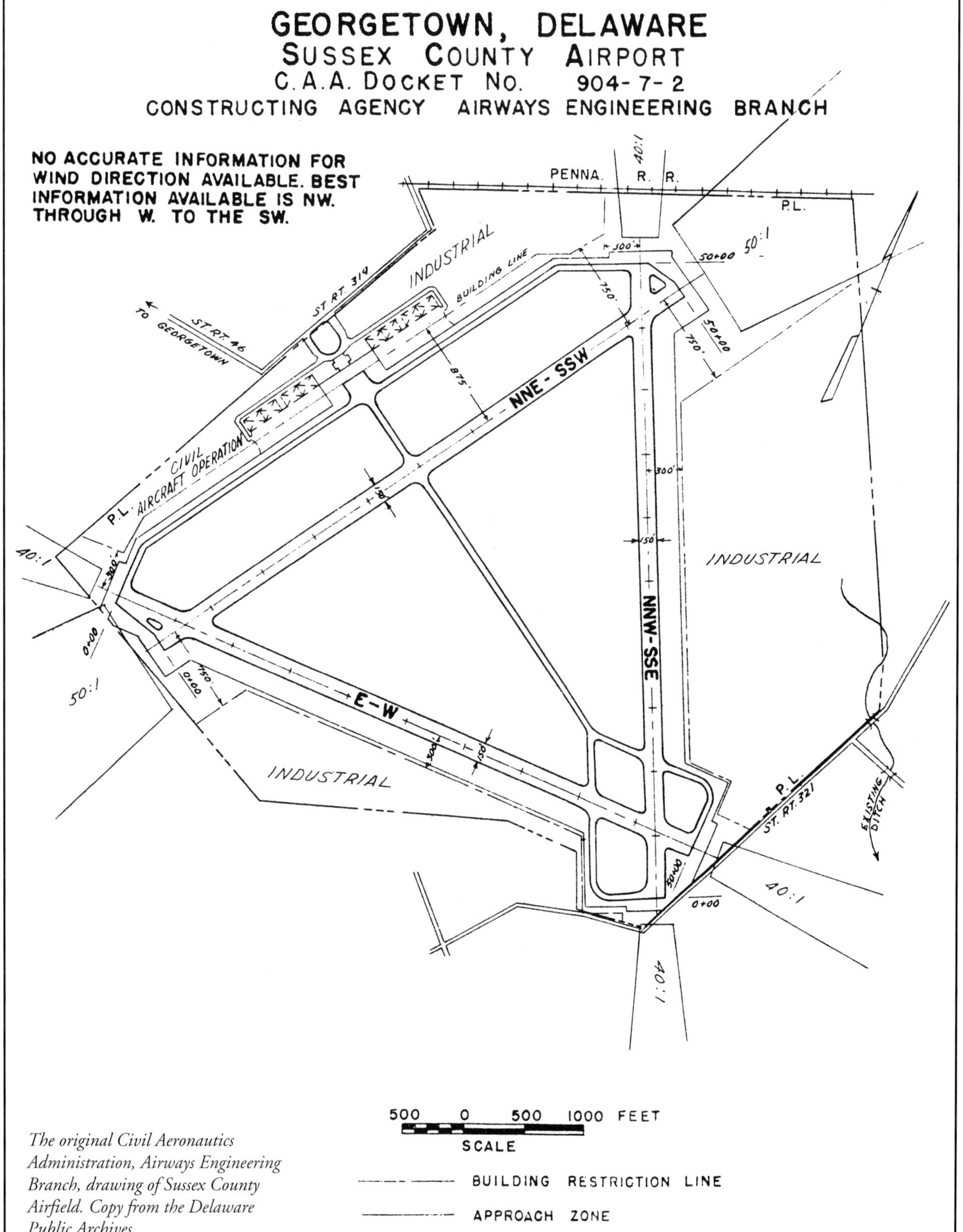

The original Civil Aeronautics Administration, Airways Engineering Branch, drawing of Sussex County Airfield. Copy from the Delaware Public Archives.

FARMERS – WOODSMEN

U R Needed

500 Acres of Timber Must be Cleared Within 90 Days for the

UNITED STATES NAVY'S GEORGETOWN AIRPORT

Sign Up For One of These Jobs
IMMEDIATELY
AT

War Manpower Commission's United States Employment Service

WEST MARKET STREET, GEORGETOWN

Poster from Delaware Public Archives.

When construction was approved, an action that would bring an outcry from the public today, it went without one complaint. On November 7, 1944, the military flew over the surrounding towns dropping leaflets which called for woodcutters to clear timber from the airfield. Although the newspapers disclosed that the wood was needed to build a simulated aircraft carrier deck, Lawrence Lind said, "This was a decoy announcement. It simply did not happen as reported." Lawrence Lind, a native of Minnesota, was one of the Navy personnel stationed at the base. During his tour of duty with the 23rd Carrier Aircraft Service Unit (CASU), he also courted and married Mildred Lee Truitt of Lewes. After the war he would work for All American Engineering. Lawrence Lind passed away in December 1997, just three months after this interview.

The Sussex County Airport would serve a different role in the war than Dover or New Castle. Sussex was used by the U.S. Navy as an auxiliary training facility for Cape May County Naval Air Station, Wildwood, New Jersey. No pilots or airplanes were ever assigned to Sussex County Airport during the war.

The first unit to arrive at Georgetown was the 23rd CASU. The complement of twenty-five men and two officers arrived in Georgetown, November 27, 1943. There were no housing facilities available at the airport, and the men found living quarters in the homes of local residents. Although the 23rd CASU would be allocated additional personnel over the years of active duty, it would be the only military unit to serve at the airport. The two airplanes listed below are the only two Lawrence Lind remembered using the airport. The TBF-1 *Avenger* is the same type of airplane flown by President George Bush during WWII.

GRUMMAN TBF-1 AVENGER SPECIFICATIONS

Engine	Wright *Cyclone*
Type	R-2600-20
Horsepower	1,750
Wingspan	54.2 ft
Length	40 ft
Height	16.5 ft
Empty weight	10,700 lbs
Gross weight	18,250 lbs
Maximum speed	267 mph
Cruise speed	237 mph
Range	1,130 miles
Crew	3

The airplane in the foreground was used by the commanding officer to fly back and fourth to Wildwood, New Jersey. The other airplanes were damaged during maneuvers. Photo by Lawrence Lind.

The Grumman TBF-1 Avenger *was a Navy carrier-based torpedo-bomber. It used the Sussex Airport for practicing carrier landings. Photo from Lawrence Lind.*

The pilot placed his parachute on the wing and awaits rescue after he failed to pull up in time from a dive and his Curtiss SB2C Helldiver *crashed in the marsh. Photo from Lawrence Lind.*

CURTISS SB2C HELLDIVER SPECIFICATIONS

Engine	Wright *Cyclone*
Model	R-2600-20
Horsepower	1,1750
Wingspan	49.9 ft
Wing area	422 sq/ft
Length	36.8 ft
Height	13.2 ft
Empty weight	10,547 lbs
Gross weight	16,616 lbs
Maximum speed	295 mph
Cruise speed	158 mph
Range	1,165 miles
Crew	2

Lawrence Lind vividly remembered the duties the 23rd CASU had at Georgetown:

> The Curtiss SB2C dive-bomber used the adjacent marsh lands for bombing practice. A large target was displayed for the pilots to use for bombing. Unfortunately, many of the pilots had little flying experience and a lot of planes crashed. One plane, with the pilot and navigator, crashed in the marsh and was entirely consumed. We also hauled wrecked airplanes back to the airport, disassembled them and shipped the parts out. Although we did simple repairs outside, we did not have any major repair facilities since there were no hangars.
>
> A simulated carrier deck was painted on the runway to outline the landing area for pilots. A mobile lighting truck was used at night to light the simulated aircraft carrier deck for night sorties. The signalman used lighted paddles for pilots of the Navy Curtiss SB2C and the Douglas TBF-1 airplanes to help guide them in. The airport was not equipped with lights, so the corners of the simulated deck were lighted with hand-held flashlights. This was very dangerous and I was surprised no one was killed. Those pilots were anything but skilled, and they were all over the place. Although our work was interesting, life and facilities were quite primitive at the airport. I had to sneak the few photos I got because taking pictures was strictly forbidden.

Top Right — The control tower was constructed entirely of wood. It was hot in summer and cold in winter. Photo from Lawrence Lind.

Bottom Left — These wrecked airplanes were recovered from the surrounding area and brought back to the airport. Photo from Lawrence Lind.

Bottom Right — Rescue and fire-fighting equipment was very limited. This photo shows the entire operational units. Photo from Lawrence Lind.

(Left) Chesapeake Airways DC-3

(Right) B-29 Superfortress *dwarfs the civil aircraft.*

(Left) B-29 Superfortress *taxies into position.*

(Right) The general aviation airplanes appear toy-like under the massive 13-29 Superfortress *wing.*

(Left) Flight of eight - B-29s.

(Right) P-47 Thunderbolt. Today, it is the rarest of WWII warbirds.

(Left) Flybys were numerous.

(Right) Row of AT-6 Texans.

These photos of the Grand Opening were taken by Bill McDaniel, of Dover.

On March 25, 1947, the War Assets Administration declared the Airport surplus and turned it over to the county. A grand reopening of the Airport was scheduled for Memorial Day weekend. The opening was supported heavily by the military under the direction of Colonel Goldsboro. The Georgetown Lions Club sponsored the two-day event.

This would be the first time our military air power would be available for public display in Sussex County. On opening day, a flight of eight Boeing B-29s, followed by a flight of eighteen North American AT-6s, and a squadron of AT-11s flew overhead. In addition, a flight of twelve, mighty P-47 *Thunderbolts* flew in for the event. The thundering military aircraft flying overhead gave the spectators an eerie feeling of what the enemy must have witnessed during the war.

The 4th Fighter Group came with the military's fastest fighter, the jet-propelled Lockheed *Shooting Star*. They demonstrated their spectacular speed by flying round trip from Georgetown to Philadelphia in 24 minutes. It would also be the first (and only) time the mammoth B-29 *Superfortress*, (the same type that bombed Tokyo) would be displayed to the public in Sussex County. In addition to these military aircraft, the B-25 *Mitchell*, Grumman F6F *Hellcatt*, Douglas C-45 *Skytrain*, Chance-Vought F4U-1A *Corsair* and Curtiss SB2C *Helldivers* were flown in for the event. Civilian aircraft manufacturers were also invited to show off their new postwar airplanes. Bellanca, Cessna, Aeronca, Ercoupe, Stinson, Fairchild, and many more, hoped for an aviation-minded public to consider flying for business or pleasure.

Many flying and parachute events kept the show alive with a breathtaking display of aviation performances. The Levy Court Commissioners and other political dignitaries wandered through the ranks looking for support for everything from re-election to new tenants. The public mood at the event was one of enjoyment and pride that Sussex County contributed to the war effort. This event was possibly the largest military/civil aviation show ever held in Delaware.

Between 1947 and 1950, several attempts were made to bring scheduled passenger and cargo service to Sussex County Airport. Chesapeake Airlines, Maryland Airlines, and All American Airways each made attempts at making the Georgetown stop profitable. Although each made a commendable effort, there was insufficient volume to continue operations. The Levy Court then sought out the Delaware Congressional delegation to use their influence to move the Delaware National Guard to Sussex County Airport. The noble effort was in vain, and the Guard remained at New Castle.

All American Engineering became the salvation of the airport when it was decided, in 1949, they would start moving all testing operations from the DuPont Airfield to Sussex County. All American Engineering would test catapults, arresting gear, energy absorbers, satellite retrieval systems, seat ejectors and other systems for the military. It was a sad day when the company was sold and the operation was moved to New Jersey in 1970.

That same year the Sussex County Levy Court felt it was time for a name change. The name "Levy Court" was established in the early days of settlement and was empowered to levy taxes on the people to provide county services. In the more modern world, court is more likely to be associated with legal action. It was for this reason its name was changed to "Sussex County Council." The Council is still an elected body and must meet the growing needs of the county.

Inactivity and neglect had aged the runways and taxiways during the All American years. Chunks of concrete, blistered from the burning summer sun and winter ice, lay in the path of landing aircraft. Lumps of grass grew up through the cracks, pushing up the concrete, creating an uneven surface. This surface was dangerous to airplanes when landing. The buildings on the airport began to show their age and gave the appearance that the airport was abandoned. With only a small contingent of privately owned aircraft there, the Sussex County Council felt the expense of maintaining three runways was prohibitive. Council voted to repair one runway to alleviate any chance of liability, and the two remaining runways were painted with a large "X", denoting them unusable.

The airport remained more-or-less dormant until Leona Hemsley, of New York, signed a long-term

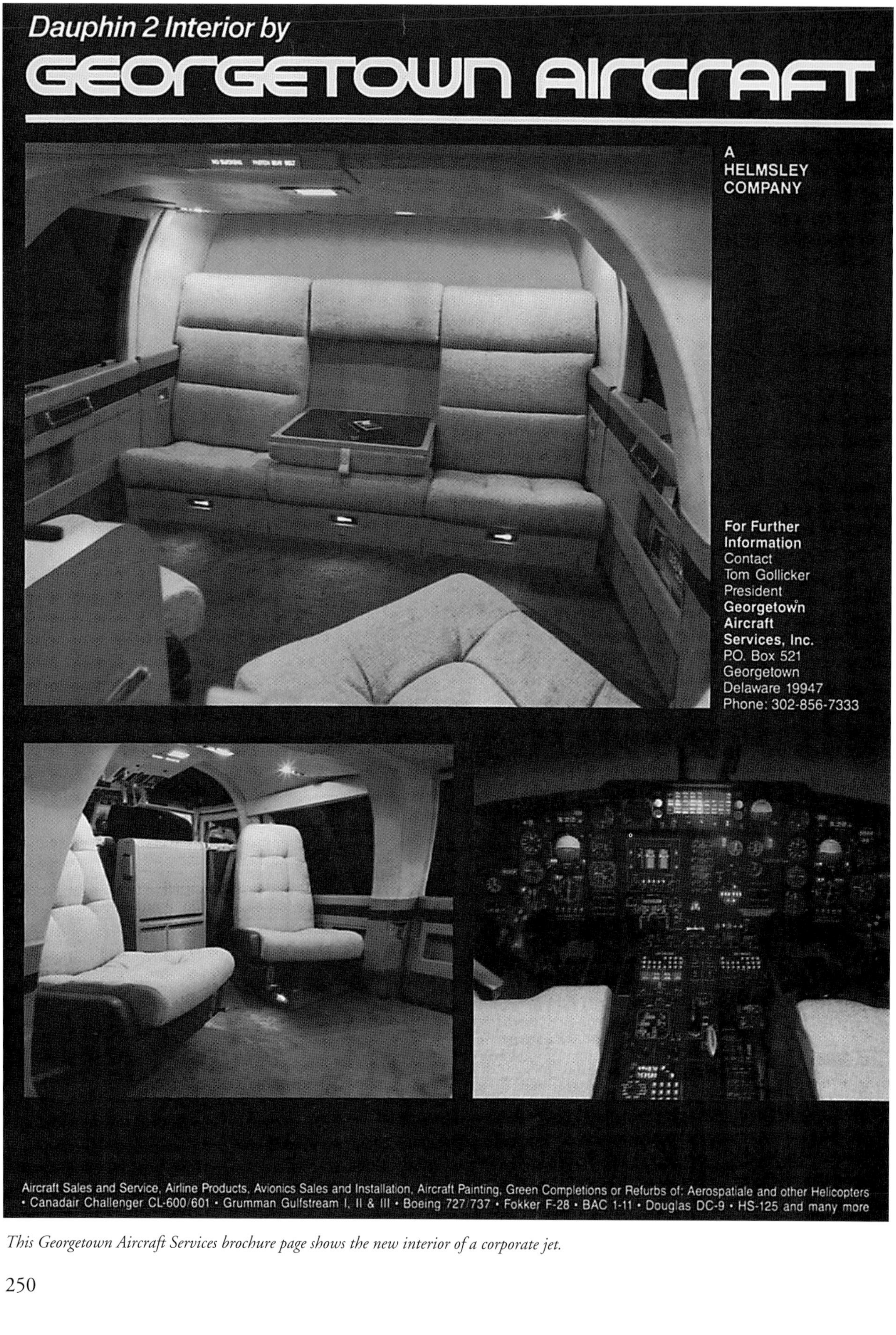

This Georgetown Aircraft Services brochure page shows the new interior of a corporate jet.

lease with the Council and created Georgetown Aircraft Services in 1982. Three large hangars were built for crews to work on aircraft as large as the three-engine Boeing 727 jet. The location of the hangars are on the industrial site bordering the east side of the airport. One of the three hangars was not used for production, while the third housed Leona Hemsley's personal Boeing 727-100 jet. In airline configuration, the 727-100 could carry 94 passengers. Georgetown Aircraft Services refurbished the airplane to an executive layout to please the boss.

Sal Scarfi, operations manager of Georgetown Aircraft Services, oversaw the complete refurbishing of corporate aircraft and helicopters. An engineering staff custom-designed elaborate interiors and exteriors befitting the most discriminating executive. Cockpit layouts were also modified to accommodate the latest in electronic navigation, instrumentation, autopilots, and safety features. Boeing initially designed the 727, in 1965, for Lufthansa Airlines, who had requested an airplane be built that could carry 100 passengers. Georgetown Aircraft Services, at the height of production, employed 125 people. The company was shut down in early 1991 under the direct orders of Leona Hemsley. Many employees were astonished at the decision, since there was plenty of work available. Six years later, Mrs. Hemsley sold the two production hangars to PATS, of Columbia, Maryland. Leona Hemsley retained the third hangar where pilots and crew stood by to fly her 727 on demand. PATS is an aircraft modification company that specializes in retro-fitting production aircraft with long-range fuel tanks. The Sussex facility will install the special fuel tanks in the

The newly surfaced runway was dedicated May 6, 1995. The ceremonies drew a large contingent of spectators, visiting aircraft and elected officials. Photo courtesy of Sussex County Economic Development.

Delmarva Aircraft FBO center and maintenance hangar, May 6, 1995. Photo courtesy of Sussex County Economic Development.

corporate version of the 737. The tanks are fabricated in Maryland and shipped to Sussex for installation.

The Sussex County Council leased the Fixed Base Operation (FBO) in 1985 to Delmarva Aircraft, Inc. The Council felt that leasing the airport to a career-orientated company would relieve them of the responsibility of operating a business in which they had little expertise. On the surface, this appeared to be a wise decision. The lease was very liberal and allowed the FBO to forge ahead in the aviation business with little government intervention. The 1998 pilot's information manual lists the airport as manned from 7 a.m. to 9 p.m. during the winter and from 7 a.m. to 11 p.m. during the summer. A non-precision instrument approach to runways 4/22 can be accomplished utilizing the Waterloo VOR (ATR 112.6). Unfortunately, the anticipated volume of aircraft activity did not materialize and Delmarva Aircraft could not expand as they had planned. With marginal profits, Delmarva Aircraft struggled to stay alive.

Effective April 1, 1991, Frank B. Calio was hired by the Sussex County Council as Director of the Sussex County Airport and Industrial Park. Two years later, the Council combined that office with the Office of Economic Development. The Sussex County Economic Development Department, under the leadership of Frank B. Calio, has made many improvements to the airport over the years. The main runway was extended to 5,000 feet and resurfaced in 1995. The buildings and grounds no longer appear neglected and unoccupied. Much of this was accomplished by requesting federal aid from the FAA through the Delaware Aeronautics Department.

The Sussex County Airport is the location for several other small privately owned, aviation-related businesses. Two aircraft maintenance operations service aircraft; Sussex Aero Maintenance, owned by Al Nace, and GCG Aircraft Services, owned by Jame Griffiths. Several individual hangars were built by Frank Anderson and leased by Professional Leasing, Inc. to airplane owners. Student pilot-training is offered by Ted Jones of Jones Aviation.

Frank Calio and his staff are very positive about the value of the airport to Sussex County. The Council saw fit, in 1997, to buy out the floundering Delmarva Aircraft, Inc. lease. This was done to allow for a restructuring of the airport with an eye to the future. New hangars, with signed leases before construction, are under construction for resident aircraft. New runway lighting is being installed, and a new FBO building is in the planning stage. Surface transportation is also being studied to improve the visitors' ability to arrive at their final destinations.

This outlines the history of all three of Delaware's major airports. However, under the provisions of transfer of title of New Castle and Sussex County airports, the military retains the right to recapture its leasehold interest in both airports in the event of a national emergency. Actually, in the interest of national security, the right of recapture has been retained in all airport disposal actions throughout the country. Inasmuch as the FAA has cognizance over the Government's interest in matters pertaining to maintenance of the airport while in civil control, arrangements for financial relief are available through the FAA. This well-thought-out plan assures that the airports will be maintained in a constant state of readiness by the counties. It is doubtful these airports will ever fall into a state of decay, or be sold off to developers for housing development. Only the capital city, Dover, and Kent County still lack a proper general aviation facility with no restrictions. ■

CHAPTER 8

Field of Dreams

The title of this chapter was borrowed from the movie that bore the same name. The Field of Dreams *in this chapter, however, is not a baseball diamond, but an airfield from which to fly airplanes. A place pilots could call home base. A place where they could teach others with visions of breaking the bondage of earth to fly. A place where pilots could congregate, tell flying tales and tinker with their airplanes. The Fields of Dreams relives those days of the local airfield, the people who made them possible and their ultimate fate. It was a life not too far removed from the corner gas station where the owner had to be open seven days a week to carve out a living and serve the public. For the most part, the airfields lasted only a single lifetime. Privately owned, public-use airfields, like the corner gas station, rarely passed on to the next generation. Until now, these airfields, and the owners who started them, were a forgotten part of the History of Delaware Aviation.*

■ *NEW CASTLE COUNTY*

✈ Biggs Airfield 1922-1957

Much of the land in New Castle was held by the New Castle Commons through a warrant issued by William Penn. These lands were set aside as public lands but more under the control of the New Castle Commons Trustees. The lands were to be used by residents for timber, firewood, grazing, and farmland. By 1764, all the timber on the common land had been cut, the removal of which created approximately 1000 acres of farmland. The Commons divided this land into farms which were worked by the tenants. Alexander "John" and Mary Biggs (Mary was John's sister) became tenant farmers on

This rare 1925 aerial photo of Biggs Airfield is representative of airfields of the period. If it were not for the white circle, it would be indistinguishable from the other farm fields. Note the Delaware River at the top of the photo. Photo from the Hagley Museum & Library.

what was known as the "Stockton Farm No. 2." The farm consisted of 134.75 acres and was leased on a year-to-year basis. Since the airfield did not occupy the entire farm, a clause stipulated John and Mary were required to pay a percentage of the crop sales to the Commons.

The actual date that John and Mary signed the first lease on the Stockton farm is lost in time. However, the *Journal Every Evening* announced to the public in 1922 that,

> John H. Biggs has given the Government air messengers, who travel by airplane from Washington to New York, permission to land in the event of trouble, or for any other purpose, on a level tract of land on his property. He has an official marker placed on the spot so it can easily be seen by the airman from afar.

The official marker referred to is the large circle of white sand or white stones placed on the airfield to distinguish it from other fields from the air. These markers can be seen in many of the early airfield aerial photos throughout this book. John and Mary Biggs, unknowingly, established the first public-use airfield in Delaware.

As aviation grew over the years, Biggs Field increased its services to include light maintenance, fuel, hangar rental, parking, and flight instruction. The airfield was listed as "Biggs" on the first aeronautical chart printed by the Coast and Geodetic Survey in 1932. It was also included in the Directory of Airports, produced in 1937 by the Bureau of Air Commerce as:

> BIGGS FIELD: One mile North of center of New Castle and Bellanca Field. Altitude 60 ft. Rectangular, clay, 1400' E/W and 1800' N/S. Two hangars and other buildings. Entire field available. Aviation fuel available day and night.

During the early days of aviation, landing facilities were not hard-surfaced. The sod airfields offered more ability to stop the airplane when landing since most airplanes were not equipped with brakes. A fixed tail skid, as opposed to a wheel, would drag the airplane to a stop. Pilots were trained to give their primary attention to landing directly into the wind to slow the touchdown speed. Since the Biggs landing field was omni-directional, pilots simply took off and landed into the wind. Although a wind indicator, called a windsock, was located somewhere on the airfield, pilots were trained to look for other signs of wind direction. Smoke was always a good indicator. Lakes and ponds another. Trees also tell a tale by bending with the wind and exposing the underside of their leaves which appear lighter in color.

The official logo of the "Noisy Birdmen." Logo courtesy of Mike Guididas.

In 1935, a colorful, organized group of flyers who flew from Biggs named themselves "The Noisy Birdmen." Although their name suggests they were a rowdy or boisterous bunch of pilots, they were not. The nucleus of the initial twelve charter members, headed by Dr. H. S. Miller of Wilmington and A. R. Lardin, an aeronautical engineer with Bellanca Aircraft, created a well-defined charter. The charter, among other things, defined "Noisy" as *strenuous efforts* to make Delaware air-conscious. The organization flourished over the years to include 53 members. It was not a criterion for members to be pilots. Applicants who were affiliated with aviation, such as mechanics and engineers, were also accepted. Although the club was active beginning in 1935, they did not officially incorporate in Delaware until April 15, 1937. Official club meetings were held the first two Mondays of each month. The first Monday was a business meeting and the second was a social gathering.

Social fly-outs would bring many members out to the airport for cross-country flying. One such event was given front-page publicity in the *Delaware Coast News* on September 18, 1936. In part, the article disclosed that the pilots departed Biggs Airfield en masse and flew to Dover where they landed and released a homing pigeon. Following this brief stop, they flew on to the unmanned auxiliary airfield at Bethany to release another pigeon. Then it was off to Rehoboth to land at Luther B. Rogers private airfield where they were greeted by members of the Rehoboth Flying Club. Before departing the airfield for a motoring trip to Rehoboth, the last pigeon was released. A late luncheon was held at the Henlopen Hotel for the combined group of forty before they returned to the Rogers airfield at 6 p.m. for the return flight to Biggs Airfield.

These jaunts, and other flying activities, were also covered extensively by the *Journal Every Evening* in a weekly feature article titled, "Along the Airways." World War II brought an end to the activities of "The Noisy Birdmen" and they disbanded to serve the war effort, never to reunite again.

J. Allison Buck, Delaware's first federally licensed pilot, also flew from Biggs Airfield. Delaware's first woman pilot, Miss Myrtle Brown (1900-1985, also from New Castle County) learned to fly from Biggs. Tom McKenna, a certified flight instructor, taught the late Frank Tallman, noted Hollywood film-maker, how to fly from Biggs Airfield. By 1938, Tallman earned his flight instructor's rating and taught flying at Biggs Field before moving to California, where he earned a national reputation supplying airplanes and pilots to film makers. Another celebrity of sorts who learned to fly from Biggs was Joe Pine, radio announcer for WILM. Joe went on to become a huge success on the West Coast as a television talk show host. Reynolds Jones, owner of Smyrna Airfield, also learned to fly at Biggs Airfield before starting his own privately owned, public-use, airfield in Smyrna. This only documents a few people from Biggs who made a career of flying, or flew for pleasure.

Although the airfield was inactive during WWII, Biggs field was still kept in operational condition in case of an emergency. As soon as the war ended, the activity at Biggs field grew with postwar aviation. The last lease available for review (1951) disclosed an annual rental of $1,250.00 to be paid quarterly. This lease also limited the number of airplanes allowed on the airfield as resident airplanes to twenty. To accommodate the aircraft, Biggs erected a row of ten "T" hangars which were rented to airplane owners.

This aerial photo of Biggs Airfield was taken in the late 1930s. Photo courtesy of Harry Van Den Huevel.

The airport came to an end in September 1957, when John and Mary Biggs decided to close the airfield and only rent five acres, which included the farmhouse and outbuildings. The Commons sold the airfield property on March 1, 1959, to developers who created the development known today as "Penn Acres." As John and Mary grew older, they moved from the Commons farm to 100 West 7th Street in Wilmington. John passed away on December 31, 1969, at the age of 73. Mary passed away October 8, 1971, at the age of 77. In contrast to the other airfields in Delaware that were started by pilots, John and Mary Biggs never learned to fly. Mary and John Biggs' *Field of Dreams* left a thirty-five year legacy by establishing the first Delaware public-use airfield in 1922 that had its beginning as an auxiliary airfield.

This aerial photo (circa 1931) shows the Air Service hangar location in the upper left corner of Bellanca Airfield in relationship to the Bellanca factory. Photo from the Daniel Kohle collection.
Inset: Randolph Holladay, Air Service, Inc.

✈ Air Service, Inc. 1928-1960

When the du Ponts sold the Spring Garden Farm to Giuseppe Bellanca for his factory in 1927, they retained the right to establish a public-use sales and service operation on a portion of the land. In aviation, this is called an FBO (Fixed Base Operation). Although many historians assert that the DuPont Airfield was a commercial operation at this time, it is in fact not true. Henry delved into various aviation interests, but enjoyed his own private haven and had no interest in opening it to the public. Henry and Francis built a hangar and operations building on the north side of Bellanca Airfield in 1928 and rented it for $250 per month to Air Service, Inc. This is confirmed by a lease located in the H.B. du Pont papers at the Hagley Museum & Library. The facility was built, not in the interest of making money, but because Henry and Francis felt a deep conviction that Delaware needed a proper airfield which would be open to the public. This conviction was born out of the visit of Charles Lindbergh and the *Spirit of St. Louis* in mid-1927. The

Air Service, Inc. hangar and operations building, circa 1931. The airplanes from left to right are; Kinner-Fleet, Krier-Reisner (KR-21), Stinson, Jr., KR-21, KR-31. Photo courtesy of Johnny Ward.

The new 1935 Air Service, Inc. hangar was larger and used many windows to take advantage of natural lighting. Photo courtesy of Joe Griegg.

DuPont Airfield could not provide a hangar for Lindbergh's airplane, which was embarrassing to say the least. It is also conceivable that Henry felt it would discourage flyers from wanting to use his own DuPont airfield.

On October 15, 1928, Air Service, Inc. was incorporated in Delaware. The official documents filed with the Delaware Secretary of State, Department of Incorporations, name J. Allison Buck, president; John H. White, vice president; and James R. Morford, secretary/treasurer. Richard (Dick) Morgan and Randolph (Ran) Holladay were hired by the board as the FBOs. Air Service, Inc., with the Stinson franchise, grew by leaps and bounds as the depression years began to fade. In 1935, the original building burned to the ground. A new building of similar design was immediately built on the same site.

Upscale aircraft such as, Waco, Beechcraft, Bellanca, Fairchild, and Lockheed were commonplace on the airfield. High profile dignitaries, such as the du Ponts, Carpenters, Clarence Chamberlin, Amelia Earhart, Wallace Beery, Anne Harding, Paul Tallman, Wiley Post and many more, were documented visitors.

Among others, Johnny McChesney Morgan (Johnny Mac) got his first airplane ride from Henry du Pont in his Buhl *Airster* open-cockpit biplane in 1927. This introduction to aviation hooked him for life. Johnny Mac, as he was called by everyone who knew him, would spend his entire life in aviation. Johnny Mac's brother, Dick, taught him to fly from the Bellanca Airfield in 1931, at the age of sixteen. Since Dick Morgan and Ran Holladay ran the FBO on

Johnny McChesney Morgan and his older Brother Richard (Dick) pose in front of the Carpenter's Lockheed. Photo courtesy of Mrs. John Morgan.

Bellanca Airfield, Johnny Mac spent all his spare time there. Over the years, he became the pilot of choice for many notables, including the du Ponts and the Carpenters while he was working for Air Service.

Air Service, Inc. would continue to prosper until the war broke out in December 1941, when it was closed to civil aviation. The airfield returned to operational status as soon as the War Department lifted restrictions in the summer of 1945. At this point Johnny Mac assumed the position of FBO until the airfield was sold to developers in 1960. Johnny Mac later accepted a job with Kip du Pont at Summit Aviation in Middletown, Delaware, until he retired. Although the airfield is closed, the hangar (on French Town Pike, Rt. 273) with its distinctive oval roof, is still standing.

✈ Buck Airfield 1928-1933

J. Allison Buck, a native Delawarean, served in the U.S. Navy during WWI. When the war ended he decided he had enough of the ocean of water, and the ocean of air became his new frontier. Victor

Allison Buck's airfield he started in 1928. Photos from the Hagley Museum & Library.

Dallin, commander of the Pennsylvania Air National Guard, taught Allison how to fly in 1922. Dallin was also a professional aerial photographer who photographed much of Delaware, Pennsylvania, and New Jersey. Allison's first airplane was a WWI surplus Curtiss *Jenny* he and his father purchased still in its shipping crate. They assembled the airplane and flew from a local field near their home in Wilmington. Once Allison felt comfortable flying the *Jenny*, he used it to do some barnstorming around the country to build flying time and earn money.

When the Department of Commerce required all pilots to obtain a license in 1926, Allison Buck became Delaware's first licensed pilot. He would also become a noted figure in Delaware aviation by serving as president of Air Service, Inc. when it incorporated in Delaware, October 14, 1928. On the very next day, Allison Buck incorporated his own business "Delaware Flying Service," located between Minquedale and Farnhurst.

Allison Buck could not shake his insatiable desire to own his own aviation business. He wanted the freedom of independence. He had earned the respect of his fellow airmen and a following of student pilots. To get him started, H.B. du Pont made him a loan to purchase four Waco Airplanes. The loan papers, located in the H.B. du Pont papers at Hagley Museum & Library, were very informal. There was no mention of interest, only that the loan would be repaid upon the sale of the airplanes.

Buck would jump at the chance to fly anything that had wings and built a solid reputation as a skilled pilot. He would perform aerobatics at air shows to thrill the public and earn extra money. He was a

Allison Buck and his Curtiss Jenny. *Photo from the Hagley Museum & Library.*

close friend of the du Ponts and taught many of them how to fly their airplanes. He was also a part-time test pilot for Bellanca and test flew many Bellanca aircraft prior to delivery.

An official opening date for the Buck Airfield could not be found. However, Secretary of State records indicate Delaware Flying Service was incorporated on October 16, 1928. Aviation journals reveal Delaware Flying Service became an agent for Waco and Fairchild airplanes and offered instruction, charter flights, sightseeing, rentals and glider towing. The single hangar roof, painted with large letters "AL BUCK" and the official white circle on the ground were the only visual signs there was an airfield below.

Allison Buck had a flair for publicity. In this photo, he displayed his Lambert MonoPrep parasol airplane in an automobile showroom on Market Street in 1931. The MonoPrep was the forerunner of the MonoCoupe. Photo courtesy of the Buck family.

April 19, 1930, Mickey Efferson, noted wing-walker and parachutist, married Mildred Hitch (center) aloft over Wilmington. Allison Buck, far right, flew the six seat Bellanca and Ralph Minker was minister. Photo from the Hagley Museum & Library.

The Delaware Glider Unit No. 1 was a club organized in 1930, and the membership used Bellanca Airfield for its base of operations. However, in 1931, it moved to Buck where there was less congestion. The officers were: Ralph Weaver, president; James O'Neal, vice president; James Scott, secretary-treasurer; and Donald Best, engineer. The board of governors included, James Morford, H. B. du Pont, Ralph Weaver, Stewart Poole, James O'Neal, C. S. Horn, Jr. and Franklin Cook. Extensive research turned up little more than the club flew gliders for their personal enjoyment.

Allison Buck's Field of Dreams lasted only about six years due to a tragic event. On July 15, 1933, Allison Buck lost his life test-flying a Bellanca CH-300 with CAB official, John Moran. The Bellanca CH *Skyrocket* was being test flown before being delivered to Richfield Oil Company in California. Investigation of the wreckage revealed the propeller blade failed in flight. The imbalance of the propeller's turning almost 2000 rpm unbalanced the engine and ripped it completely out of the mounts. Eyewitness reports confirmed the airplane and the engine fell separately. The airplane crashed into the Delaware Bay killing both occupants. Al Buck had been flying since 1920 and was only 36 years old. He was also a member of the elite QB (Quiet Birdmen) organization, and he was buried at the Mt. Salem Cemetery. He had been divorced from his wife Sarah Smith Buck, who lived with her parents, at 705 Van Buren Street, Wilmington. They had an eight year-old daughter Barbara Ann Buck.

Howard Turpin, a free-lance flying instructor, took over the airfield operation until the estate could be settled. After about eighteen months, Howard Turpin moved to Skycroft Airfield north of Wilmington, and Buck Airfield became history. The property is now the site of Gracelawn Cemetery.

✈ Skycroft Airfield 1932-1937

In 1932, Maurice Paschall, owner of Paschall Coal & Oil Company, decided he would take up flying. His homestead, north of Wilmington, included enough land for a landing strip and hangar. He named the small airfield Skycroft and soon allowed several other local pilots to join in the fun. Johnny Ward and Joe Barkley, Sr., kept their Taylorcraft in the hangar on Skycroft. The airfield was quite small, covering only about 40 acres on land that is known today as Carrcroft. Bound on the west by Weldin Road and on the south by Shipley Road, it was a bit challenging for pilots. The short 1300-foot-long runway had trees and electric wires on both ends. One large hangar, partitioned into bays, had the capacity to house five airplanes and an operations office.

A thorough search was made for an actual opening and closing dates of Skycroft. Since it was originally opened to accommodate Charles Paschall's own airplane, it evolved with little fanfare. The best scenario is that a few locals kept their airplanes on the airfield for convenience sake. The airfield never grew beyond that point. In contrast, the Paschall home still stands as this is written. Skycroft Airfield is shown on the second edition of the Coast and Geodetic Survey aeronautical chart printed in 1933. If it were not for this evidence, it would probably have been lost in time. It appears Skycroft lasted until 1937. Howard Turpin, a gypsy flight instructor who taught flying at Buck and Skycroft, accepted a job at the Dover Airfield as FBO in 1937. Skycroft Airfield was no longer listed on aviation charts in 1938.

✈ Cherry Island Airfield 1938-1941

The DuPont Airfield, Bellanca Airfield, and Dover Airfield, with their new hangars and fresh paint, stood in stark contrast to the Cherry Island Airfield located at the end of Wilmington's 12th Street. What Cherry Island Airfield lacked in creature comfort, it made up in determination. If you didn't know where it was, you would most likely not find it from the ground, nor from the air for that matter. Driving east from Wilmington on 12th street, you would cross the railroad tracks, provided a freight train was not blocking the way. If it was, you just went back from where you came and tried another time. A rutted, single-lane, dirt road wound through high marsh grass to all points of the compass before the lone operations hangar and yellow Piper *Cub* came into view. If you were wealthy, you would make a quick retreat before getting out of the car. On the other hand, if you had limited funds and a burning desire to fly, you were in the right place.

William and Johnny Ward were poor folks who were afflicted with flying fever at a very young age. In 1938, the brothers rented the land for their *Field of Dreams* for $25.00 monthly, and they were in business. They scrounged used lumber to build the framework for their self-engineered hangar and lean-to operations building located on the very edge of the marsh. Much of the siding was made up of discarded Hercules Powder Company resin barrels which were manually flattened on the ground with a hammer. If the structure could claim any fame in history, it probably was the first one built in Delaware entirely from recycled materials. The landing area was simply a mowed field of sorts. (Stubs of marsh grass can hardly be called grass.) When it was dry, the land looked like a parched, dry riverbed. When it rained, the airfield became a sea of mud. When the rain stopped, the water would run between the dry mud cracks, and a few hours later the

The combination hangar and operations office built by the owners. Photo courtesy of Johnny Ward.

An unidentified woman watches while Johnny Ward holds the gas hose and his brother, William, fills the tank of a Piper Cub *using a large funnel and a chamois to trap any dirt or moisture. Photo courtesy of Johnny Ward.*

field was usable again. Since there were no designated runways, pilots simply looked at the lone windsock located on the hangar roof and took off (or landed) into the wind. With an elevation of seven feet above sea level, high tide and heavy rain could temporarily turn Cherry Island into a seaplane base.

This spartan facility was born out of the desire of William and Johnny Ward to teach flying under the government-sponsored Civil Pilot Training Program (CPTP) and make a living. The CPTP paid flight instructors to teach flying to young men in case of war. One Piper *Cub* was purchased with a small down payment followed by many backbreaking monthly payments. Pilot instruction was $2.50 per hour. The Piper *Cub* rented for $6.00 per hour. A manually operated gas pump dispensed 80-octane aviation fuel at $0.23 per gallon. In the beginning, this was the only airplane on the airfield. Later, several DuPont employees started a flying club and added two more Piper *Cubs.*

Cherry Island Airfield was typical of the early privately owned, public-use, flying facilities in Delaware and across the nation. The depression offered little in the way of extra money for flying. The brave ones carved out landing facilities by cutting trees, digging stumps, and mowing fields. A hangar of any kind was a luxury most could ill afford. For the most part, the airplanes were tied down out in the open weather. More important than the amenities, was their *Field of Dreams* from which they would fly.

A storm in 1941, which blew the roof off the building, failed to discourage the brothers. Part of the roof can be seen next to the gas pump in the bottom photo on the preceding page. Once again, they scrounged materials to make the repairs while continuing operations. Cherry Island, like the other airfields throughout Delaware, was closed for the duration when WWII broke out in December 1941. If an airplane remained on any airfield, the owner was required to make it inoperative. This was usually done by removing the propeller and one aileron and taking them home. The remainder of the airplane sat forlorn, waiting for a long manmade war to end before it could return to its element, the sky. The Ward brothers answered the call to serve their country and the airfield was returned to marsh grass and wildlife, never to open again. Cherry Island was later chosen by the city as a landfill. What was once the lowest land in the area, is now one of the highest in the state.

✈ Baker Airfield 1950-1960

The topography of upper New Castle County consists of gentle rolling hills and meandering streams that flow into the Christina River. It is beautiful country, rich in heritage. The Swedes were the original European settlers in 1638. The town of Wilmington, with its deep-water port, prospered over the years to become the most populated city in the state. Gracious estates, with manicured landscaping, suggest this land is the choice of many who have excelled. This upper end of Delaware is fast-paced, and high profile.

In stark contrast, the lower end of New Castle County, south of the Chesapeake and Delaware Canal, contains pockets of farmland. The lifestyle slows dramatically. The small towns of Odessa, Townsend, Delaware City, and Middletown are reminiscent of rural America. The lifestyle of the people is a laid-back, unassuming one. It is a lifestyle of commitment to a balance of physical hard work and relaxation rather than a show of wealth. In the 1950s an airfield in this location would not have been considered worthy of creating a fuss. Today it would be an entirely different matter.

Aviation, and airfields in particular, after WWII, appeared to grow throughout the state like dandelions in the spring. By 1953, the aviation sectional chart showed a total of twenty aviation facilities in Delaware. Around 1950, Franklin B. Baker, Sr., owner of Baker Petroleum, purchased a 209-acre farm six miles north of Middletown to serve as the family residence and headquarters for his business. The farm provided additional space for equipment, and a barn to keep a horse for his sons, Ralph and George. A vain attempt was made to grow crops on the unused portion, but Frank was not a dedicated farmer. He then gave serious thought to flying for fun. He and his sons took flying lessons from Mr. Lovett, who owned a small airfield just across the

This aerial photo of Baker Airfield was taken in 1959. Photo from the du Pont collection.

Delaware line in Maryland. They earned their pilots licenses and decided, in March 1953, to use the unused portion of the farm for a private airfield, which would eliminate traveling to/from the airfield in Maryland. No thought was given to turning the airfield into a public-use facility. It was a place for father and sons to keep an airplane and go fly. It was their private *Field of Dreams.*

As time passed, the airfield became a favorite place for airmen to fly to and nest awhile. The hospitality was always congenial, and the flying tales were endless. Soon the airfield began to take on the look of a public-use facility. On November 4, 1955, Stewart Poole, Delaware Aeronautics Commission secretary, approved Frank Baker's request to license the facility for a commercial, public-use airfield.

Although the drawing would not be acceptable in today's world of government regulation, it was commonplace in the middle 1950s. The drawing was attached to a single-page form supplied by the Commission and mailed. Stewart Poole wrote Frank Baker on November 4, 1955, indicating his application was approved, and he would stop by in Middletown for him to sign the papers and collect the $1.00 fee. The entire transaction was completed in thirty days. This was all that was needed to enter into the aviation business in those days.

The facility was appropriately named "Baker Airfield." Frank erected a hangar/operations office near the farmhouse, installed aviation fuel pumps and, over the next three years purchased 3 Aeroncas, 1 Ercoupe, and a Piper PA-12. Aircraft rental and a

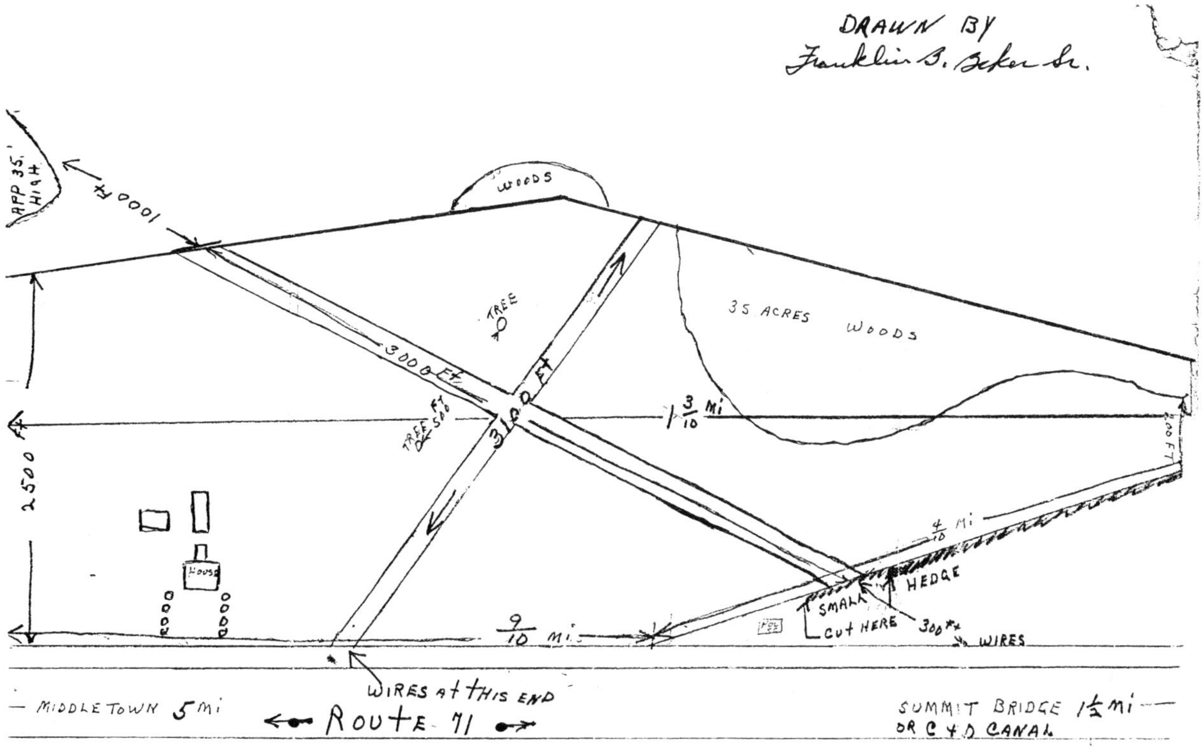

This rough sketch was drawn by Franklin Baker, Sr. and submitted to the Delaware Aeronautics Commission for approval in 1955. Art from the du Pont collection.

The operations building and hangar on Baker's Airfield circa 1959. Photo from the du Pont collection.

Richard C. "Kip" du Pont, Jr., (second from right to left) with fellow model airplane enthusiasts at St. Andrews School, 1951. Photo from the du Pont collection.

flying school were offered to the public as well as tie-down parking. The flock of birds began to grow, and a full time airport manager/flight instructor was hired to keep things in order. Frank never intended for the airfield to achieve any great level of prominence in aviation. It was a hobby that served as a break from his oil business.

✈ Summit Airpark, 1960

There is no scientific medical evidence which confirms that *flying fever* is hereditary or contagious. However, if one took the time to list all the du Ponts that flew airplanes, this might bear further investigation.

Richard C. "Kip" du Pont, Jr., son of the late Richard C. du Pont, founder of All American Aviation, caught flying fever at a very young age building model airplanes. He attended Tower Hill, Fay School, and graduated from St. Andrews in Middletown. He learned to fly from H.B. du Pont's airfield in 1953, at the age of sixteen. After graduation from St. Andrews, he enrolled in the University of Miami

Aviation Management & Airline Administration Program and Embry-Riddle Aeronautical School in Miami. During the summers, Kip worked for All American Engineering where he flew personnel to/from Sussex Airport from the DuPont Airfield. When he completed his Miami schooling, he started flying for Mohawk Airlines and joined the Delaware Air National Guard to do what he liked best, *fly.*

In 1959, a series of aviation-related events took place which changed many people's lives, including Kip du Pont's. Mohawk Airlines stewardesses went on strike which grounded their pilots. Kip, although satisfied with flying for Mohawk, did not relish the idea of being grounded in a dispute that was not of his doing. To return to flying, on December 11, 1959, Kip du Pont started his first flying business, *Manor Aviation, Inc.,* which offered charter flights from New Castle County Airport. A twin-engine Beechcraft model 18, fitted with a refreshment bar, lavatory, and executive seating for five was available for $0.60 per mile. In the spring of 1960, Kip added aerial spraying of agriculture from Baker Airfield in Middletown.

Johnny McChesney (Johnny Mac) Morgan. Photo courtesy of Mrs. J. McC. Morgan.

Manor Aviation's Beechcraft-18, N830K. Photo from the du Pont collection.

Also in 1959, Piasecki Airport, on the former Bellanca site, had been mired in union-management litigation for several years. Rumors persisted that Piasecki was going to sell to developers and leave Delaware if he lost the legal battle. If this were to happen, Air Service, Inc. would need to relocate or cease operations. Air Service, founded in 1930, and now headed by Johnny McChesney Morgan, and Ed "Smitty" Smith, service manager, had a long list of aviation followers including the du Ponts and the R.R.M. Carpenters. There were also a number of dedicated owners' aircraft parked on the airport that would need a new home base. Kip saw a window of opportunity to start his own airfield. If Piasecki sold to developers as was rumored, New Castle County Airport would become the sole airport serving the County, except for Baker Airfield, in Middletown, where a small group of fliers nested their birds on Frank Baker's farm.

The aviation industry recession after the war had started its recovery by 1959. Cessna and Piper had new designs on the drawing boards and were eager to introduce their new models. Cessna would offer a new high-wing, two-seat trainer called the 150. Piper followed later with a new, all-metal, low wing trainer called the *Cherokee.* Both of these aircraft offered tricycle landing gear, which allowed easier ground control and visibility. The same companies also offered four-seat versions for business and pleasure flying. Unemployment was low and the people were coming back to flying. The good news was that factory orders were beginning to create a backlog.

At the age of twenty-three, Kip sensed the recovery coming and entered into negotiations with Frank Baker. He leased Baker airfield, with an option to buy, for a period of five years to start Summit Aviation, Inc. On May 6, 1960, Summit Aviation, Inc. was registered as a subchapter "S" corporation in Delaware. Kip du Pont was president, Frank Baker, vice president, M. Satterthwaite, secretary, and Walter Timm, treasurer.

The first ten-unit steel T hangar at Summit was erected by Pete Bryce in June 1960. Photo from the du Pont collection.

The first order of business was to develop a long-range business plan for the airfield. From this plan, the first objective was to bring more business to the airfield. Kip was aware that the majority of airfields throughout the state offered airplane owners makeshift wooden hangars, designed and built by the airfield owners. Many had dirt floors, sagging doors, leaking roofs, and were anything but burglar proof. Airplane owners still had nightmares of the many hangared airplanes destroyed during the 1954 hurricane *Hazel*. The decision was made to construct quality steel T hangars, with concrete floors and electricity, which would overcome all these shortcomings. It was hoped that airplane owners would move from their present locations to a more secure environment for their airplanes.

Ground was broken for the first ten-bay, steel T hangar in June 1960. This was none too soon, for Frank Piasecki made good his threat when he lost his legal battle with the union. He sold Piasecki Airport to developers and gave notice to Air Service that they had thirty days to vacate the premises. Air Service did not choose to relocate their business and ceased operations. Kip du Pont purchased all the special aircraft maintenance tools and moved them to Summit. Johnnie "Mac" Morgan rented an office in the NCCA terminal and continued to fly for the Carpenter family until 1964, when he went to work for Summit. Al Bowen, Ralph "Bucky" Buckingham and Edward "Smitty" Smith became Summit employees on January 1, 1961.

The decision to construct the hangars first paid off. As soon as the first hangar was completed, it was necessary to construct an identical second hangar. Two bays of this unit would serve as a temporary operations office and maintenance facility. The remaining eight sections would be leased to airplane owners for storage of aircraft. It was a tornado-like beginning with most of the hangar spaces rented by the time they were completed. By the fall of 1960, Summit Aviation on Baker Airfield, gave the appearance of being a rock-solid aviation business.

In spite of all this frenzied activity, Kip found time to court and propose to Caroline MacLaren Johnstone. Caroline admits that when she answered, "yes" it included a promise to learn to fly and obtain her pilot's license. Caroline soloed from Baker Field three days before they were married on May 26, 1962. Caroline was not inflicted with *flying fever* at the same level as Kip and let flying take a back seat to other matters. Caroline recalled with a broad smile;

> Kip asked me one day if I had gotten my pilot's license. He said he was flying to Wichita for a Cessna dealership conference and I was not going unless I had obtained my pilot's license. I thought he was kidding and let it slide as I had done for the past couple of months. He was not, and left me home alone! By the time he returned, I had flown every day until my instructor signed me off to fly to Laurel Airfield to have John Reed approve my private pilot's license. I was never left home alone again!

This photo shows a new Cessna-172 in front of the new hangar. Note the sign on the hangar reads, "Summit av—Bakers Field." Photo from the du Pont collection.

Harold Gorrell, Cessna factory representative, (in airplane) shows the fine points of the new model 172 Skyhawk *to Kip du Pont. Photo from the du Pont collection.*

Charlie Raysor, was named sales manager in 1962. Photo from the du Pont collection.

The Cessnas temporally shut down traffic on Route 13 at Harrington until they were safely parked in the display area. Photo from the du Pont collection.

The focus of Kip du Pont, from the very beginning, was to build a modern, well-planned, full-service, general aviation facility. A Cessna Aircraft franchised dealership covering the Delmarva Peninsula, parts of Pennsylvania and southern New Jersey, was added in 1961. A firm believer in advertising and promotion, Kip made guest appearances at local civic clubs to promote flying as a business tool and a sport. He also invited the DuPont Aero Club to drive, or fly in for a personal tour of the airfield. Richard (Dick) Zebley was hired to start an FAA-approved flight school and be chief flight instructor. To draw student pilots, new Cessna 150s would be used for flight instruction. Charlie Raysor was hired to oversee the new aircraft sales department. To further promote the beautiful Cessna airplanes, Charlie Raysor had two Cessnas on display at the Harrington Fair Grounds in 1962. To accomplish this feat, the airplanes landed on the practice dirt race track, taxied out onto route 13 and through the main gate to the display area. The airplanes stood in stark contrast to the farm tractors and combines displayed in the same area and drew crowds of spectators.

In 1962, Kip sold his interest in Manor Aviation to his cousin, Richard S. du Pont. Manor's only asset was the Beechcraft model 18. To keep costs down Richard employed part-time qualified pilots for the charter operations. If additional aircraft were needed, Summit made them available.

Every venture did not end in a success story. Manor Aviation expanded their services May 29, 1963, to include scheduled, nonstop flights from Wilmington to Rehoboth Airfield. The flights would be limited to weekends and holidays during the summer. The timetable was listed as follows;

	FRI.	SAT.	SUN.
Leave Wilm.	3:30 pm	9:00 am	3:30 pm
Arr. Rehoboth	4:05 pm	9:35 pm	4:05 pm
Leave Rehoboth	4:30 pm	10:00 am	4:30 pm
Arr. Wilm.	5:05 pm	10:35 am	5:05 pm
Leave Wilm.	5:30 pm		5:30 pm
Arr. Rehoboth	6:05 pm		6:05 pm
Leave Rehoboth	6:30 pm		6:30 pm
Arr. Wilm.	7:05 pm		7:05 pm

Air fare (one way) – $9.00 plus Federal Tax

Charlie Knapp was named pilot of the new six-passenger Cessna 205 airplane, supplied by Summit Aviation. Backed by an extensive advertising campaign, the Vacationer, as the flight was called, carried only 142 passengers the first year of operation. This was insufficient volume to continue, and the flights were cancelled the following year. The problem was not in the flying, but the inconvenience of the traveler not having an automobile to use at the destination.

A couple of years later, Richard sold Manor Aviation to Ed Byrd (a former pilot for Manor) and Sam du Pont. Due to an in-flight, structural failure around 1967 on the Beechcraft, Manor Aviation was closed down permanently.

The new, twin-engine, Cessna Skymaster *at Summit. Photo from the du Pont collection.*

Cessna notified dealers they were going to introduce a revolutionary new twin-engine airplane in 1963, that would be heralded the safest twin-engine airplane ever built for general aviation. This new airplane would be called the *Skymaster* with one engine up front pulling and the second in the rear pushing. In aviation terminology, this is called center line thrust. In this design, if either one of the engines quit, the flying characteristics of the airplane would not change. It simply would fly a bit slower. Summit

Summit flight operations were only halted on two short occasions during the runway surfacing operation in 1963. Photo from the du Pont collection.

Note the name change on the hangars to "Summit Airpark" and the small line shack on the right for line service personnel. Photo from the du Pont collection.

Aviation was one of the first dealerships in the country to have one on display.

The five-year lease with Frank Baker expired in 1964. In February, Frank agreed to sell the airfield to Kip and stay on as a consultant for ten years. By the end of 1964, the "Baker Field" signs on the hangar were removed. The facility was renamed "Summit Airpark." Summit Aviation became the fixed base operation (FBO). Although this was sensational news, it was overshadowed by Caroline giving birth to their first child, Richard C. du Pont, III.

An all-season runway, taxiway, and parking ramp became the most important (and expensive) construction project to date. In 1964, plans were made to have Melvin Joseph Construction Company, of Georgetown, hard-surface the main runway (35-17) 35' wide and 3400' long, with lighting. Although narrow by today's standards, it would allow Summit all-seasons' operational capability. To allow for continued flight operations during construction, the second grass runway (29-11) became the active runway. With a vision to the future for Middletown,

1964 Lear Jet at Summit. Veteran USAF jet pilot, Colonel Joe Anastasia was named chief pilot. Photo from the du Pont collection.

Kip purchased an additional twenty-seven acres of land, adjacent to Summit, from Charles and Sarah Goodwin in 1965. Over the years, the runway would be extended, widened, and a separate taxiway installed.

✈ Summit North

With encouragement from John Rollins, Kip decided to acquire the Lear Jet dealership for Delmarva in 1965. At the same time, it was decided to expand the exposure of the Lear Jet and Cessnas at an extension of Summit Aviation on New Castle County Airport. The location at NCCA was given the generic name of Summit North for convenience sake. "Johnny Mac" Morgan was named vice president and general manager. The new Lear Jet would be available for charter by Summit Aviation and operate from NCCA. To accommodate this expansion, eight additional acres were leased for twenty-five years from the County, and a new 96x80' service and office building was erected. Steel T hangars were also erected, which could accommodate thirty aircraft, for lease to airplane owners.

In two short years the Summit North program ceased operations. The venture proved unprofitable, and the Lear dealership was dissolved. Summit North moved back to Middletown. Although the Summit Aviation Charters and Del/Air operations remained at NCCA, the scheduling, bookkeeping, and accounting were moved to the Middletown site. With no longer a need for the leased eight acres of the New Castle operation, the lease was transferred to Atlantic Aviation. Johnny Mac moved back to Middletown as vice president and general manager of Summit Aviation. Kip's cousin, Richard S. du Pont flew some charter flights and helped oversee Summit's Charter Department.

This same year, Kip was asked to serve on the board of directors of All American Engineering. All American Engineering was founded by his father in 1938. This honor must have given Kip an incredible sense of pride since the du Ponts had divested all their holdings in the company many years before. He served on the board until the company was sold in 1970. Not to be outdone, Caroline, on November 10, 1966, delivered Virginia Hemingway du Pont. Three years later, Caroline would repeat the event by delivering a second daughter, Allaire. The family resided on a farm near Chesapeake City called "Great House Farm."

Richard "Kip" du Pont became a director of All American Engineering in 1965. Photo from the du Pont collection.

The decision to consolidate all of Summit Aviation's operations in Middletown created a need for additional space in 1968. This expansion would become the largest building program undertaken by Kip du Pont for Summit. The complex would consist of new construction added to the south side of the existing buildings. The design called for a larger service hangar, ramp service center, avionics department, new corporate offices and lounge to be constructed. Outside improvements included landscaped gardens, and an extended paved parking area for aircraft and autos.

The 1968 renovated flight training center (formerly the original Baker Airfield buildings). Photo from the du Pont collection.

This new glass-enclosed ramp service center was also part of the 1968 construction. Photo from the du Pont collection.

The roadside view of Summit's new general aviation complex. Photo from the du Pont collection.

Kip du Pont and one of three Short Skyvans *purchased by Del/Air. Photo from the du Pont collection.*

The flight training center, which now boasted one hundred and fifty students, also needed larger quarters. To accommodate this department, the original structure on the south side of the airport, used by Frank Baker back in 1950, was completely renovated. The reconstruction included separate offices for flight instructors, a Link simulator instrument trainer and an audio-video room for film presentations. Flight training programs were also established at Middletown H.S. and Tatnall schools. To increase the utility of the Airport, an FAA-approved ADF/ VOR instrument approach was added May 24, 1971.

✈ Del/Air

Del/Air was another company founded jointly by Kip du Pont, George Strawbridge, and Bob Crompton. In the beginning, Del/Air utilized two single-engine Cessna 206s and a Cessna 172 to provide rapid deployment of critical cargo from remote airfields to Philadelphia International. Del/Air occupied offices on the north side of NCCA until 1972. Although small in the beginning, Del/Air, under the leadership of Joe Wear, would grow over the years to service most of the entire Atlantic states. Del/Air's growth required larger aircraft and three Short *Skyvans* were purchased to fill the need by 1972. At this point Del/Air left NCCA for Philadelphia for more convenient connections. Summit Aviation would provide all the service for these cargo carrying, multi-engine aircraft. This required maintenance personnel to attend special schools provided by the Short Brothers Aircraft Company. To accommodate the Short *Skyvans* and other large transient air traffic that were using *Summit Aviation* during this period for service, the runways at Summit were lengthened to 4500 feet with 500 foot grass overruns. By the mid-1970s Del/Air serviced Philadelphia, Charlotte, Atlanta, Dayton, and several other Midwest points.

Del/Air showed excellent growth, and additional locations were added throughout the east coast to meet customer demands. Atlanta became a mini-hub to serve Savannah, Birmingham, Huntsville, and Chattanooga. Under the direction of Joe Wear, Del/Air grew from a handful of employees to over 650. It was in the mid-1970s that a deal was reached between Delta and Del/Air to haul all of Delta's oversize cargo which could not be accommodated in Delta's DC-9s or 727s. The only condition was that Del/Air must change their name to eliminate any possible connection with Delta. This was agreed upon when Kip du Pont allowed Del/Air to change the name to Summit Airlines. Although Kip approved the use of the name Summit Airlines it was not connected with Summit Aviation in any way.

The debt/equity ratio suffered when expansion rose to serve 46 cities in 22 states. Additional aircraft were needed to meet this demand and L-188 and DC-9 freighters were leased to meet the need. Surface transportation was also expanded to pickup and deliver heavy freight. It finally came to the point that additional funds were needed to meet the expanding operations. By 1981, fifty-one percent of Summit Airlines was purchased by E. F. Hutton. Four years later, E. F. Hutton divested all interest in the operation and sold it to a New England-based group. At this point, Kip divested all his holdings in Summit Airlines. The existing management, including Joe Wear, conflicted with the new owners and left the company. Two years later, the airline was put into Chapter 11 bankruptcy and never recovered.

Summit Aviation's board was always well-structured with capable people in their related fields. Although members of the board changed on occasion over the years, by 1972 the board consisted of Kip du Pont, president; Walter E. Timm, vice president; Charles Raysor, vice president; Johnny McC. Morgan, vice president and general manager; J.B. Willis, secretary; Albert J. Morris, assistant secretary and treasurer. Although some members of the board changed from time to time, the policy of appointing the best was always adhered to. By 1975, Summit converted their subchapter "S" corporation to the standard "C" corporation. All of the stock in Summit Aviation was privately-held by the du Ponts.

It was quite obvious Kip du Pont liked almost all airplanes. What is interesting is the wide range of airplanes that caught his fancy. His personal collection of airplanes was quite varied. These included an *extremely slow,* (36 hp, 60 mph) 1936 Aeronca C-3 antique; a (550 hp, 190 mph) North American AT-6

The ultimate in luxury was Kip's executive DC-3. Photo from the du Pont collection.

The ultimate in speed was Kip's former WWII fighter, the North American P-51D Mustang. *Photo courtesy of Joseph Greigg.*

The ultimate "flying turtle" was Kip's 1936, Aeronca C-3. Photo from AOPA PILOT *March 1981, cover.*

Texan; a *very fast* (1,695 hp, 400 mph) 1945 North American P-51D *Mustang*, and a *very posh* (two, 1400 hp engines, 180 mph) classic Douglas DC-3 executive airliner. Kip did not store the airplanes in a hangar and never allow them out. They were flown frequently. The Aeronca was featured on the cover of the *Aircraft Owners & Pilots Association (AOPA)* magazine in March 1981. Kip flew the airplane himself for aerial photos used in the feature story. The P-51D *Mustang* could be seen at many of the air shows throughout the East Coast. The Douglas DC-3 was used to fly throughout the country and the Caribbean. In 1977, Kip flew the DC-3 on a rare trip to Cuba. On board were Kip du Pont, pilot, Richard S. du Pont, co-pilot, Caroline du Pont, Mr. and Mrs. Atwood and their two daughters Jan and Susie, who were unofficial representatives of president Jimmy Carter. Although the flight was termed "unofficial" as far as the State Department was concerned, it was blessed by the president. The objective was to meet with Fidel Castro, in a non-State Department atmosphere, to query how Castro felt about the chief executive. This flight was the first private airplane to land in Cuba since the Bay of Pigs.

Jan Churchill, noted aviation pilot, writer/author, and Caroline du Pont entered in the last All Woman Transcontinental Air Race, Powder Puff Derby, in 1977. Competitive events involving gender were being tested in the courts. Rather than subject the organization to possible litigation, the historical event was terminated in 1978. The duo piloted Jan's Cessna 182 *Skylane* (sponsored by Summit Aviation) from Palm Springs, California, to Tampa, Florida. One hundred and forty starters were registered for the 2,200 mile flight. Jan and Caroline made a great showing by finishing fourth in the proficiency flight.

Summit Aviation was blessed with quality people throughout the years. Their dedication to aviation was rewarded by the continued growth of the company. However, as some employees entered their senior years and retired, their skills would be difficult to replace. Of the many who retired, two icons stand out. Johnny Mac Morgan, who soloed from Bellanca Airfield in 1930, retired from Summit Aviation in 1980. His vacancy on the board was filled by Caroline du Pont. Edward "Smitty" Smith followed four years later in 1984.

Unforeseen Turbulence

There is no way a business can predict world events. In 1973, America supported Israel in its war with the Arabs. This action angered the Arab nations whose primary attraction in the Persian Gulf is oil. To complicate matters, Iran is Shiite Muslim, Iraq is Persian and Saudi Arabia, Arab. These factions historically never accepted each other's religious beliefs in harmony. In many cases they engaged in war with each other in futile attempts to resolve issues. In spite of their dislike for each other, they are united in the area of oil distribution, their only major export. Their organization was called OPEC.

In a display of anger over the U.S. involvement in the war, OPEC decided to show the world how important their oil really was. The combined agreement of the oil-producing nations under OPEC curtailed the amount of oil to be pumped by a huge margin and escalated the price per barrel from $2.90 to a staggering $11.00. Our dependency on foreign crude oil had been well-documented, but never taken seriously. The U.S. government was totally unprepared for such a rapid depletion of our oil supplies. The end result of the embargo was a severe shortage of fuel throughout the country. Energy costs created unparalleled inflation rates never seen before in this country. Aviation fuel also felt the wrath, and FBOs attempted to limit their meager fuel supplies to flight training and established customers. Transient aircraft pilots were told no fuel was available. Kip and Caroline were in Florida with the DC-3 on a business trip and waited days to find enough fuel to return to Delaware. Cessna Aircraft canceled their 1973 EXPO in the fear that their visitors would not find fuel to return home. Many other major events throughout the country followed suit.

The oil crisis was hardly over when another unforeseen aviation disaster struck which would virtually destroy the entire single-engine airplane industry. Product liability lawsuits were filed against Cessna, Piper and Beechcraft citing accidents involving aircraft over twenty-years-old and older. Ancient product liability laws had been enacted with no time limits, which left opportunity for liability. By the early 1980s, Aircraft product liability lawsuits reached an all-time high of $209.7 million dollars. Cessna,

Piper, and Beechcraft ceased building single-engine aircraft. In 1979, the general aviation manufacturers collectively produced 17,000 aircraft. By the mid-1980s new general aviation aircraft soon declined to a paltry 1,085 units. To make matters worse, the lawsuits also covered new spare parts. Parts became scarce and escalated in cost to unacceptable levels for many consumers to tolerate. Serviceable, used parts from salvage yards did little to resolve the problem. Many aircraft service centers simply refused to use them in fear of future litigation. Fuel costs had tripled. Sales of new aircraft were almost non-existent. Insurance rates for operating student aircraft ballooned out of site. Rental of airplanes was equally affected. The aviation industry was taking a terrible economic beating.

Tragedy does not play favorites. When it strikes, it delivers a severe blow. Although tragedies often destroy the material things we have worked hard for all our lives, it cannot compare with the loss of a loved one. No one was prepared for the untimely death of Richard "Kip" du Pont on March 31, 1986, at the age of forty-nine. His passing, at his home, from a severe heart attack was not anticipated by anyone. A shock wave crashed down on the family and the aviation world. Kip loved aviation and all the people who were involved in the profession. The pilot who flew a primitive Piper *Cub* was received in the same manner as the pilot of a jet. Caroline and the family had barely recovered from the realization that Kip was gone when two years later her only son, Richard, was killed in an automobile accident.

The Recovery

The loss of Kip and Richard was a stunning blow. The banks and trustees suggested Caroline sell, pay off the loans, and get out of aviation. (What they were really saying was they felt she could not step in, replace Kip, and keep the company solvent.) Investment speculators began appearing like vultures in the sky in hopes of taking advantage of the situation. Suppliers also became nervous and skeptical. They wondered if their bills would be paid promptly, or at all. Lending institutions felt that making loans to a company in a depressed aviation market which was mired in litigation was highly risky.

Caroline du Pont, Chair of the Board
Photo from the du Pont collection

If Caroline decided to retain the business, a reorganization of the officers and directors would be the first step. Caroline became Chairman of the Board and Dennis Einstein assumed the position of general manager. Kip du Pont's vacancy as president was filled by his long-time friend and airline pilot, Patrick (Pat) J. Foley, who had served Summit as a consultant since 1971. Kip and Pat became friends when they attended aviation schools together in Miami in the mid-1950s. Pat's management style and personality were very forceful and strong-willed. Caroline recalled that Pat pressed her hard, on many occasions, to make unpopular but necessary decisions involving the entire staff. Caroline recalled,

> The reorganization included laying off many very capable people who had worked for Kip for years. Summit Aviation was run like a family business, and this was the most heartbreaking and difficult task I have ever faced in the business world.

Former airline captain Patrick (Pat) Foley took over the position of president of Summit Aviation. Photo from the du Pont collection.

Major changes were made in Summit Aviation's commitment to general aviation. Kip had established a highly diversified market and was the driving force of the entire operation. In a robust economy, all business accepts a reasonable amount of fat. In the event of a business depression of this magnitude, the fat had to be trimmed immediately or suffer the consequences. The consequences were total failure of the business.

The Cessna sales division was closed. The Summit Air-taxi certificate (Charter flights under FAA part 135) was not renewed. The student training program was leased, but later closed permanently. Summit Aviation's business would be reduced to leasing hangar space to aircraft owners and maintenance.

Under FAA rules, all aircraft are required to have an annual inspection. Since the only sizeable aviation market left was parts and maintenance, this would become Summit's niche in the market. The remaining staff was restructured and retrained in their newly created positions. Summit Aviation secured authorization to service Sikorsky, Bell, and Boeing helicopters, as well as all general aviation piston and jet aircraft. In addition to airframe maintenance, the avionics shop was upgraded to include installation and service of navigational equipment, transmitters, receivers, stormscopes and radar. In the end, Summit emerged stronger than ever. In fact, the transition was so well-planned and implemented, Summit was honored with "Delaware's Blue Chip Enterprise Initiative" for 1991. This award, sponsored by Connecticut Mutual, Nations Business and the Chamber of Commerce is given to the company who has shown outstanding leadership in face of seemingly insurmountable adversity.

With an eye toward the future, Summit purchased two adjacent properties known as Manor Farms (78 acres) and Mt. Pleasant Farms (115 acres) to accommodate larger aircraft. Summit Aviation, in 1998, consists of 540 acres and employs over one hundred aviation technicians. Allaire du Pont, Caroline's daughter, has her private pilot's license and is working on additional ratings to improve her skills. She is also a graduate from Mary Washington College with a BA degree. Allaire was elected to serve on Summit's board in 1998.

Caroline, and the team at Summit, are dedicated to keeping Kip's dream alive and well. Although he is gone, his *Field of Dreams* remains a monument to his dedication to aviation.

✈ Piasecki Airport 1956-1960

Frank Piasecki was born in Philadelphia in 1919. During the late 1920s and 1930s, the golden age of aviation inspired Frank to pursue aviation as a career. In 1943, Piasecki had organized the PV Engineering Company in Morton, Pennsylvania, where they developed the PV-1. The P stood for Piasecki and the V for his partner, Venzi. When WW II broke out, the military endorsed helicopter development as a rescue vehicle. With military financial support, Frank and his team of engineers continued to improve the design. Although many improvements were made in the design during the war years, no helicopters entered into regular service during that period of time. The helicopter, although capable of taking off and landing vertically, was considered by many to be unorthodox, ungainly, and dangerous. A. Felix du Pont, who served for a time on the Piasecki board of directors, felt otherwise. Although Felix did not fly, he was an avid supporter of Piasecki and his flying machines.

In 1946, the name of the company changed to Piasecki Helicopters Corporation and introduced the HRP-1, the first practical tandem rotor helicopter. By 1952, Piasecki had created such a name in helicopter development that he was chosen by the U.S. Chamber of Commerce as one of the top ten outstanding men in America.

On April 29, 1955, Frank Piasecki was removed as Chairman of the Board of his company by the influence of Laurence Rockefeller. Not one to take a move of this nature lightly, Piasecki formed a new company he named "Piasecki Aircraft Corporation." The similarity of the two names caused a court action and the name was changed on March 9, 1956, to Vertol Aircraft Corporation. The name VERTOL stands for Vertical Takeoff and Landing.

The experimental, single-seat, Transcendental Aircraft Company TR-1 Tilt-rotor in flight on Bellanca Airport. Note the Air Service hangar in the left background. Photo courtesy of Mario Guerrieri.

On October 23, 1956, Piasecki entered the Delaware scene by purchasing the land, buildings, and remaining military contracts from Giuseppe Bellanca. The name Bellanca Airfield was immediately changed to Piasecki Airport to capitalize on the nationally-recognized Piasecki name. Frank Piasecki had also purchased the right to the four-seat *Ercoupe* and the twin-engine *Ercoach,* with the idea of going into production in the Ballanca plant. Piasecki also inherited the UAW Local 840 union contract in the deal. Apparently, Piasecki signed the purchase agreement not knowing the union contract came with the deal. The Bellanca-union contract stated the agreement was binding upon *successors and assigns of the parties involved, and that no part of the pact is to be changed by any sales or mergers.* Adding fuel to the fiery issue, Piasecki locked out union members and refused to allow them in the plant and replaced them with non-union employees. The union complained in court that this action was in complete disregard for the contract which called for seniority roster hiring practice.

Years of litigation ended in December 1960, with a defeat for Piasecki. During the years of litigation, Piasecki completed the remaining Bellanca contracts and a few helicopter components. No complete helicopters are known to have been built in Delaware. Newspaper accounts confirm there were 137 employees involved, each earning $125 weekly. This could cost Piasecki somewhere near $17,000 every week the employees were not working for Vertol. Although he now owed years of back pay to former employees, the actual cash settlement was never announced. Frank Piasecki then sold the 400-acre site July 1960, to Capital Realty of Montreal, Canada, and joined forces with Boeing Airplane Company. The Delaware plant was shut down and all production equipment was moved to Morton, Pennsylvania.

Boeing established a new division appropriately named Boeing-Vertol. The name VERTOL would not return to Delaware until the 1980s when BOEING-VERTOL moved to New Castle County Airport to assemble and test the V-22. Before the venture got underway, the U.S. Government required Boeing to join Bell Helicopters to share in the development. The name of the project now changed to Bell-Boeing V-22.

However, history records that the V-22 was not the first in the development of the *Tilt Rotor* aircraft in Delaware. In the early 1950s, Mario Guerrieri and Robert Lichten, both aeronautical engineers, designed and built a single place version of this type of aircraft in a hangar on the northeast corner of NCCA. The prototype utilized flexible drive shafts to drive the rotors from an air-cooled engine mounted in the fuselage. This design produced excessive vibration, and the flexible shafts were replaced with solid drive shafts and universal joints. Once completed, all test flights were made from Bellanca Airport.

Since the design was very radical for the time, a pit was dug to protect the operator who used a remote-control flying apparatus for initial flight testing. Utilizing umbilical-type control cables, the operator hunkered down in the pit during the hovering test flights. Ballast was used to compensate for the absent pilot. Transition from hover to forward flight was accomplished by a pilot after the hover testing was completed.

Much of the history of this early attempt at tilt-rotor design is lost in time. Mario, who retired in Florida, was alive and well and in his 80s, and did confirm the aircraft flew on many occasions. He recalled that they were struggling for funds for continuing the project when they sold the design to William Cobey of Glen Riddle, Pennsylvania. Mario moved to the West Coast to work for Hiller Helicopter, and Lichten was employed by Bell. Cobey established the Transcendental Aircraft Corp. and named the aircraft the TR-1. Flight testing continued with a variety of engineering changes to improve flight performance. The final flight of the TR-1 took place along the edge of the Delaware River, when the propeller pitch began to change. Confused, the pilot allowed the aircraft to descend into shallow waters on the edge of the Delaware River. The pilot was unhurt and the TR-1 was removed from the river with little damage.

Cobey later received a contract from the Army Air Corps, Wright Field, to construct a new machine for further development. Cobey cannibalized the TR-1 to build the TR-2. The second model was test-flown and later abandoned for reasons unknown. The remains of the TR-2 were parked beside a barn on Cobey's property for a number of years before they vanished. No other aircraft were built or flown. Although attempts were made to locate William Cobey, they were not successful.

■ *KENT COUNTY*

✈ Dover Airfield, 1933-1950

The grand opening of the Dover Airfield, 1933, drew a huge crowd. Photos by Herman C. Taylor.

Barnstormers during the 1930s found Dover a friendly place to offer airplane rides. From the air, the obvious field of choice was Maple Lane Farm located on the east end of Division Street at what is now known as Route 13. A barnstormer would circle the town a few times to announce his arrival and land to offer rides. Prices varied with the times but were generally in the neighborhood of $5.00 per ride. Business was good, and Daniel M. (Miff) Wilson, the owner of Maple Lane Farm, actually encouraged the activity. Not all farmers were hospitable to barnstormers. Many drove the flyers off because of concerns about liability and damage to their crops.

In time, Miff Wilson decided that his farm was an ideal location for a community airfield. Aviation was growing, and government-supported air mail was not available to Dover because of its lack of aviation facilities. Another reason was that in New Castle the Bellanca Airfield had been operational for five years and was reaping international publicity with record-breaking Bellanca-built airplanes.

On January 26, 1933, Miff Wilson, with Ernest V. Keith, esq., who acted as counsel, formed a corporation and sold preferred stock in "Dover Airport, Inc." at $5.00 per share. The following persons bought into the corporation.

Cert. Date	Name	No.	Shares
January 26, 1933	Daniel M. Wilson	1	200
January 26, 1933	Richard E. Wilson	2	200
January 26, 1933	L. D. Shank	3	5
January 26, 1933	W. Ross Jones	4	2
January 26, 1933	U. S. Corporation	5	5
January 26, 1933	W. M. Hinkle	6	2
January 26, 1933	Melvin Hopkins	7	2
January 26, 1933	H. V. Holloway	8	3
January 26, 1933	H. H. Hanson	9	1
January 26, 1933	Walter T. Massey	10	2
January 26, 1933	James C. Wickes	11	2
January 26, 1933	G. Leslie Gooden	12	2
January 26, 1933	Betty Gooden	13	2
January 26, 1933	George E. Dawson	14	2
January 26, 1933	M. A. Hartnett, Jr.	15	1
January 26, 1933	Max Terry	16	1
January 26, 1933	M. A. Hartnett	17	1
January 26, 1933	Howard Buckson	18	2
January 26, 1933	Cahall Sipple	19	2
January 26, 1933	John B. Hutton	20	2
January 26, 1933	Norman P. Morris	21	2

January 26, 1933	John Metz	22	2
January 26, 1933	William F. Cook	23	1
January 26, 1933	J. Allen Frear, Jr.	24	2
January 26, 1933	Wilbur D. Burton	25	1
January 26, 1933	George Voshell	26	1
January 26, 1933	Edward F. Evans	27	2
January 26, 1933	Ruby M. Helm	28	2
January 26, 1933	C. F. Millman	29	1
January 26, 1933	Fred A. Townsend	30	2
January 26, 1933	Grover C. Frear	31	2
January 26, 1933	E. V. Kemp	32	3
January 26, 1933	F. H. Shakespeare	33	2
January 26, 1933	J. Oscar Plummer	34	5
January 26, 1933	E. H. Trone	35	3
February 6, 1933	George Hill	36	5
February 6, 1933	B. F. Johnson	37	2
February 6, 1933	A. L. Leaden	38	1
February 20, 1933	Morris Simon	39	1
February 20, 1933	Irving Buckson	40	1
February 20, 1933	Graham Stewart	41	2
February 20, 1933	Herbert Cluley	42	2
February 20, 1933	Fred Bell	43	2
February 20, 1933	M. A. Frazier	44	10
February 20, 1933	N. M. Newton	45	5
February 20, 1933	W. G. Newton	46	2
February 20, 1933	Wallace Handy	47	2
February 23, 1933	B. F. Simmons	48	2
February 23, 1933	Charles D. Allee	49	2
February 23, 1933	Ralph J. Wilson	50	4
February 23, 1933	William Smith	51	2
February 23, 1933	Walter Warrington	52	1
April 1, 1933	John A. Custer	53	2
April 5, 1933	W. L. Harris	54	2
May 4, 1933	Richard E. Wilson	55	100
May 22, 1933	Alfred S. Biggs	56	5
February 5, 1934	D. M. Wilson	57	80

Few will argue that an airport is not a vital part of a community transportation plan. However, when it comes to funding, you will have to burn the woods and sift the ashes to find financial support. Historically, it appears this problem was present whether the airport was government-funded or privately funded. In any event, Miff Wilson was more than surprised by the lack of community support for an airfield to serve the capital city. His original concept was to sell 6000 shares of preferred stock at $5.00 per share to raise $30,000 for construction. The amount did not appear to be unrealistic since Wilmington had raised over $100,000 to finance the Bellanca Airfield just five years earlier. The money would be used to build a service hangar, rental hangars, operations office, fuel tanks, tie-down parking, and auto parking. Miff had also contacted the University of Delaware's Department of Agriculture for information on planting the proper grass for runway use. Any leftover funds would be used for start-up purposes.

Only 705 shares were sold. Of the total shares issued, the Wilson family purchased 580 shares for $2900. This left a paltry 125 shares, or $625 invested by the community. In spite of the reduction of anticipated funds, Miff Wilson was determined to complete the project. Journal entries indicate that W. M. Newton was the contractor for the single service hangar. The hangar was 50'x100' and built of concrete block and steel. The 50' width utilized clear-span trusses for the roof. The hangar door was 45'x13' which allowed access for aircraft storage and service. Morris Wheeler provided the steel, and Strohman painted the building and "DOVER" on the roof. When it was completed, the hangar would hold ten airplanes.

An application to the United States Department of Commerce, Aeronautics Branch, disclosed that the airfield's entire surface was suitable for landing. Since no lighting was provided, the airfield was for daylight use only. Airport license number 227 was issued by the Department on May 24, 1933.

Dover Airport, Inc. was owned by the stockholders. The first stockholders' meeting was held February 20, 1933. At that meeting, Miff Wilson was elected president, Richard Wilson was elected vice president, Ralph C. Wilson secretary, G. Leslie Gooden treasurer, with Herbert Cluley, Ernest Keith, and J. Allen Frear Jr., as directors. However, the airport's business activities, under the name "Dover Municipal Airport" were under the control of Richard Wilson. One of Richard's first decisions was to hire J. B. Purnell and C. W. Clements, both of whom held air transport pilot ratings and instructor ratings, to oversee the daily operations.

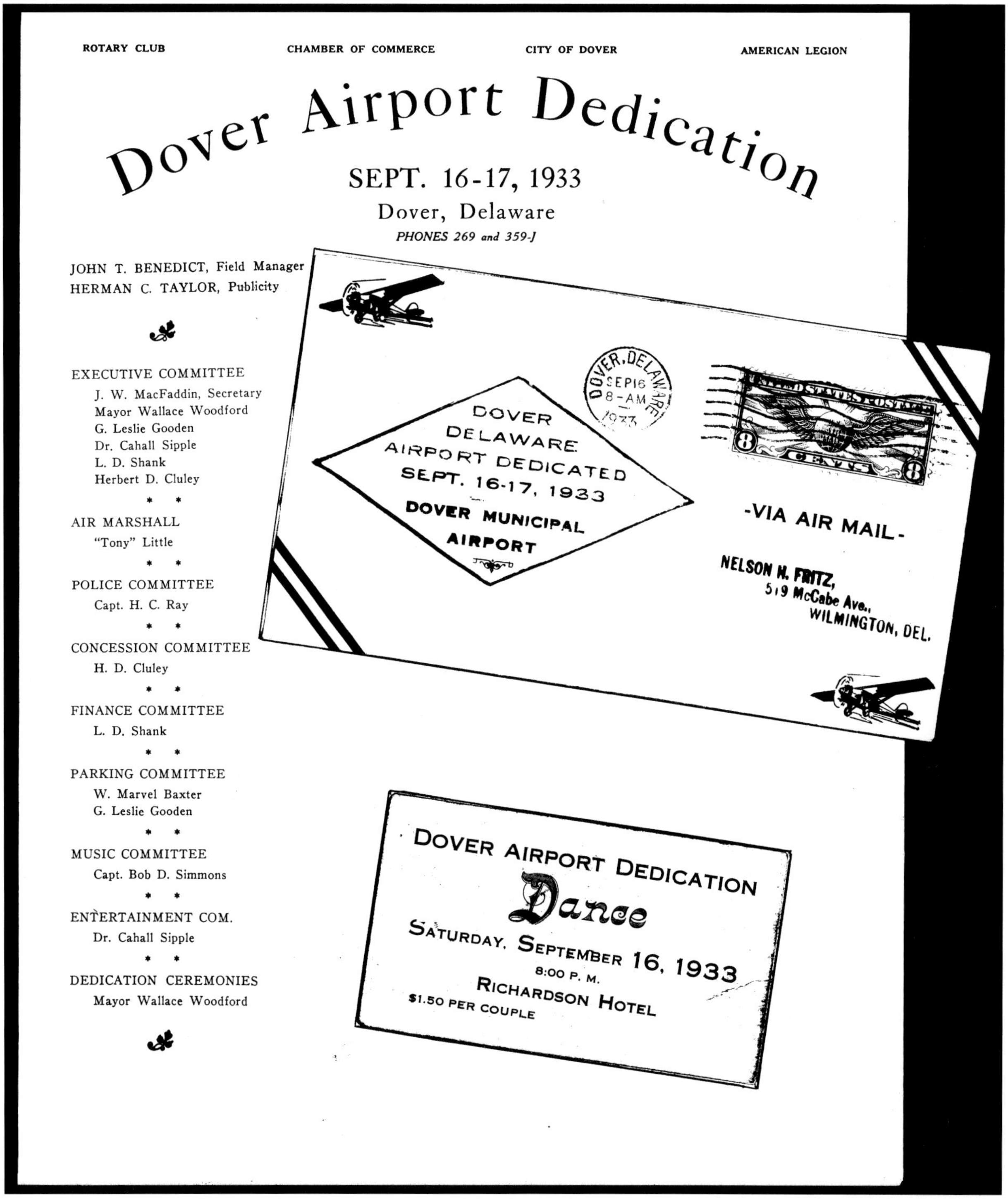

ROTARY CLUB | CHAMBER OF COMMERCE | CITY OF DOVER | AMERICAN LEGION

Dover Airport Dedication

SEPT. 16-17, 1933

Dover, Delaware

PHONES 269 and 359-J

JOHN T. BENEDICT, Field Manager
HERMAN C. TAYLOR, Publicity

EXECUTIVE COMMITTEE
J. W. MacFaddin, Secretary
Mayor Wallace Woodford
G. Leslie Gooden
Dr. Cahall Sipple
L. D. Shank
Herbert D. Cluley

AIR MARSHALL
"Tony" Little

POLICE COMMITTEE
Capt. H. C. Ray

CONCESSION COMMITTEE
H. D. Cluley

FINANCE COMMITTEE
L. D. Shank

PARKING COMMITTEE
W. Marvel Baxter
G. Leslie Gooden

MUSIC COMMITTEE
Capt. Bob D. Simmons

ENTERTAINMENT COM.
Dr. Cahall Sipple

DEDICATION CEREMONIES
Mayor Wallace Woodford

DOVER DELAWARE AIRPORT DEDICATED SEPT. 16-17, 1933 DOVER MUNICIPAL AIRPORT

DOVER, DELAWARE SEP 16 8-AM 1933

-VIA AIR MAIL-

NELSON H. FRITZ,
519 McCabe Ave.,
WILMINGTON, DEL.

DOVER AIRPORT DEDICATION
Dance
SATURDAY, SEPTEMBER 16, 1933
8:00 P. M.
RICHARDSON HOTEL
$1.50 PER COUPLE

Although the dates on this material indicate September 16, 17, inclement weather postponed the event until the 23rd and 24th.

The original airfield records indicate that local socialite and aviatrix Alma Horsey Fooks became the first tenant in April 1933. Alma owned two airplanes; a coal-black, open cockpit Fleet biplane, and a Stinson "K" cabin monoplane. Open-cockpit flying is very enjoyable in the summer. Winter flying is another story entirely. It appears Alma enjoyed both worlds of flying.

At the second meeting of the board of directors, plans were discussed about a grand opening or dedication to officially open the airfield. It was decided that members of the city council should be invited to serve on the committee since the airport was built to serve the community. The August 18, 1933 edition of the *Dover Republican* reported that the committee consisted of John W. MacFaddin, secretary; Herman C. Taylor, publicity; Mayor Wallace Woodward, dedication ceremonies; Cahall Sipple, entertainment; L.D. Shank, finance; G. Leslie Gooden, parking; Bob Simmons, music committee; and Herb Cluley, concessions. The newspaper report also confirmed the committee hired Tony Little as Air Marshall to provide the air show as part of the dedication program. Mr. Little was from Norristown, Pennsylvania, and owned Patco Airport and Pylon Racing. The event was scheduled for September 16 and 17, 1933. In addition to the flying, a dance would be held at the Richardson Hotel on Saturday night. Admission would be $1.50 per couple. The last event on Sunday would consist of an automobile race. The race was named, "The Depression Race," and any car entered could be claimed for $15. It was anticipated the program would draw 30,000 people. When pilots heard about the dedication, scores pledged to fly their airplanes to Dover and compete in the events. The Dover Post Office was contacted to offer philatelic collectors a "first day issue" cancellation stamp to capture the great moment for all time.

With an anticipated crowd of 30,000, additional support was needed. The Dover Rotary Club, the Dover Chamber of Commerce, the City Council, and the American Legion pledged their support. The committee did a fantastic job of publicity and everyone looked forward to the event.

(Left to Right) Charles Clements, James T. Benedict, and Mickey Efferson. Photo by Herman C. Taylor.

A temporary wooden cross and flagpole bearing the American flag were marked by members of the Smyrna American Legion. The man with the tie is Henry Carey. The others are unknown. Photo by Herman C. Taylor.

A CROSS IN THE MARSH

August 19, 1933, was a typical hot summer day in Dover that included roving thunderstorms which seemed to appear out of nowhere. Harold McMahon, a transport pilot from Oyster Bay, New York, had left Roosevelt Field in a Lockheed *Vega* for a flight to Atlantic City, then to continue on to Washington, DC. Harold was an accomplished pilot, employed by Skyloft, Inc. of Washington, DC. During his career he had made many friends that included Charles Lindbergh, Clyde Panghorn, Al Williams, Frank Hawks, Clarence Chamberlin, and Giuseppe Bellanca.

McMahon departed Atlantic City alone and headed east across the Delaware River. The hot summer sun created considerable haze which made for poor visibility, normal for August flying. The *Vega* sped along at two and one-half miles a minute and entered into a thunderstorm near Leipsic before the pilot realized he was caught in a fatal trap. Lightning flashed all around him as he fought the airplane in the extreme turbulence. Newspapers reported that lightning probably struck the ill-fated plane. That appears unlikely since the Lockheed *Vega* was constructed of wood. A more viable conclusion is that the extreme turbulence associated within the thunderstorm caused structural failure of the aircraft. Local residents heard the unfamiliar roar of the engine and looked in horror as the airplane, with one wing missing, plunged from the sky straight down into the soft marsh. The marsh consumed the entire remains of the plane with the exception of a few small scattered pieces of fabric and wood. The impact created a seventy-five-foot-wide crater, twenty feet deep, which filled with water immediately.

On Sunday morning, Herbert McCarty and Andy Surini, who were pilot friends of the McMahon family, flew from Roosevelt Field to Dover to see if they could recover the body. Hundreds of people showed up at the site to offer assistance, to no avail. By noon on Monday, McCarty telephoned Mrs. McMahon to advise her that the recovery of her husband's remains was not possible. Reluctantly, she agreed to leave the body where it had fallen.

Aerial view of the grand opening of the Dover Airfield, September 23-24, 1933. Photo from the Hagley Museum & Library.

Since the site was almost inaccessible by land, McCarty and Surini planned an aerial funeral service for Saturday, August 26. In the meantime the Careys, on whose farm the airplane crashed, erected a large temporary wooden cross and a pole for an American flag to mark the site. A total of 15 airplanes, three of which were Army planes from Mitchell Field, New York, arrived on Saturday for the 2:30 p.m. funeral services. When the airplanes were ready to depart, they were led to the site by John T. Benedict, manager, flying Alma Horsey Fooks' Stinson. Also in the Stinson were Miss Alma Fooks and Miss Cornelia H. Taylor. Clyde E. Pangborn, who had recently flown a Bellanca nonstop from Japan to the U.S., flew the lead airplane once the site was located. Also on board with Panghorn were Mrs. McMahon and Dr. William Hayes, of New York. Dr. Hayes, a minister, read the ritual of the Episcopal Church while flying over the site. As Dr. Hayes pronounced the committal to the grave, red roses were dropped from the airplane by Mrs. McMahon. Following the lead airplane, 14 other airplanes, all piloted by friends of Harold McMahon, flew in an aerial procession and dropped flowers over his final resting place.

Since Harold McMahon was a reservist, the American Legion's Walter L. Fox Post #2, Dover, and the David Harrison Post #14, Smyrna, and the Delaware National Guard, Dover, all paid tribute to their fallen comrade. Three volleys of rifle fire were discharged by the National Guard, which echoed across the marsh. A Legionnaire played taps while the American flag was lowered slowly from the makeshift wooden pole and folded into a triangle. At the end of the ceremony the flag was taken to Dover and presented to the widow, Mrs. Amanda McMahon.

The dedication of the Dover Airfield that was planned for September 16 and 17 was cancelled on September 15, when it became obvious that inclement weather would prevail. Frantic committee personnel notified the newspapers, radio stations and manned telephones to notify everyone that the date would be changed to the following weekend, September 23 and 24. Printed programs were discarded and new ones printed by the *Delaware State News.*

A thorough search of historical documents, old newspapers, and interviews did not reveal who was responsible for renaming the airfield. The second printing of the program states, "Harold E McMahon Field" followed by, "Dover Airport Dedication." It appears that some were deeply moved by the airplane accident that had occurred a month earlier and taken the life of Harold E. McMahon. On Saturday, September 23, 1933, the Dover Airfield was filled with airplanes from far and near to pay another tribute to Harold E. McMahon and name the airfield in his memory.

An unknown mason, with trowel in hand, makes the finishing touches on the permanent monument of Harold E. McMahon. Photo by Herman C. Taylor.

In the absence of Mayor J. Wallace Woodford, Chairman L. D. Shank, member of the Airport Dedication Committee, presided over the dedication ceremony. He outlined the history of the airport since the project was first undertaken, and then introduced Mrs. Amanda McMahon, widow of the deceased pilot.

> We are gathered here today that we may pay homage to a real flyer. That we may perpetuate his name—that we may continue in his work. In his passing may we find inspiration to strive for that goal of perfection in his chosen calling. Let us strive to emulate the best in him and carry on with a renewed vigor for still a youthful industry. It is the spirit of the air that those brave ones who go ahead shall not have perished in vain. We have our lesson to learn in the nature of their passing and when our time comes, as it surely will, then we have overcome every element. It is our prayer, she concluded, that no accidents may ever mar this beautiful landing field and I hereby dedicate it the Harold E. McMahon Airfield, and happy landings.

Following Mrs. McMahon's address, the first day of the two-day program commenced with a Dover High School Band Concert. A full day of outdoor activities ended at 8:30 in the evening. Saturday night a dance was held at the Dover Country Club.

The next day local residents began purchasing airplane rides to see Dover from the air. At noon a concert was given by the Kent County Band followed by a luncheon on the field for the visiting pilots. At 1 o'clock the flying events started with a spot landing contest which was won by pilot Summerville of Philadelphia. The "Bring 'em back alive" contest was won by J. Wesley Smith of Philadelphia. Fifteen teams participated in the relay race which was won by the "White Team." This team consisted of Walter Dean, Philadelphia; George C. Brinkerhoff, College Park, Maryland; J. Wesley Smith, Philadelphia; and James E. Kries, Washington.

Mickey Efferson, Wilmington, prepares to depart the plane for his exhibition parachute jump. Photo by Herman C. Taylor.

The first prize for aerobatics was awarded to Al Williams of New York. Charles Clements of Dover came in second. The free-for-all race was won by J. Wesley Smith, Philadelphia. Charles Clements from Dover came in second. The final car race (14 cars entered) was held at 4 p.m. and was won by Charles Paradee of Dover.

A spectacular parachute jump from 10,000 feet was performed by five jumpers, led by Mickey Efferson. Mickey, from Wilmington, a former member of the Royal Air Forces, was invited to jump at virtually every airfield opening throughout the area. The passing of a hat for donations could earn him as much as $8.00, which was a week's work for many. The conclusion of this event was followed by the presentation of awards. The dedication ceremonies could be considered an overwhelming success when the newspapers estimated that 30,000 people attended the two-day affair.

Dr. Cahall Sipple, chairman of the entertainment committee, and his wife entertained invited guests at their home, "Silver Lake Manor" after the afternoon exercises. Among the honored guests were Mrs. McMahon and Lt. Commander E. F. Stone, copilot of the Navy NC-4, the first airplane to cross the Atlantic Ocean in 1919. Stone was also responsible for saving the life of the only survivor of the Akron dirigible which had crashed only a few months earlier off the shores of New Jersey. Lieutenants W. A. Burton, Deacon Edge, and George Bowerman were ordered to accompany Commander Stone to Dover and return to Cape May, their duty station. The U.S. Army was represented by Lieutenants McKenna and Merchant who flew their Curtiss biplanes in from Bolling Field, Washington, D.C. Following the Sipple party, a banquet was held at the Hotel Richardson followed by a dance at Maple Dale Country Club to honor visiting pilots.

The original wooden cross placed at the crash site was far from adequate to mark the final resting place of Harold McMahon. A permanent marker was funded by donations from Mrs. McMahon, fellow pilots and Skyloft, Inc. Records from the Recorder of Deeds office, dated November 29, 1933, confirm that Henry and Earl Carey sold the 25-foot-square parcel of land to the widow for $1.00. Logs were cut and a temporary path was built across the marsh to the site. Henry, Earl, and John Carey later drove twenty-foot pilings into the marsh and poured a concrete slab two feet above the high water mark for the base of McMahon's permanent grave marker. The concrete base was designed to hold a nine-foot concrete cross with a bronze plaque inscribed, "Reserve Corps, Harold E. McMahon, Born June 1, 1897 — August 19, 1933." At the base of the cross is a headstone which is inscribed, "Harold E. McMahon, California, Born June 1, 1897, Staff Sgt. Air Corps - August 19, 1933."

The sign "DOVER AIRPORT" on the building was never changed. It continued to be displayed on the roof and on the front entrance. This was in contrast to the official 1934 aeronautical chart issued to airmen that listed it as "McMahon Airfield." However, this listing only lasted for one year as the 1935 edition reverted back to "Dover." It cannot be explained, only assumed, that the owners felt "Dover" to be a more appropriate listing for the capital city airfield.

The lane to the Wilson Farm connected to Kings Highway is now severed by Route 13. The road running east-west near the hangar is North Little Creek Road. U.S. Army Air Corps photo, 1935.

Stockholders Come To An End

Letters were sent by Ralph C. Wilson to each stockholder on December 27, 1935, as follows;

> At a meeting of the stockholders on December 6, 1935, a motion was unanimously passed as follows, to wit:
>
> "Now, therefore, be it resolved that in the judgement of the Board of Directors of Dover Airport, Inc., it is advisable and most for the benefit of the Dover Airport, Inc., that the same shall forthwith be dissolved, and to that end and in conformity with the laws of the State of Delaware, it is requested that all of the stock-holders having a voting power of consent in writing to a dissolution of said corporation upon the payment to each of the said stockholders by Daniel M. Wilson the sum equal to the par value of each share of stock plus ten percent (10%) thereof, it being the understanding that he, the said Daniel M. Wilson, is ready and willing to make said adjustment."
>
> This resolution was adopted after Mr. Keith, a director of the Corporation, explained, "that so far the main responsibility for the operation of the airport has been upon Mr. Daniel M. Wilson and his son, Richard E. Wilson; and that, as it is quite unlikely that the Dover Airport, Inc., will be as valuable commercially as it was intended and hoped for, it seemed only fair and right under such circumstances that Mr. Daniel M. Wilson and others holding fee simple title to the land used by the airport should have the same restored to them; particularly in view of the fact that most of the cost of constructing the airport was borne by the said Daniel M. Wilson."
>
> If you will please sign and return the vote appearing at the bottom of this letter, together with your stock certificate, the amount due you will be forwarded immediately. Please have your signature on Certificate of consent witnessed."

Wilson's displeasure with the community's purchase of stock was quite evident with this note scrawled on an envelope (copy of envelope below).

> "Records of First Dover Airfield—Dissolved D. M. Wilson paid all stock holders what they paid in and 10% of their investment - which in most cases was damn little."

Although the corporation was dissolved, the Wilsons continued to operate the airfield and kept it open year-round. A letter written to Jim Flood, editor of the *Dover Post,* by Delaware artist John C. "Jack" Lewis in 1994, describes his first airplane ride from the Dover airport.

> As a child in the middle 1930's, I lived on Kings Highway, and the area stretching for a considerable distance east of the DuPont Highway and situated between East Division Street and Maple Parkway. As you know, the Dover Post was then the hangar.
>
> An event forever etched in my mind occurred at the Dover Airport about 1937. The world-famous aviator, Clarence Chamberlin, landed there with his Curtiss Condor plane and offered rides over Dover for $5.00! It seemed the whole town had turned out to welcome him. How my parents could afford it during the depression I don't know, but my brother,

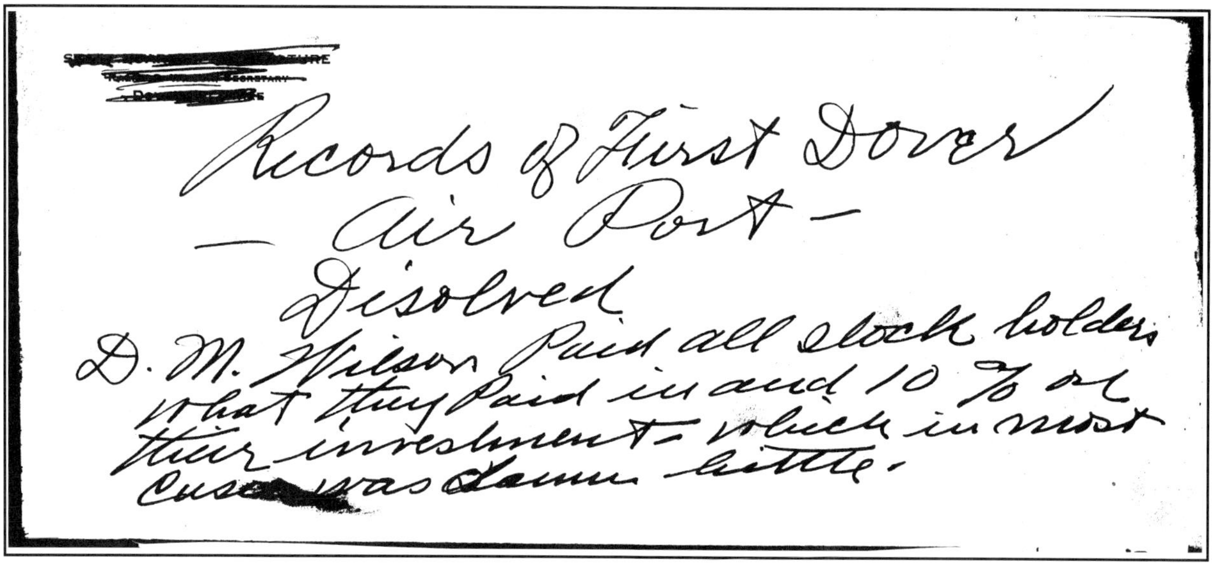
[redacted]
Records of First Dover
— Air Port —
Disolved
D. M. Wilson Paid all stock holders
what they Paid in and 10 % of
their investment - which in most
Case was damn little.

Clarence Chamberlin flew this Curtiss Condor *to Dover after the opening of the Rehoboth Airfield in 1938. Photo by Herman C. Taylor, from George Jones collection.*

Fred, and I, ages 10 and 11, prevailed upon them to let us go up for our first airplane ride.

I remember that our fellow passengers included Charles Terry, later Governor, and Dr. Isaac MacCollum, later Secretary of State. I remember Dr. MacCollum because he delivered my two brothers and me (so I was told later).

Another reason I remember it was that the plane parked close to the fence along Maple Parkway and later taxied over to the hangar. The pilots had hung a set of wheel chocks on the fence, which were used when they parked the plane. When they arrived at the hangar they realized they left them hanging on the fence. Being right up in their faces as a nosy kid, I heard them talking about retrieving the chocks. I immediately volunteered to get them, and did so, running all the way. In appreciation, one of the pilots gave me a quarter - big money in those days!

I was also given a postcard at the time, with a picture of the plane and a facsimile of Chamberlin's signature, which I still have.

The Dover Airport touched many people in Kent County with memories that have lasted a lifetime. When a person takes his first airplane ride, it is never forgotten, even if he did not like it. The depression years offered little excitement for those who remember those hard times. But, an airplane ride broke through that barrier with a short escape from the harsh reality of poverty.

Daniel "Miff" Wilson and his granddaughter pose next to a Douglas DC-3, circa 1938. The Douglas had no problem operating from the Dover Airfield. Photo by Herman C. Taylor.

This 1979 photo was taken by Jim Flood, Sr. prior to purchase for the home of The Dover Post. *Photo courtesy of Jim Flood, Sr.*

The original hangar can hardly be identified in this 1998 photo taken by the author.

During WWII the airfield was closed to flying operations. The tenants either sold their airplanes or placed them in storage. Flying would not resume again until late 1945. The postwar economic boom kept the airfield viable until the early 1950s when Miff Wilson sold the property to developers for a shopping complex and housing development named "Edgehill."

The original hangar survived and was occupied by several businesses, which included a warehouse and roller rink. In 1981 it was sold to Jim Flood, Sr. for the Dover Post Publishing Company, whose offices occupy the building as this is written. The original home of Miff Wilson also survives nearby and still stands, majestically surrounded by huge oak trees.

✈ Smyrna Airfield – 1946 -

The bright orange windsock flops lazily in the warm summer breeze. The light wind rattles the leaves of corn that grow heartily nearby. The smell of freshly cut grass fills the air. Under the sole shade tree a group of pilots sit on a picnic bench swapping flying tales that are as old as flying itself. In spite of the aging hangars, the airfield is clean and well kept. If someone placed a few antique airplanes of the 1930s there, one could imagine being in a time warp. This quaint airfield lies just one mile east of Smyrna on Woodland Beach Road just beyond the Delaware Route 1 overpass. The camaraderie is infectious. The pilots who keep their airplanes here love this airfield. Smyrna Airfield is a flyer's paradise at its best.

Reynolds Jones, 1997

Reynolds Jones, the owner, took his first airplane ride in a Rearwin *Sportster* at Miff Wilson's Dover Airfield in 1936. As this is written, he still proudly displays a photo of a Rearwin *Sportster* under the glass top of his office desk. That single flight would convince the young Reynolds Jones that flying would be part of his life. The Jones family had a history of farming, and for Reynolds to break from farming to venture into flying was considered frightening by his parents. Airplanes were for daredevils and maniacs with a death wish! They fell out of the sky like raindrops! They made such a racket you would lose your hearing in short order! In spite of the negative comments, Reynolds took flying lessons at Biggs Airfield and paid for them from his allowance. At Biggs Airfield, he purchased a one-half interest in an Aeronca *Chief* in 1942.

The military restrictions imposed during WWII forced them to move the airplane out of the no-fly zone to Lancaster, Pennsylvania. Several others followed suit, and they would all pile into a car and drive to Reading on weekends to fly. The war would not take away their fun of flying! This went on for four agonizing years until the war was over in 1946.

Reynolds' father and mother purchased the 170-acre farm east of Smyrna on Woodland Beach Road in 1946. Reynolds was delighted when he was given permission to use a small portion for an airfield. He jumped on the tractor and mowed his *Field of Dreams* for a 1200-foot runway. Since Reynolds was still responsible for farming to make a living, the airstrip was used mostly on weekends. He sold his half interest in the Aeronca *Chief* and purchased a Piper J3 *Cub.* When Wells Faries, a local businessman, purchased a new Ercoupe from the factory in Riverdale, Maryland, and wanted to keep it on the airfield, Reynolds applied for a license. At that time, the State Aeronautical Department required a public-use airfield to have two runways with a minimum of 1800 x 300 feet each. Once the airfield was operational, Wells Faries purchased a prefabricated hangar from the defunct Red Lion Airfield in New Castle and moved it to the Smyrna Airfield to house the Ercoupe. Reynolds then hired a part-time instructor to teach flying in his Piper *Cub* to help offset expenses.

Reynolds never gave any serious thought to making

This photo of Smyrna Airfield was taken about 1960. Photo from Reynolds Jones.

a living off of flying. This was fortunate, for those who felt otherwise failed at an alarming rate. Farming still took up most of his time. Flying was simply something he enjoyed. Reynolds' next decision was one he never regretted. He explained, "I sold my Piper *Cub* and bought an engagement ring for my wife to be, Barbara Thompson, in 1950." Reynolds continued with a wry smile, "By the time I found out how all these kids came along we had five." Although the airfield stayed open, it would be ten years before he would own another airplane and go back to flying.

A handful of local pilots kept the airfield open by tying down their airplanes and renting the single vacant hangar. The CAP was also a welcome tenant and kept their airplanes at Smyrna from 1954 until the mid-1990s. The airfield was now reduced to a single east-west runway. In an effort to expand the airfield services, Reynolds earned his commercial pilot's license in 1958. Reynolds admitted that once you own your own airplane, and you sell it, there is a definite void in your life. By 1960, Reynolds found a two-seat Cessna 140 and bought it for his personal airplane. He also acquired a part 135, FAA-approved air-taxi certificate to use the airplane for charter flying. By 1966, he sold the 140 and purchased a four-seat Cessna 170. About this same time, Reynolds acquired a commercial pilot's license and flew for the State of Delaware, Division of Fish

& Wildlife (formerly the Federal Wildlife Service). Their requirements were to fly an employee over the marsh where he would count ducks, geese, and muskrat houses. He provided this service to the department for over fifteen years.

In 1968, Al Belcher, a DAFB C-141 pilot, asked Reynolds if he could use the airfield to teach flying if he bought his own airplane. An agreement was reached, and Al purchased a new Cessna 150 for instructing. Unfortunately, Al was transferred to McGuire AFB two years later. Reynolds bought the Cessna from Al and passed his FAA exam for an instructor's rating. He finished training all of Al Belcher's students and took on more between farming. Reynolds kept an exacting log of the over 160 people he taught how to fly. Reviewing the log, some prominent names stood out: Leroy Sayres, Ron Sayers, Kevin Fleming, Dave Henderson, John Murphy, Mike Jacob, and Dave Deakyne. Leroy Sayres was probably the wildest of the bunch. He was never satisfied flying around in a perfectly good airplane. He enjoyed jumping out! He made 92 parachute jumps before his appetite was satisfied.

The last entry for instructing students in Reynold's log is dated 1993, when he retired from flight instructing. In 1966, Reynolds sold his Cessna 170 and purchased a 1964 Cessna, four-seat, 182. He used this airplane for "air taxi" operations for a number of years. He still owns the airplane as this is written.

The Jones family never had ambitions to become wealthy. In 1974, Reynolds decided to quit farming altogether. Reynolds added on a small operations office and became a full-time FBO on his own airfield. The kids were grown and off on their own, reducing the need for the extra income. The airfield had also grown over the years to accommodate more airplanes. Hangars had been built to house seven aircraft for owners who desired inside storage. Outside parking could accommodate fifteen more. It was now time to slow down a bit. None of his four children shared any of his interest in flying. They have all chosen other professions.

As Reynolds and his wife Barbara look back over their lives, they say in earnest, "We would not change one thing." Their way of life has provided lifelong friends. Like other airfields that are privately owned, Smyrna Airfield will some day become a part of Delaware aviation history. The manmade birds that have used the airfield all these years will have to find another place to roost. None of the Jones family has any interest in committing itself to operating an airfield open to the public seven days a week. The family has decided to turn the airfield back to farmland or sell it off to developers. It is not the first nor the last time this will happen to airfields that are built on dreams.

✈ Jenkins Airfield 1946-

A visit to the Jenkins Airfield is much like strolling through an automotive bone yard. Carcasses of broken airplanes are scattered among narrow winding trails throughout the property. Salvageable parts, like wings, stabilizers, and rudders, are stacked on end in open racks awaiting an overjoyed buyer who needs the part desperately. A scattering of privately owned, flyable airplanes are tied down, or housed in hangars which can barely protect them from the elements. Hangar doors are nonexistent. Birds, rabbits, and other animals of nature roam freely and have readily adapted their lifestyle to the JJJ-Y Ranch (Joe Jenkins Junk Yard).

Joe Jenkins was another young man of the 1930s who fell in love with airplanes and flying. Since he was too young to drive, he rode his bike from Camden-Wyoming to the Dover Municipal Airfield to get his first airplane ride in 1934. Howard Turpin bestowed the honors when Joe took his first airplane ride in a prewar Porterfield. Joe went on to take flying lessons from Turpin for more than a year. He decided to change instructors because Turpin kept all student log books in his possession. Joe acknowledged, "I never really knew where I was when I couldn't review my flying time." It can be assumed this practice kept students from wandering off to other flight instructors during their training period.

By 1939 Joe Jenkins had joined the Rehoboth Beach Flying Club by purchasing a share in their Aeronca 058-B at Rehoboth Airfield. He flew the Aeronca

This aerial photo of Jenkins Airfield was taken in 1998 by the author.

until he received his draft notice on December 31, 1940. He reported for basic training on January 7, 1941, at Fort Dix, New Jersey. His first assignment was to the motor pool where he drove staff cars and motorcycles. Later, he went on military maneuvers to South Jersey, North Carolina, Virginia, and back to Fort Dix. After the maneuvers, he stayed at Fort Dix for three weeks, when the entire division was transferred to Alexandria, Louisiana, where he took his flying cadet exam. Joe recalled, "Most of the ones that took the flying exam did not do it for the love of flying. It was more because of the hate of walking."

From Alexandria he was transferred to Fort Lewis, Washington State. Joe was there when the Japanese shelled March Field from a submarine off shore. He remembers that this episode shook everyone up, and he was immediately assigned to motorcycle patrol from Fort Lewis to Vancouver, British Columbia. That shelling made many believe there was a real possibility the Japanese would invade America. Security was intensified and at this point was at an all time high.

Three months later Joe was transferred to Santa Anna, California, for preflight training. After completing this, he was assigned to Ryan Field, Tucson, Arizona, for primary flight training. After primary training he went north about sixty miles to Morana for basic flight training, then another sixty miles to Phoenix for advanced flight training. At Phoenix, he completed his training in the Lockheed P-38 *Lightning* fighter. Joe's next assignment was Orlando, Florida, where he was given two months of extensive training in night fighting. After completion of this assignment he was sent to Salines, California, to await further instructions. In Salines he was checked out in the Lockheed RB-37, which was later used to fly his entire outfit to Algeria, North Africa.

In North Africa the unit was split up, and Joe was assigned to the Army Air Force and the 414th NFS (Night Fighter Squadron). At this time America did not have a night-fighter aircraft available, so he flew fifty missions in the British-built *Beaufighter.* When the American-built Northrup P-61 *Black Widow* was provided to the 414th NFS on December 20, 1944,

Joe Jenkins stands proudly beside the Northrup P-61 First Nighter *which was assigned to him in North Africa during WWII. Photo courtesy of Joe Jenkins.*

it was assigned to Joe Jenkins. By this time they were stationed at Elmas, Sardinia. Next they'd go on to Alghero, Borgo, Corsica and finally to Pisa, Italy.

The *Black Widow* that Joe Jenkins flew was paid for by Northrup employees through a war bond drive. This was followed by a contest which allowed the bond buyers to name the airplane. The winning selection was made by Albin A. Johnson who suggested *First Nighter.* The name was boldly painted on the nose in large letters before it left the factory. However, one of the 414th ground crew painted out the original lettering and redesigned new nose art incorporating a Vargas Girl along with new lettering which was more in keeping with the times.

The first P-61 was fitted with Pratt & Whitney R-2800 engines which produced 2000 hp each. The early version carried only a pilot and navigator. The top gun turret had a patch covering the opening since turrets were in short supply. The early night fighters did not carry a rear gunner for the same reason. Later models were equipped with superchargers, top turrets and tail gunner. With a gross weight of 28,000 pounds, 66-foot wingspan, and 48-foot length, it was considered a huge airplane for the crew of two. With a fuel capacity of 640 gallons, the P-61 had a speed of 227 mph and a range of 1400 miles at 10,000 feet. Joe had flown the *First Nighter* on thirty-eight night missions by the time the war ended in May 1945. The P-61 carried three 1000-lb bombs which were generally used on railroad facilities, bridges or supply depots. The bombs painted on the side of Joe's P-61 represented bombing runs.

After the war, it took awhile for Joe Jenkins to get adjusted to civilian life, and he pondered several options. One was a lucrative offer to fly for TWA. After flying eighty-eight military missions, he considered flying people all over the country boring, and he turned the offer down. He finally decided to start a crop-dusting business. He went to South Carolina and purchased a Piper *Cub* which was converted for crop-dusting. On the return flight he landed in his uncle's farm field just outside Camden-Wyoming, Delaware. His uncle promptly purchased about a half ton of lime and Joe sprayed it on his field the same day.

Joe purchased a farm near Camden-Wyoming in 1949, which would serve as his base of operations. This airfield could support the many activities he would engage in during subsequent years — flight instruction, storage for private aircraft, crop-dusting and airplane salvage. As a flight instructor, Joe taught many local people how to fly from two grass runways which are still in use today.

Joe's first involvement in airplane salvage was accidental. Before the war he had read an article in *Popular Aviation* about a man in California who bought and sold all kinds of used airplane parts. This stayed in the back of his mind until hurricane Hazel mangled many airplanes in its fury in 1954. Two Piper *Cubs* which were damaged at Sunglo Airfield near Dover were offered for sale. Since he used Piper *Cubs* for crop-dusting and flight instruction, Joe felt

A bone yard by any other name is still a bone yard. These photos are representative of Joe's collection of airplane carcasses. Photos by the author.

they could be used for parts. There were also a couple of airplanes heavily damaged at Smyrna and their owners no longer wanted them. Joe bought these airplanes as well. As time passed, Joe bought more damaged planes, and he was placed on various aviation insurance underwriters lists for bidding on wrecked airplanes. In many cases these airplanes were purchased as-is, where-is on the crash site. Now Joe and his son Joey trek into virtually every state in the union picking up wrecks and loading them on a trailer to bring home to sell or salvage.

Joe continued in the aerial application business and flight instruction for about ten years until his salvage business grew to the point he decided this would be his niche in aviation.

Some of the airplanes that are purchased are rebuilt and sold in flying condition. Others are sold as "project" airplanes to those with the time and ability to make them flyable again. Then there are others which are stripped and sold for parts. Joe has shipped parts all over the country and, on a few occasions, overseas. More often than not, customers will show up with a trailer and load the parts, or entire airplane, at Joe's airport site.

The most remarkable thing about Joe Jenkins' operation is his memory. There are no computers or elaborate inventory lists to review with appropriate prices. With literally hundreds of thousands of parts, and many wrecked airplanes scattered over many acres, the inventory is kept in his personal memory. The telephone will ring and the customer will ask for a part and, in most cases, receive an immediate reply. If Joe does have to check for a part, he knows exactly where to look.

(Left to Right) Al Nace, Sussex Aero, Joe Jenkins and Joe's son Joey, were swapping many old aviation tales when this photo was taken by the author in 1997.

Today, Joe Jenkins' flying log lists a host of civilian and military airplanes. His favorite airplane? There are two; the English *Beaufighter* and the Northrup *Black Widow.* Portraits of the original airplanes he flew hang proudly in his office/den.

Joe Jenkins is a very laid-back, easygoing man. It is more than evident that he is happy with his life. Did the airplane salvage business make him wealthy? If Joe is wealthy, it is all in inventory. But if wealth is measured in happiness, then Joe Jenkins is indeed a wealthy man. He was captivated by airplanes as a young man and that fascination never waned. There is no doubt that Joe Jenkins found and lived his *Field of Dreams.*

✈ Harrington Airfield 1946-1950

The pilot brought the airplane across the airfield at a mere ten feet off the ground. A wooden plank wall was directly ahead of him. If he did not pull up soon he would crash directly into the wall at eighty miles per hour. The crowd gasped as the pilot pulled up at the last second to avoid a crash. Then he made a second pass and repeated the same heart-stopping near miss. Everyone was sure that he was going to smash the small airplane into the wall on the second run. Now he was going to make a third pass. Earl Newton, the pilot, had thrilled the spectators twice with his near misses. This time, it was going to be for real. On this pass he would not pull up and deliberately crash the airplane nose-first into the wall. The impact caused the wings to crumble as the fuselage ground to a halt in the dirt. Earl Newton, stunt pilot, flying a 65 hp Funk, had completed the highly publicized final event at an air show on the Harrington Airfield June 12, 1949.

The day was filled with acrobatic flying, a delayed parachute jump from 5000 feet, and a ribbon-pick-up while hanging from the landing gear of the air-

Daniel V. Link, treasurer of Harrington Airfield, Inc. and a Piper J-3 Cub. *Photo courtesy of Ronald Link.*

plane. Another hair-raising event included shutting off the airplane engine at 1000 feet, climbing out of the airplane, and restarting it by hand. This was accomplished by holding on to the airplane with one hand, with both feet placed firmly on the landing gear strut and using the free hand to pull the prop through from the back side. Another stunt called for Earl Newton's partner, Walt McGinnis, to climb out of the airplane, hang from the landing gear struts and pick up a ribbon from the runway. Although they received a small payment for their efforts, it was the thrill and excitement that drove the men in this world of acrobatic entertainment. Earl Newton drove a bus and Walt McGinnis was a factory worker, which they depended on for their livings.

The Harrington Airfield became an operational business after WWII when Atlee Link took flying lessons from Sewell McIllavine in Federalsburg, Maryland. After obtaining his flying license, Atlee started to look for land suitable to start his own *Field of Dreams.* Carl Goslin, who owned a farm about one mile east of Harrington on Route 14, became interested in joining Link in the venture. The duo soon grew to five when John Andrews, Daniel V. Link, and Daniel R. Link joined in and incorporated Harrington Airfield on March 7, 1946. After acquiring an Aeronca dealership and purchasing two new airplanes, they applied for government approval to operate a Primary Flight School under the GI Bill. This approval was issued on February 4, 1947. It also allowed veterans to take forty hours of flight instruction, paid in full by the government. Three T hangars were built of wood, with sheet metal roofs, to shelter the airplanes. A 1000-gallon fuel tank was also installed to store the 80/87 octane aviation fuel. On July 17, 1947, N.M. Whitten, from the CAA, inspected and approved the airfield for public use.

INCORPORATED UNDER THE LAWS OF THE STATE OF DELAWARE

No. 5

100 Shares

Harrington Airport, Inc.

Capital Stock 500 Shares
Without Nominal or Par Value

This Certifies that Daniel V. Link is the owner of One Hundred Shares of the Capital Stock of Harrington Airport, Inc. Fully Paid and Non-Assessable transferable only on the books of the Corporation by the holder hereof in person or by Attorney upon surrender of this Certificate properly endorsed.

In Witness Whereof, the said Corporation has caused this Certificate to be signed by its duly authorized officers and its Corporate Seal to be hereunto affixed this 26th day of March A.D. 1946

Daniel V. Link Jr. SECRETARY

Atlee B. Link PRESIDENT

SHARES WITHOUT PAR VALUE EACH

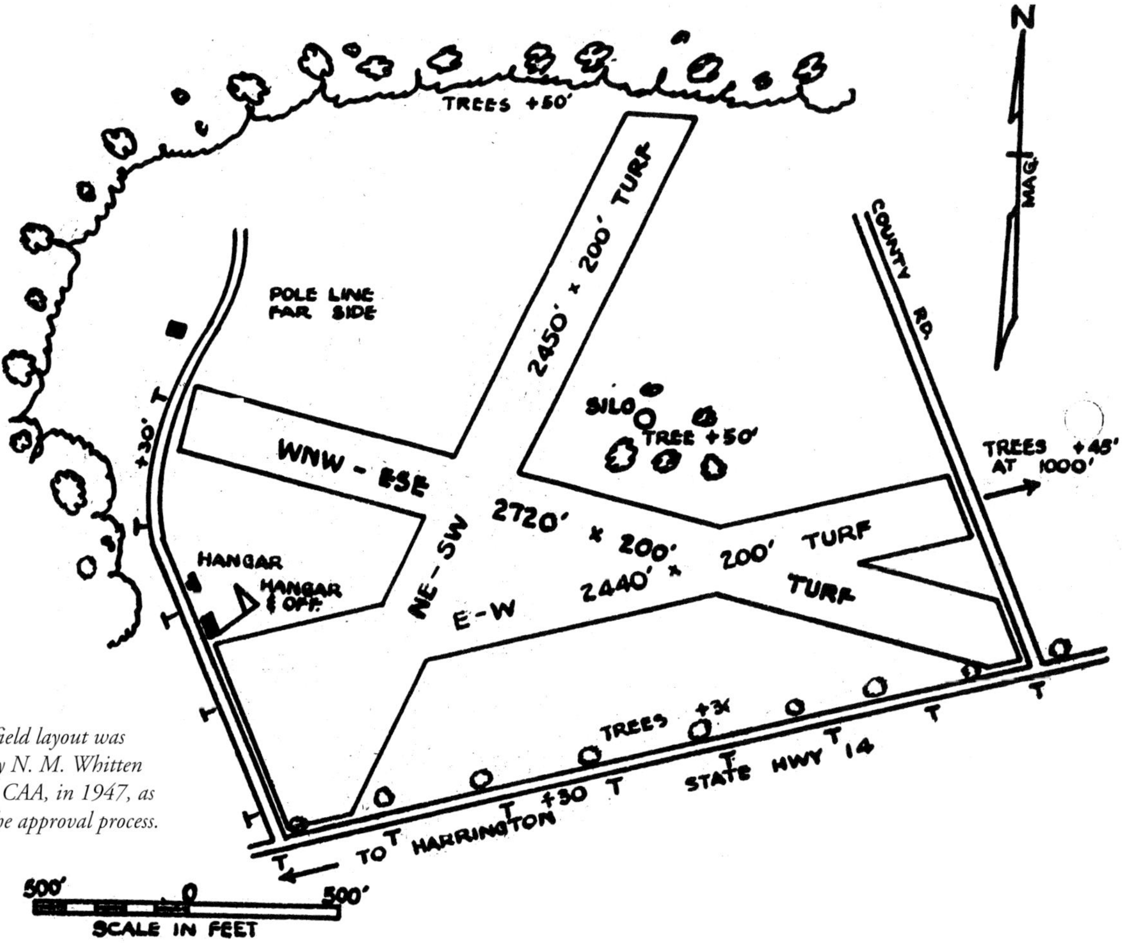

This airfield layout was drawn by N. M. Whitten from the CAA, in 1947, as part of the approval process.

Business was brisk, and Atlee Link could not keep up with the volume of student instruction. In time, it became necessary to hire Joe Jenkins and Jerome Howard as flight instructors. Joe would fly down from his airfield in Camden-Wyoming to teach in the evenings and on weekends. He would also deliver the Harrington *Journal* weekly newspaper by air to the small outlying towns. Joe recalls;

> I threw the marked bundles in the airplane and took off. When I came to the town, I would open the door and throw the bundle out to the paper boy below. I never bothered to land, that would take up too much time.

Some of the local residents who learned to fly during this time were Jimmy Pearson, Fred Greenly, Chipman Minner, Calvin Minner, Lloyd Carter, Franklin Butler, James Hobbs, and Lloyd Wilcutts. Jimmy Pearson would be the only one to make flying a career. He became a professional pilot and flew for the airlines before moving into an airline management position.

By 1948, the group purchased a new Bellanca *Cruisair,* four-seat airplane to add a charter service. The business grew so rapidly there was no time for pleasure flying. The fun was now replaced by seven-days-a-week of hard work and commitment to customers. The next significant move by the Links was to buy out Andrews and Goslin around 1949.

The government-funded GI program came to an end in 1950. Without the government footing the bill, student flying came to an abrupt end, and the Harrington Airfield began to struggle to stay in existence. Like many of the other Fields of Dreams which depended on the government program for survival, Harrington Airfield fell victim to the times.

The airfield came to an end on March 16, 1956, when Atlee Link sold the farm to David Ryan, who used the land for farming. Years later, Ryan sold off parcels of the land which were developed into homesites. Apparently David Ryan had a nostalgic feeling about the heritage of the airfield and appropriately named the development *Airport Farms.*

✈ Milford Airfield 1947-1989

Willard Jackson was a native of Milford. After graduating from Milford High School, he moved to New York City to study diesel mechanics. After finishing school in New York, he found a job with the U.S. Naval Academy, Annapolis, Maryland. At the Academy he was responsible for the maintenance of the Academy's diesel engines on submarines and other naval vessels. When WW II broke out, he wanted to join the military. However the academy would not release him because of his vitally important skills. A persistent man, he continued to request active duty and eventually was granted permission to join the U.S. Army engineers. Warrant officer, JG. Willard Jackson was sent to Belgium and France and served aboard minesweepers to help clear the harbors of mines.

Helga and Willard Jackson in 1984. Photo courtesy of Helga Jackson.

After the war, the War Assets Administration started selling off surplus equipment to servicemen still stationed overseas. Willard decided to purchase a Jeep, for personal transportation. However, he was not satisfied with the military version and promptly cut it in half, extended the wheelbase and installed fenders fabricated from aircraft auxiliary fuel tanks. At the motor pool where he performed this surgery, he met a friend who flew small airplanes. Although Willard had no flying experience, he was talked into buying a surplus Piper L-5 *Grasshopper.* Willard purchased the airplane for $300, crated it and his Jeep and shipped them to the DuPont Airfield in Wilmington. Mechanics at the DuPont field reassembled the airplane, painted it, and licensed it for flying in the U.S. He then had the airplane flown to Milford to a private airfield owned by Bill Drummond. It was from this field that Willard was taught how to fly by Bill Manlove and Bill Drummond.

Willard met his wife to be in Bremerhaufen, Germany. Helga was a U.S. citizen who worked on the staff of General George C. Marshall. General Marshall was known for the rehabilitation of Germany which was to be known as "The Marshall Plan." Helga was fluent in German as well as English, and she worked as a translator on top-secret documents. After Willard returned home, Helga was transferred to Washington D.C. to continue her service with the government. Once Willard obtained his pilot's license, he would fly to Washington in his Piper L-5 to see Helga. Although courting a lady by flying is quite unique, this was not a first in aviation history.

Willard Jackson was intrigued by the government's offer to train pilots under the GI Bill. His high school teacher, Mrs. Ross, informed Willard that the farm across the road from her on the outskirts of town was for sale. It consisted of 80 acres, a farm house, barns, and a sawmill. After walking the entire property lines, he arranged for financing and bought the farm.

Ross Carruthers, Willard's nephew (left), and Willard Jackson sit in the summer sun in 1947, pondering a lifetime of aviation dreams. Photo courtesy of Helga Jackson.

This aerial photo was taken after hurricane Hazel. Two airplanes can be seen in the ruble of the demolished hangars on the right. Photo courtesy of Helga Jackson.

Willard used the sawmill to cut trees and turn them into lumber to build his first hangar. Fortunately, all but 11 acres of the farm was cultivated for growing crops, and there was enough clear land to create two runways. Willard hired his nephew, who was an airplane mechanic, to service and license the airplanes. Since neither one was a qualified flight instructor, he leased the airfield for a flying school to David Kenton who taught flying. He also rented out the farmhouse and the portions of the airfield that could be farmed. He later sold several acres of property on the west end to the Delaware Mosquito Control Commission. The combination of Willard and Helga, who kept the books, proved to be a successful business team.

Willard and Helga were married in 1949, in Milford. They lived with Willard's parents until their new home was built on the airfield property. They had three children; one boy, Randall and two daughters, Barbara and Susan. None of the three Jackson children would be allowed to drive an automobile until they knew how to fly an airplane first. "It was a Willard Jackson requirement," Helga stated factually. Helga went on to explain his rationale; "Willard knew it required a lot of self-discipline, judgment, coordination, and respect for a piece of equipment to fly an airplane. In his mind, this would teach the children important values when they drove a car." Barbara Jackson soloed on her sixteenth birthday. The event was well covered by the local press. It was quite unusual for a teenager in the 1960s to be able to fly an airplane, but unable to drive a car. It was especially newsworthy since women rarely sought flying as a profession. Barbara would be the only one

Willard Jackson built and rented this fly-in, drive-in, restaurant which was quite popular during the late 1950s. Photo courtesy of Helga Jackson.

of the children to continue flying by becoming an airline stewardess for TWA.

Willard never did like the idea of instructing students and never obtained a flight instructor's license. However, he did obtain commercial and instrument ratings which allowed him to do aerial spraying for farmers as well as create an air taxi service. He also supplied flights for a Millsboro parachute club. The parachute club drew many local residents out to the airfield for weekend jumping festivities. Willard also passed all the FAA tests to become an AP/IA mechanic. With these ratings he could perform service and license his own airplanes. As an airfield operator, Willard owned many airplanes over the years. However, his pride and joy was his Cessna 180 *Taildragger.* Willard and Helga were members of the Cessna 180 club and flew their airplane to many club meetings. They also purchased property in a residential airpark on the Keys in Florida to which they would fly in the winter months.

Violent storms are the curse of any airfield owner. Milford airfield was not to escape the fury of hurricane Hazel in 1954. Hangars were ripped apart and airplanes were tossed about like toys. The airfield looked like a war zone.

Willard felt that he could increase revenues if he built a fly-in restaurant facing the newly constructed dual highway on the east end of the airfield. In 1949 he began construction, and once again he went to the woods and cut the timbers to build what was probably the first fly-in, drive-in, restaurant in Delaware. When the building was completed, he rented it as an operational business. Pilots would fly in and locals would drive in to have breakfast or lunch. The combination of parked airplanes and parked automobiles around the building was unique, to say the least.

General aviation's economy took a major decline in the 1970s when insurance rates soared and product liability claims created easy money for claimants and lawyers. When pilots sold their airplanes because of rising costs, the Milford drive-in restaurant business declined and was shut down. The building remained empty for some time until it was finally leased as a liquor store. As the town of Milford continued to grow and real estate became more valuable, the site was sold and the building torn down.

After serving the community for forty years, Willard Jackson passed away at the age of 70 of prostate cancer. Helga tried desperately to keep the airfield open.

This 1973 aerial view clearly shows the newly constructed dual-lane highway adjacent to the Milford Airfield. Photo courtesy of Helga Jackson.

Their son, Randall, died in a motorcycle accident and both daughters were married with family commitments. It was far too much work for one person, and Helga closed the airfield permanently in 1989. The remainder of the real estate is being sold to developers.

Will Milford ever replace the airfield to serve the community? That is highly unlikely due to the large amount of real estate an airport consumes. The permits, regulations, site approval, environmental impact studies and public approval are almost insurmountable to overcome. The cost of constructing the infrastructure is also a huge financial commitment. The Jackson family created a *Field of Dreams* in Delaware aviation history which in all likelihood will not be repeated. Sussex County Airport will serve Milford for the foreseeable future.

✈ Dover Airpark, 1948-1964

Thomas W. Murray, Sr., moved to Dover from the Ocean City/Salisbury, Maryland area in 1926, looking for work. He and his family barely carved out a living in Dover selling Freuhauf trailers and working at anything else that would pay the rent and put food on the table. Tom rented a small garage on North Street, where he would rivet brake linings

Thomas W. Murray, Sr. Photo from the Thomas W. Murray, Jr. family.

After WWII broke out in December 1941, Tom Murray's trailer business clientele changed. A War Department order proclaimed all Freuhauf trailers be allocated to the war effort. That decision put an end to trailer sales to dealers and any sales income to Murray, as small as it was. Gas rationing reduced the use of automobiles, and his repair business dropped off as well. Most orders issued by the War Department touched everyone directly or indirectly.

Things looked mighty bleak until Tom remembered that he hauled Richard du Pont's gliders on trailers. He hit upon the idea that the military would need large aircraft sub-assemblies transported to assembly plants throughout the country. They would also need to pick up downed aircraft. The more aircraft the government built, the more parts would have to be shipped, and the more likely accidents would happen. He felt there was an opportunity to start a transportation business that specialized in hauling oversized loads.

onto brake shoes and provide light maintenance to cars and trucks to supplement the meager trailer sales. Survival was a week-to-week challenge living in a small rented home on Governors Avenue.

Tom had become friendly with the du Ponts in upper New Castle County where he would haul Richard du Pont's gliders on weekends. This is where he saw the custom-designed trailers Richard du Pont used to haul his gliders to soaring meets. Unlike most trailers that hauled airplanes with truck-tractors, du Pont hauled his behind the family car with ease. Being in the trailer business, Tom looked the trailers over closely and marveled at their ability to haul the bulky aircraft. Little did he know that his interest in du Pont's trailer would spark an idea that would make him a very wealthy man.

Tom talked to Richard du Pont about his idea of starting a business which specialized in hauling airplane sub-assemblies and moving aircraft. U.S. manufacturing plants were already supplying aircraft to our allies in Europe. Airplanes were built in parts by various subcontractors and assembled in factories around the nation. The opportunity was there to move the sub-assemblies to assembly plants and pick up downed aircraft. Richard agreed with Tom that the timing was right for such an operation.

Tom Murray left that day with a recommendation from Richard du Pont to take with him to the War Department. Tom drove to Washington and met

Tom Murray would buy any car that would run to pull his trailers. This 1936 Packard sedan was one of the more prestigious ones selected. Photo courtesy of Thomas W. Murray, Jr. family.

with the War Department officials and outlined his plan. With their support, the next step was to acquire the necessary Interstate Commerce Commission permits. Murray once again outlined his plan and presented the endorsements from du Pont and the War Department. In a bold request, he asked for ICC-MC permits to operate in all forty-eight states and Alaska. The head of the ICC thought for a moment and responded, "Give him the permits. After the war is over they won't be building any more airplanes anyway."

Little did he know the driving force of this man named Murray. With the permits in hand, Murray went into action. He named his new company "U.S. Aeroplane Carriers." U.S.A.C. was issued its first government contract to move aircraft parts from Grumman Aircraft in Long Island, New York, to a Texas assembly plant. Tom hand-crafted and designed his own trailers that could be pulled behind cars or trucks. These units were fabricated in Miff Wilson's Dover Airfield hangar. All the parts needed to construct the trailers were secured from automobile junk yards. New parts were impossible to purchase without government approval. Since those approvals were too bureaucratic for Murray to contend with, he solved the problem by using salvageable parts. Since gas rationing restricted most pleasure driving, owners were willing to sell their cars cheaply rather than store them for an unknown number of years.

Another contract called for the transportation of radar units to Alaska from the East Coast via the new Alcan Highway. Alaska was considered vulnerable to invasion by the Japanese, and this was given the highest priority. Tractor-trailers were used for this mission. Dual fifty-gallon drums of gasoline were mounted to the front of the tractors to supply the needed fuel to traverse the highway. The trucks ran in pairs on the highway as a safety measure, since no service facilities were available anywhere.

This May 4, 1944, photo shows C-46 wings ready for shipment. Note the steel bar to deflect any low-hanging tree branches. Photo courtesy of Thomas W. Murray, Jr. family.

This battered 1936 Buick delivered six C-46 fins to the assembly plant. Photo courtesy of Thomas W. Murray, Jr. family

As time passed, U.S.A.C. drivers could be any place in the 48 states plus Alaska, and Tom felt he needed rapid transportation in the event of a problem surfacing. A driver might become ill and need to be replaced. A truck, car, or trailer might require a mechanic or a vital part to get back on the road. He decided to buy a Beechcraft *Staggerwing* to support the men and machines on the road. The airplane was purchased for a fraction of its cost due to the rationing of aviation fuel and government restrictions. These restrictions did not apply to U.S.A.C., which was now considered vital to the war effort. Pete Maslanka, a certified flight instructor, was hired as the official company pilot. He would teach Tom Murray how to fly so he might obtain his license. The airplane turned out to be an excellent investment and performed many rescue operations.

Tom Murray's 1939 Beechcraft Staggerwing, *Serial #304, N2422 on Dover Airfield. Photo courtesy of Thomas W. Murray, Jr. Family.*

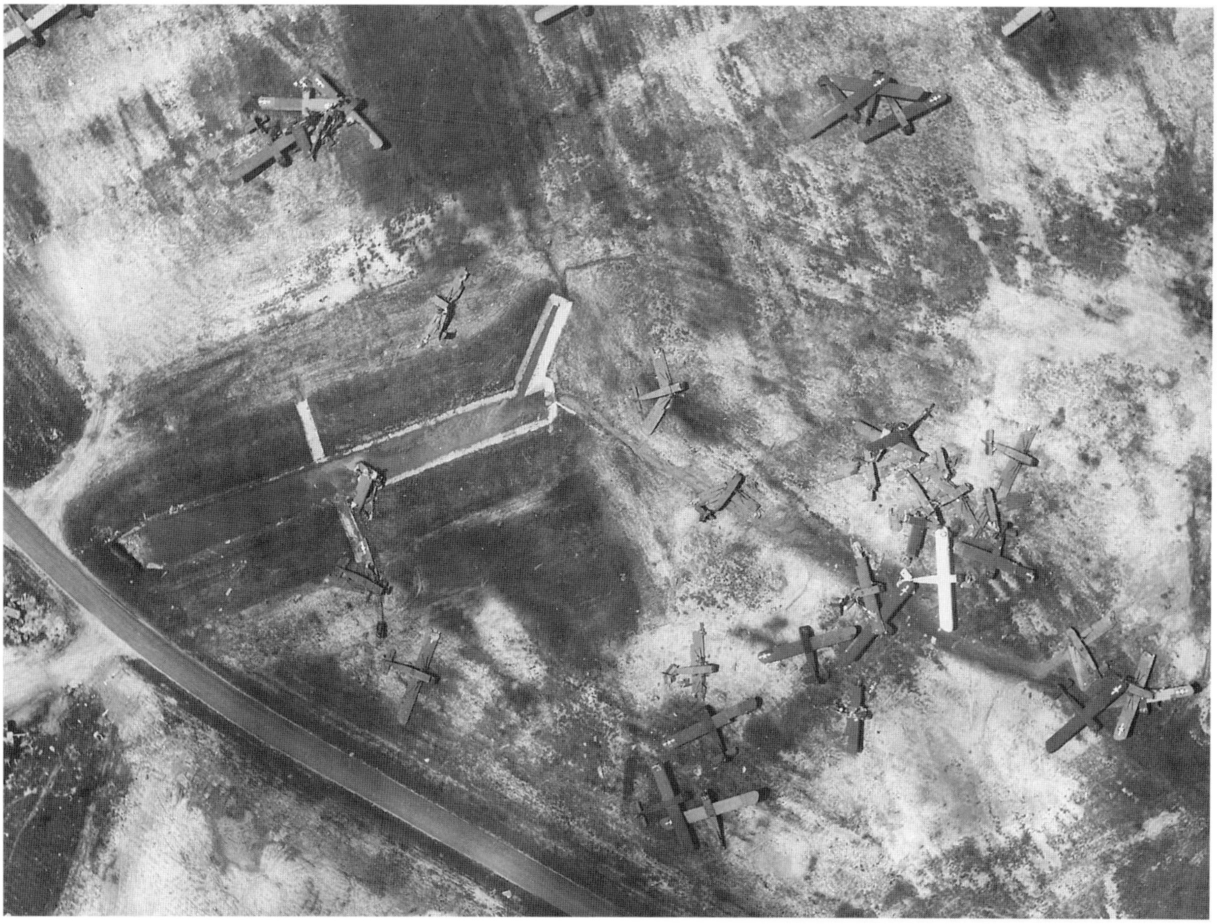

This 1944 aerial photo shows the destruction the storm inflicted upon the gliders on Maxton AAFB. Photo courtesy of Thomas W. Murray, Jr. family.

Three loaded U.S.A.C. trucks, loaded with glider wings and control surfaces in downtown Pittsburgh. Photo courtesy of the Thomas W. Murray, Jr. family.

Tom Murray, Sr. helps steady an outer wing panel as it is removed from the glider. Dismantling the gliders was the responsibility of U.S.A.C. Photo courtesy of the Thomas W. Murray, Jr. family.

A glider fuselage is carefully loaded on a trailer which was being pulled by a Packard coupe. Photo courtesy of the Thomas W. Murray, Jr. family.

A devastating hurricane raged through Maxton AAFB, North Carolina, in September 1944, destroying hundreds of military troop-carrying gliders. The gliders had to be dismantled and returned to the Heinz Company in Pittsburgh for rebuilding and salvage. Tom Murray and Pete Maslanka flew the Beechcraft *Staggerwing* to Maxton AAFB to evaluate the scope of the project. The damage was massive. Murray's mind worked feverishly on how he could remove all the gliders to Pittsburgh without inflicting more damage. He took accurate dimensions of everything; wing length, cord, rudder, stabilizers and fuselage. Pete Maslanka was visibly shaken when Tom Murray told the Base Commander U.S.A.C. could do the job. At this point Tom did not even have one proper trailer to haul the massive gliders. They returned to Delaware with an approval to move the gliders with the utmost speed.

Thomas W. Murray was a ramrod. If his mind was set on a problem to be solved, just stay as clear of him as you would of a whirling propeller. Freuhauf would not sell trailers to Murray because of the war allocation in spite of Tom's pleas to the government. In desperation, Tom immediately assembled a crew of welders to fabricate trailers of his own design which would haul the oversized loads. His crew built trailers which could be towed behind automobiles or trucks. His designs paid special attention to making them low to the ground to avoid overhead obstructions as well as providing maximum protection to the fragile cargo. There were no interstate highways, and the gliders would be transported through many small towns where electric wires crossed the road. One trailer design would carry four glider wings. Another would carry the fuselage while the interior of the glider fuselage was utilized to carry the tail assemblies on a special roof rack.

Murray's ideas paid off handsomely. Besides being a ramrod, Tom was a visionary and a workaholic. He knew that when the war was over he would be back

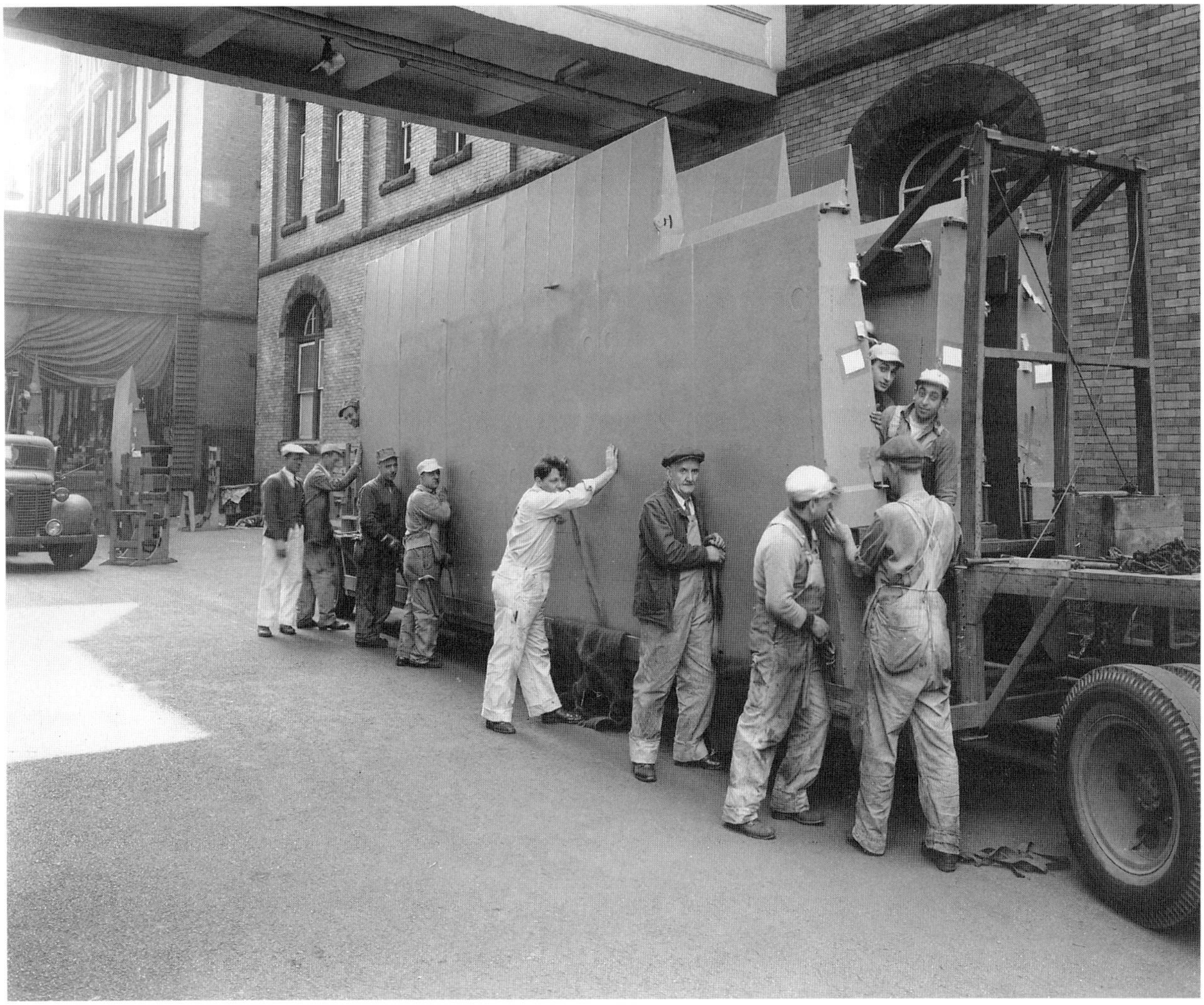

The size and weight of the wings can be appreciated in this photo. Eleven men are preparing to unload one from a trailer. Photo courtesy of the Thomas W. Murray, Jr. family.

in the pilot's seat of his Beechcraft *Staggerwing*, looking to expand operations. The Dover Airfield was owned by the Wilsons and Tom Murray was uncomfortable with that association. Now that he had money, he purchased the Needle's Nursery Farm north of Dover (now Dover Downs) which fronted Route 13. The farm totaled over 300 acres, which was more than enough to create a new airfield with three runways and to construct a new building to support his growing operation. Never one to miss an opportunity, Murray managed somehow to get the government to release German prisoners of war being held in a camp outside Leipsic for use in clearing the land for his future airfield.

Prior to the war, Harold Turpin was the Fixed Base Operator (FBO) at the Dover Airfield. When the War Department closed the airfield for the duration, Harold moved inland to Hagerstown where civil flying was not restricted. After the war, Harold once again took over the responsibilities of FBO and flying at the Dover Airfield. At the same time, Tom Murray began moving to their new location. After Murray moved, the Dover Airfield closed and Harold Turpin left the area.

U.S.A.C complex located on Route 13, circa 1955. Note the multiple grass runways and hangars behind the buildings. Photo courtesy of the Thomas W. Murray, Jr. family.

This aerial photo was taken in 1998 to illustrate the changes in the property over the last 50 years. Photo by the author.

After the war, many companies that depended solely upon government contracts shut down when they could not find work to fill the void. Tom Murray was an exception. He expanded his operations. The runways were leveled and seeded at Murray's north Dover Airfield location in 1947. A new operations building was built facing Route 13 during 1947-48. Besides modern offices and a diner, the new building had a fabrication and welding shop which would be used to build and maintain the special-bodied trucks and trailers that hauled oversized loads. Hangars were also built to rent to local airplane owners. Tom also traded in his faithful Beechcraft *Staggerwing* for a new Beechcraft *Bonanza.*

The new U.S.A.C. site on Rt. 13 was second in size only to Playtex in Dover during this period. An Air National Guard jet is loaded for transport. Photo courtesy of the Thomas W. Murray, Jr. family.

This 1951 Martin PBM amphibian fuselage illustrates the size of the loads U.S.A.C. handled on a regular basis. Photo courtesy of the Thomas W. Murray, Jr. family.

A complete center section of a multi-engine aircraft hangs over both sides of the trailer in these two photos. Photos courtesy of the Thomas W. Murray, Jr. family.

Naturally, Murray needed aviation fuel, vehicle fuel, and lubricants for the new site. He soon signed a franchise with ESSO (now EXXON) and became a dealer. The franchise allowed him to sell the products at retail, while he used them in his business at wholesale prices.

When the war was over, there was more prosperity for Murray. Since the Summit and St. Georges bridges were still not constructed at this time, Douglas DC-3's hauled perishable farm products to major cities. This airfield was Dover's only public-use aviation facility capable of handling large aircraft. Fresh corn, tomatoes, beans, and squash were flown to Pittsburgh, New York, Baltimore, Washington, and Philadelphia from Murray's new facility.

To say Tom Murray's idea of specializing in hauling oversized loads was an excellent idea is almost an understatement. From his headquarters in Dover, Murray established a network of more than ninety locations with representatives throughout the country. If an airplane went down, U.S.A.C. picked it up. If an oversized load needed to be moved, U.S.A.C. moved it. If a driver got sick, or equipment needed repair while en route, Murray loaded a relief driver or mechanic in his personal airplane and flew him to the rescue.

The ICC officer was dead wrong about airplanes becoming extinct after the war. U.S.A.C. had the permits to operate in all forty-eight states and Tom Murray put them to good use. U.S.A.C. was a gold mine.

After the war, Murray changed the name from U.S. Aeroplane Carriers to U.S.A.C. Transport. The word "transport" was added to the name in an effort to convince customers the company would haul other oversized loads besides aircraft. U.S.A.C. Transport

The long wooden pole, used to deflect power lines, was standard equipment on all runs. Photo courtesy of Thomas W. Murray, Jr. family.

The M-29 Weasel was sold or leased with a trailer. Properly sealed, the Weasel was capable of forging rivers or lakes. Photo courtesy of the Thomas W. Murray, Jr. family.

hauled Chris-Craft yachts from the Salisbury, Maryland factory. Helicopters, TV towers, jet engines, storage tanks, and guided missiles were also standard cargo. At this point his son, Tom Jr. joined the company and conducted most of the internal operations of the company while Tom Sr. focused on expanding the customer base. They concentrated heavily on the war surplus equipment market by buying blocks of Boeing/Stearman PT-17 *Kaydets* and Piper L-4 *Grasshopper* airplanes and reselling them to airport operators as trainers.

With the war now over, the War Assets Administration (WAA) was formed to sell off the bulk of the military inventory. Many of the aircraft that were not purchased ended up as scrap metal. The world was at peace and the cost to maintain the equipment far outweighed the cost of selling off the huge inventory. The funds could be used to help pay off the huge war debt. There were thirty locations across the country designated as WAA surplus centers. To purchase an airplane (or group of airplanes) all you had to do was walk in, inspect the airplane of your choice, look at the log books provided, and answer a few questions, none of which included, "Do you know how to fly this airplane?" Next, you would sign your name a couple of times, pay for it "as-is-where-is" and fly it away with a complimentary tank of gas.

Pete Maslanka ferried many Piper L-4 *Grasshopper* and Boeing PT-17 *Kaydets,* purchased by Murray, into the Dover Airfield from WAA sights across the country. The PT-17 *Kaydet* was nicknamed the "washing machine" because of the vast number of pilots who "washed out" in flight training trying to master the art of flying this huge biplane. The twin-seat, open-cockpit aircraft was known for its strength, and the 220 hp Continental radial air-cooled engine provided excellent acrobatic training. It was designed as a trainer and was not noted for its speed. It weighed 2800 lbs and flew about 125 miles per hour.

The Piper *Cub* on the other hand was very slow at 85 miles per hour. However, it was a forgiving airplane of only 65 horsepower that weighed a mere 1220 lbs. The military version of the Piper was designated the L-4 *Grasshopper.* No other airplane of the day offered more economical flight instruction or ownership. Tom Murray bought a bunch of them and sold them just as fast.

Tom Murray, Sr. poses next to Eva Braum's German Horsch Cabriolet. Eva was Adolph Hitler's mistress and the car was brought to America by the U.S. Government. It was declared surplus at Aberdeen proving grounds and sold for $700. Tom later sold it to Arthur Godfrey. Photo from the Thomas W. Murray, Jr. family.

The team of Tom Murray, Sr. and his son Tom, Jr. also saw a market in the M-29 *Weasel,* manufactured by the Studebaker Automobile Company during the war. An off-the-road vehicle built for military use, the wide, endless tracks allowed the vehicle to romp over marshlands, sand, and deep snow leaving only a light footprint. They bought *Weasels* by the hundreds and sold them to sportsmen, hunting clubs, beach patrols, and power companies to survey and maintain their remote power lines. Tom and his son Tom, Jr. also sold and serviced surplus Jeeps and heavy trucks. Converted to meet civil lighting requirements, these vehicles were sold in domestic and foreign markets.

There was no limit to what the Murrays would buy. They also purchased Herman Goering's special-built white Mercedes. John Marsh, a contractor in Lewes, remembers buying surplus barges from Murray to build the Rehoboth Canal. They even found markets for surplus Navy PT boats. When the Tucker car was announced in 1948, Tom Murray had one in Dover for demonstration. There were only fifty-two Tucker cars built, so one can imagine how powerful and persuasive a man Thomas Murray had become to be able to have one brought to Dover.

If there was a deal to be found, you could bet the Murrays would be in the thick of it. Tom Sr. also purchased *Woodburn,* a residence on Kings Highway, Dover. He remembered it had a history in the days of the underground railroad. It became his primary residence for several years, but he later sold it and moved to Marathon, Florida. In the 1970s

Woodburn was purchased by the State of Delaware, restored, and renamed The Governor's House. It has been the official residence of the Governor ever since. The Murrays also purchased several hundred acres of waterfront property at Bar Harbor, Maine, where they spent summer vacations. Naturally, this scenic property had an M-29 Weasel for exploring.

The final curtain came down on Tom Murray, U.S.A.C., and Dover Airfield, when he was approached by David P. Buckson, then Attorney General for the State of Delaware, to sell. Buckson represented a group of investors who included John Rollins. They envisioned the idea of a multi-sports racing complex for the property. When complete, the complex would include flat track horse racing, harness racing and NASCAR stock car racing. Tom was now in his senior years. He had burned the candle at both ends for most of his life and the realization that he could retire a wealthy man was very appealing and he agreed to sell. Thus ended an era of a man who came from Fort Worth, Texas, in 1926 with hardly a dime to call his own. He drove himself endlessly to build an empire that was known in every state in the union.

Dover Airfield and all the buildings were bulldozed to the ground and replaced by Dover Downs in 1968. Thomas Murray, Jr. retained the surplus military Jeep and heavy truck business and moved to a new site on Persimmon Tree Lane in Dover. Tom Murray, Jr., his wife Nancy, and their son, John, continue to buy and sell surplus military vehicles and parts in a world market.

✈ Delaware Airpark 1956

NASCAR race weekends bring in many out-of-state visitors to Delaware Airpark. 1996 Photo by the author.

Working the farm fields was a way of life for many people in Kent County during the great depression. Families were bound together to share the responsibilities of planting and maintaining the fields to produce food. It was a simple life with few luxuries. Most farms were devoid of electric or indoor plumbing. The more affluent farmers drove model A Fords. Others drove a buggy or walked.

The Durham family were wealthy farmers. Not monetarily mind you, but rich in family values. Their entire family was structured around being responsible for family and one's own actions. The parents' policy of, "if you want something, you will have to work for it" was inflexible. Others who did not share this belief fell to hard times and could not cope.

Floyd Durham was born in Kent County on a small tenant farm east of Dover, February 28, 1918. There were seven children in the family. Floyd was born in the middle, with two sisters and one brother older, and two sisters and one brother younger. World War I was over, and the country started to slide into a depression. Those who were fortunate enough to live on a farm fared much better than those who tried to survive in a metropolitan environment. The typical farmer during this period grew crops for feeding the family and for barter. There was little cash available to buy anything that was not essential to survival. Their lifestyle was spartan, yet overflowed with the nonmaterial things of life. Farmers tend to be independent people. This independence, by association, carried over to Floyd Durham. He has a strong-willed personality. Obstacles in his life were considered nothing more than challenges. He carries a sense of pride in being a self-made man. In spite of not being able to secure a formal education, he excelled from hands-on experience.

Prior to WWII, Floyd tried to join the Civilian Pilot Training Program (CPTP) but failed the physical examination because of an early shoulder injury received in a farming accident. After failing the physical, he thought he would never be able to fly and

put the dream out of his mind. But he never became despondent over the decision. He simply sought another profession. After learning the skills of plumbing, he started his own business. On July 16, 1943, he married his childhood sweetheart, Doris Morgan. The family grew to include a son, Harlan, and a daughter, Muriel.

The plumbing business was operated from a small building in the back yard of the family residence near the town of Cheswold. Doris became a teacher, in the Capital School District, which contributed immensely to the family. It was a much improved lifestyle compared to his farm life which had lacked electricity and the other amenities that are taken for granted today.

Sunglo Airfield, South Dover circa 1946. The race track is on the Konshak farm. The airfield, with its three hangars (note the arrow), was part of the same farm. Photo courtesy Bill McDaniel.

It was Lee Remley, a Cheswold private pilot in 1950, who encouraged Floyd to try to pass the civilian pilot physical, which was more lenient than the military CPTP examination. Floyd's dream of someday flying his own airplane came one giant step closer to reality when he passed the civil aviation medical examination. Once he had passed the physical, Lee told Floyd that Perk Evans, a Dover pilot, was going to quit flying and sell his hangar located on Sunglo Airfield, south of Dover. Floyd thus bought the hangar for $250 before he owned an airplane or knew how to fly one.

Now that he owned a hangar, all he needed was an airplane. Still not a licensed pilot, Floyd purchased an Ercoupe, similar to Remley's, from the ERCO factory in Riverdale, Maryland. Lee Remley flew Floyd to Riverdale to close the deal with Bob Sanderson, the president of ERCO Corporation. Much to Floyd's surprise, a young lady on crutches was assigned to fly him back to Delaware. Floyd vividly remembers,

> She leaned her crutches against the building and virtually crawled to the airplane. She refused help of any kind. She had learned to fly and got her license in the Ercoupe which was built with no rudder pedals. After delivering Floyd's Ercoupe to Sunglo Airfield, Lee Remley flew her back to Riverdale.

Sunglo Airfield was a part of the Konshak farm that had typical grass runways of the era. Located in Dover on South State Street Extension, the airfield was only 30 feet above sea level with a 1700-foot-long sod runway. Lee Remley and Ed Biter, Sr. kept their airplanes in adjoining hangars. Floyd, with only two and one-half hours instruction, soloed the Ercoupe. Lee Remley, Eric Lash, and George Garrison each owned Ercoupes and regularly assembled to fly in formation to flying events. Now that Floyd joined in, the flight of four identical airplanes became well known throughout the area.

Everyone was settled in and happy at Sunglow Airfield until hurricane Hazel decided to rearrange everything in 1954. Hazel inflicted major damage to all the airplanes and hangars, which closed the airfield permanently. The hangar collapsed on Floyd's Ercoupe which caused considerable damage. The same fate befell Lee Remley's Ercoupe. Floyd and Lee brought their airplanes home to Cheswold and put them in their garages for repair.

The same year Floyd brought his wrecked airplane home, some unimproved real estate adjoining the Durham property became available through the County Orphan's Court. Floyd purchased the property with the intent of creating a private airstrip. Although the property would only allow enough room for 1800 feet of runway, he felt it ample and it became his *Field of Dreams.* He started clearing and grading the land and, when he finished, proudly named his creation *Little Acres.* His dream was now fulfilled. He had a loving family, a plumbing business, a hangar with a private airstrip in the back yard, and an airplane to fly.

When David Buckson and John Rollins purchased Thomas Murray's Dover Airfield for the site of Dover Downs Raceway, it once again left Dover without an airfield. Jay Hinyub and Armond "Rocky" LaRocque were flight instructors who operated from the Dover Airfield. They both had students who had not completed their training and needed a base for operations. Jenkins Airfield located west of Camden-Wyoming and Smyrna Airfield were the only available sites. Joe Jenkins was heavily engaged in the airplane salvage business, and the flight instructors thought the mangled airplanes might leave a negative impression on students, passengers, and family members. Smyrna was another choice, but Reynolds Jones, the airfield owner, was already operating a flight school of his own. An additional problem was Wayne "Pappy" Brubaker, the local airplane mechanic who serviced the airplanes at the Dover Airfield. He faced relocation and needed a workshop for his work. There appeared to be no place to go.

Their search for an airfield then focused on Floyd Durham. Floyd's *Little Acres* lacked the infrastructure to support a public-use airfield. Adequate as a privately owned facility, it had no facilities for flight instruction and maintenance. In addition to two flight instructors and a mechanic, their airplane owners now needed to find a new home. Some sold their airplanes and quit flying. Some moved to Floyd's airfield and the remainder scattered.

The Durham family accommodated these flying nomads and allowed them to keep their airplanes at *Little Acres.* After all, the best part of flying is the camaraderie. The attic over the family garage was turned into a one-room operations center, complete with plumbing, heat, and a window air conditioner. Floyd's plumbing shop was cleaned out and moved into the family garage so "Pappy" Brubaker had a place to perform maintenance. Floyd's airplane occupied a shed with an open front, while all other aircraft were tied down outside. The entrance to the airport was the family driveway. A 250-gallon aviation fuel tank was installed to service the airplanes. At the time, that is probably as far as the Durham family felt the airfield would be developed.

The operations building was the loft over the family garage in back of the Piper Super Cub. *The Durham residence is on the left. May, 1964, photo by the author.*

An antique "Church Midwing" *being restored by "Pappy" Brubaker. In the left shed-hangar is the author's 7AC* Aeronca *and right hangar houses Floyd Durham's Piper* Tri-Pacer. *Photo dated 1964.*

The first hangar designed and built by Floyd Durham at Delaware Airpark circa 1963. Photo taken by the author, 1998.

This aerial photo was taken July 1963. Note the trees are not completely cut down through the wooded area and the runways are not hard-surfaced. Photo by the author.

It does not matter how things are, people always want things better. That is human nature. It was not long before everyone, including the Durhams, acknowledged that improvements to the airfield would be an ongoing challenge. Airplane owners wanted shelter for their airplanes. The outside elements are not favorable to the longevity of an airplane. The first hangar was of Floyd's own design. His objective was to be sure it could survive a hurricane should it ever happen. He still felt the pain of hurricane Hazel. Floyd designed, and welded together, three-inch water pipe for roof trusses. Telephone poles were buried deep into the soil to hold the metal roof. Although it was an open hangar with no doors or walls, it would protect the airplanes from the sun, rain, and snow. When it was completed it accommodated eight airplanes.

The approach to landing on runway 27 was always a challenge with 50-foot-high trees to fly over. They also seemed to fill the windshield quite rapidly when taking off eastbound on runway 9. The 1800-foot-long runway looked more like the deck of a small aircraft carrier in the middle of the Atlantic. Taking off to the west was not as much of a challenge since a farmer's fence was the only obstacle. Pilots watched the wind and the weight of their airplanes carefully before departing. No one had to tell Floyd the runway was a bit short. It was more than obvious when anyone standing outside judged every pilot's landing and takeoff. If you could fly in or out of *Little Acres,* you could handle any airfield. Many less-qualified pilots would simply avoid the airfield altogether.

Never in his wildest dreams did Floyd Durham think he would spend most of his life owning and operating his own public-use airport. Privately owned, public-use airports consume large amounts of real estate which make it more than difficult for those who are not wealthy to financially afford. Over the years, Floyd managed to buy five more

Harlan Durham provides all the aircraft maintenance. He holds an FAA commercial pilot's license, with multi-engine rating, as well as an AP/IA mechanic-inspector's license. 1997 Photo by the author.

parcels of real estate to meet growing demands. The first of these would be in 1960, to create a separate entrance and eliminate the traffic on his private driveway. Four additional parcels were purchased later to lengthen the runway on the east end. For Floyd Durham, Delaware Airpark was an evolution, not a vision. The evolution was a result of unforeseen community growth which was impossible to predict. In the end, that may have been a blessing in disguise, since he was not in a financial position to accomplish the goal in one transaction.

It turned out that buying the property was the easy part. Clearing the property and turning it into usable runway was a major task. Floyd purchased a chain saw to start cutting down the trees and selling them off for lumber. The saw rarely got a rest. Everyone at the airport chipped in to cut the trees down. At one point the pile of brush seemed to grow as high as the trees that had been cut down. The brush was burned, and a bulldozer operator was hired to remove the stumps and level the runway in 1963-64.

In spite of limited funds, Floyd's endless determination finally turned what was once forest into usable runway. The extra 1800 feet brought additional activity to the airfield. Unfortunately, the additional activity brought additional problems. The sod runway could no longer maintain quality grass growth. The spinning propellers blew the topsoil away from the root structure and the runway became clots of

Yvonne Coker poses with Diamond Aviation's new Piper Cherokee *on Loockerman Street for Dover's 250th Anniversary in 1967. The airplane was taxied from the airport with a police escort. Photo by Floyd Durham.*

The World Aerobatic Champions Pitts S2A Specials *for 1972 at Delaware Airpark. Photo by Bill McDaniel.*

grass. The uneven soil surface was also hard on the airplanes' fragile landing gear. Worse was the mud when the spring thaws arrived leaving the airfield unusable. When it was dry, the propwash created a dust storm from departing aircraft. The airfield was growing and once again needed more money spent on the infrastructure.

In these early days of private aviation, banks felt those who flew small airplanes had suicidal tendencies. That perception made loans to aviators more than difficult to obtain. The bank made the loan to build a new operations building and more hangars after the Durhams committed everything they owned for collateral. A couple of years after the new buildings were occupied, a hard-surface runway, runway lights, taxiway, parking ramp, and service hangar were installed. In spite of the expense involved in hard-surfacing the runway, Floyd never charged a landing fee to anyone who landed.

Little Acres was virtually unknown outside Delaware in 1967. Now was the time for a name change. *Little Acres* had grown beyond Floyd's expectations and a name change was now in order. *Delaware Airpark* became the name of choice. Diamond Aviation would be the name for the Fixed Base Operations (FBO), which also included a Piper Aircraft dealership. Floyd's son, Harlan, earned a commercial pilot's license, multi-engine rating, with AP/IA mechanic and inspectors license. In 1970, Harlan would assume the challenge of maintenance of the aircraft and their inspection. The maintenance shop, built in 1973, was attached to the operations building. With the improved facility, aircraft began to fly in from around the globe. Many business people flew in to do business in Dover and Kent County. Aircraft from as far away as Canada, England, Alaska, and Mexico are a few which come to Floyd's mind.

The World Aerobatic Championships were held in Salon de Provence, France, July 18-31, 1972. The U.S. contenders for the men's division were Charlie Hillard, Gene Soucy, Tom Poperenzy, Art Scholl, and Bill Thomas. The women's aerobatic team consisted of Carolyn Salisbury and Mary Gaffaney. Eleven countries and 59 pilots, including the USSR, would compete for the championship. On June 26, the U.S. Aerobatic team flew their six open-cockpit, bi-wing, "Pitts S2A Specials" into Delaware Airpark. They used the airport as a stopover before flying into DAFB to be loaded on a C-5 *Galaxy.* Al Hedgecock, then station manager for WDOV-Dover and Sug Adlam opened the morning program by intro-

This transient Douglas DC-3 dwarfs the operations building. Note the service hangar has not yet been built. Photo by Floyd Durham.

This Air Alaska turbo-prop Fairchild F-227 dwarfed the operations building when it landed in 1978. Note the service hangar is now finished. Photo by Floyd Durham.

ducing Mary Gaffaney. Al remembers, "Mary looked like a sweet matronly lady who had just taken a batch of cookies out of the oven."

The same day, she accompanied Hillard, Poperenzy and Soucey in a wild demonstration of their aerobatic capabilities at Delaware Airpark. After their demonstration, Al remarked, "If she could bake cookies the way she flies that Pitts, they would be the best in the world." Little did he know at the time that Mary Gaffaney would bring home the cookies. Mary Gaffaney won the title of *Women's World Aerobatic Champion.* The men's team also showed they had the *Right Stuff* when Charlie Hillard won the *Men's Aerobatic Championship.* This was the first time Americans had ever won both titles. On the return flight to Dover, via C-5, they once again stopped at Delaware Airpark where they were greeted by local pilots and hailed for their accomplishments.

The weekends at Delaware Airpark were filled with family events. Fly-outs to breakfast or lunch were commonplace. Late Sunday afternoons were reserved for family cookouts under the shade trees. Dr. Winfred Mroz started a sailplane club with two sailplanes, and parachutists engaged in skydiving. Others created flying clubs where they jointly owned shares in an airplane. Activity on the airport was enhanced by "Penny-A-Pound" rides to the public twice each year.

Floyd continued to look for ways to improve Delaware Airpark. An instrument approach was requested from the FAA. This required considerable engineering since it would operate in conjunction with Dover Air Force Base (DAFB) radar and approach control. Delaware Airpark was the first, and only, public-use airport in Kent County with a VOR instrument approach. The FAA minimums were set at 400-ft ceiling and 1-mile visibility. The instrument approach was approved in 1970.

Delaware State University, under the direction of Tom Knight, operates a flight school from Delaware Airpark. The University offers ground school in airport management, weather, and air-traffic control. In addition, they offer flight training in various single-engine and multi-engine aircraft. Federal Aviation Administration (FFA) ratings include private, commercial, instrument, and multi-engine. Delaware Airpark is located four miles from the DSU campus. The job placement rates for graduates of this program are extremely high.

In 1995, Floyd invested in a dedicated telephone line and antenna that would allow pilots to activate flight plans from their airplanes. This "clearance delivery" (125.95) is another customer service that pilots use free of charge at Delaware Airpark. Effective October 1996, Delaware Airpark had the first FAA-approved GPS (global positioning system) instrument approach in Delaware.

Floyd Durham has achieved in Delaware what is considered nearly impossible in the entire field of aviation. He built a privately owned, public-use airport, complete with hard-surface runway, taxiway and ramp, instrument approach, operations building, service hangar, storage hangars and lighting entirely from personal family income, and he survived.

Bill Elliott (left) and Floyd Durham (right), NASCAR race weekend, June 1972. Photo by the author.

Most privately owned businesses do not pass on to the next of kin. This is especially true of airports that require long hours of daily operations. In 1998, at age 80, Floyd Durham can still be found at his beloved *Field of Dreams* overseeing the daily operations and instructing students. Although he has made plans for succession of *Delaware Airpark* to his children, his only concern is the unforeseeable future.

Aerial view of Chandelle Estates, 1998. Photo by the author.

✈ Chandelle Estates - 1967

Route 9 is one of Delaware's most scenic two-lane routes. Starting in New Castle County, it meanders through lush farmlands, tidal waters, narrow bridges and marshlands flush with cottontails. In Kent County, two small villages on Route 9, Little Creek and Leipsic, survive from families engaged in fishing as a way of life. This route is also used for the entrance to Bombay Hook, one of the Federal Government's most renowned wildlife areas in the east. The refuge becomes a haven for ducks and Canada geese during the migration seasons. It is hardly the place one would look for an airport.

The 103-acre Holt farm, with an 1830's farmhouse, two miles east of Dover, borders on Delaware Route 9 and Silver Leaf Lane. The Holt Estate heirs placed this property up for public sale to be held in February 1967. The farm had been abandoned for several years and had fallen to decay. In spite of the broken windows and its general run-down, shabby appearance, a number of interested parties appeared on the day of sale. Although the turnout for the bidding created considerable interest, the property did not bring a high enough bid to satisfy the heirs.

Several weeks after the auction, Josh Twilley and Jack Wich, co-owners of National Enterprises, Inc., thought the property could have another potential beyond farming. Josh and Jack were both pilots and often viewed potential investment properties from the air. With the Holt farm they felt they could get a return on their investment if the property could be developed into a residential airpark. This, they felt, could be accomplished without sacrificing the integrity of the farm. The idea of a Residential Airpark was not a new one. Many such airparks have been developed successfully throughout the country. Owners build their homes and hangars on their properties and share a common runway from which to fly their airplanes. The concept is similar to a boat

Josh Twilley and Jack Wich. 1998 photo by the author.

owner's desire to own a piece of waterfront property where he has access to his boat.

Josh Twilley wore several hats, which included being an attorney, landlord, investor, and a private pilot. Josh was a native of Kent County and possessed a keen knowledge of land values and their potential for resale or investment. Josh felt Jack had an untapped talent of being able to sell almost anything. Both had their strengths, but in different areas. With Josh Twilley's financial backing and expertise in real estate transactions, combined with Jack's ability to sell, National Enterprises proved quite successful. At its highest point, it held title to 180 properties.

Jack has always loved airplanes and flying. At age eight he started building flying models and flew in his first airplane at twelve. When WWII broke out, he was seventeen. He enlisted in the USMC as an aircraft crew chief. By the time he was twenty, he had earned his private pilot's license and promptly joined the U.S. Army Air Corps. After twenty years of military service, Jack retired from the military service in Dover.

After some discussion and planning, Josh and Jack made an offer to the Holt heirs and National Enterprises purchased the farm in March 1967. Jack and his wife, Ann, had decided the old farm house could be restored to become their residence, and they purchased the site from the corporation. It can be seen on the right side of the aerial photo at the beginning of this series. The properties facing Silver Leaf Lane were divided into residential sites to be sold. Approximately twenty acres was allocated for the runway, hangars, and parking ramp for visiting aircraft. Jack and Ann immediately started the long, laborious job of renovating the eighteenth century farmhouse. They moved their Piper *Tri-Pacer* from Delaware Airpark and parked it in the back yard. Josh followed later with his Ercoupe.

During the twenty-one days of runway construction, a problem of dust arose when the graders continually exposed the sandy soil to wind and sun. Jack said, "The dust was so deep, you could drop a hammer on the runway surface and it would be lost." The contractor brought in a watering truck which promptly ran out of water. In desperation, Jack installed high volume water pumps and connected several thousand feet of hose with sprinklers to control the dust. He recalled, "Actually, it worked too well. The runway became a sea of mud." Jack decided it needed a good crop of grass before the runway could be used to any extent. When he found that the grass seed would cost $4,000 the idea was rejected. Forlorn, he admitted he had built an airstrip that was unusable in its present condition.

Time, in many cases, solves more problems than people. In a few months nature grew a crop of grass that would be the envy of any golf course. Soon pilots began to visit the airfield to review the restoration progress of the Wich's 1830 farmhouse. Jack and Ann started a log book for pilots to enter their names as visitors to remember. It has become a family keepsake.

The name "Chandelle" is a French word used to describe an aerial maneuver where an airplane makes a quick and simultaneous climb and turn using the airplane's momentum to increase the rate of climb. Josh and Jack picked the name because of its WWI aeronautical and French origin. Chandelle Estates would be the first residential airpark ever proposed in Delaware.

Josh and Jack never planned Chandelle Estates to become a commercial airport. The idea was to make the venture profitable for their company and provide a place for their own airplanes. Outside influence changed that concept. By 1968, a local aero club asked for permission to use the airfield. The club managers not only wanted to use the airfield for parking their airplanes, they wanted to teach new members how to fly. If Jack and Josh accepted the proposal, the airfield would have to change its designation from "Private" to "Commercial." After much thought, they decided to file the necessary papers with the Delaware Aeronautics Commission. With nothing more than a visit to the premises, the application was approved by the State Aeronautics Commissioner.

The change to commercial status influenced the decision to make further investment in the property. Josh and Jack built the first of two steel T hangars on the airfield to rent to airplane owners. These hangars would hold a total of sixteen airplanes. One hangar bay would be used as a workshop for Wayne "Pappy" Brubaker, who serviced airplanes. Pappy, the most loved airplane mechanic in lower Delaware, was located at Delaware Airpark and a closed hangar was not available to him at that time. Working on an airplane inside was Pappy's dream come true. A double-wide mobile home was also erected to serve as an operations building for students and instructors. This was followed by aviation fuel and oil to service the planes on site.

All this activity created the same problem other commercial airfields experienced. The grass runway would not support the increase in operations through four seasons. The soil would not support the airplanes in wet weather and snow removal was at the will of the weather. Once again, Josh and Jack summoned a construction company to hard-surface the runway. The grass runway was scalped, then select soil and gravel installed in its place before the final surface was applied. Jack recalled that there was an endless stream of huge dump trucks filled with hotmix. To obtain the optimum surface, four passes were made, each pass consisting of material 1.5 inches thick. By the end of the day the 28-foot wide, 2675-foot-long runway was done. Runway lights, parking pads and anchors for tie-downs were also installed to accommodate tenants and visitors.

Jack and Ann, by this time, had moved into the renovated farmhouse. The next morning it was a cloudless sky, and a perfect morning to fly. Jack and Ann climbed into the Piper and taxied to the end of the runway. Ann took the controls for the inaugural takeoff. As they flew over the airport, they began to realize how far they had come and how much money National Enterprises had invested in what was initially conceived as a residential airpark.

The economy was stable and their airport began to demand more time and attention. Josh and Jack decided to acquire the North American franchise for the four-seat *Lark.* At the peak of their involvement, they would own six airplanes. Most of the flying activity fell on weekends and holidays, depriving them of family activities. These days, which were once considered days off, now turned into *work days.*

Josh and Jack never had any aspirations of becoming airport managers. They began to wonder if the enterprise had not gone beyond their initial dream, which was to keep everything simple and uncomplicated. With the new runway, hangars, and a flight school, more activity followed. Jack and Josh realized their *Field of Dreams* had become a burden. The burden of dealing with the day-to-day activities of the airport was more than they wanted. In spite of their tremendous financial investment, the airport was not profitable. The liabilities for injuries at the airport alone were enough to cause concern.

The aero club principals were summoned to a luncheon meeting at the farmhouse. In simple terms, it was stated that National Enterprises wanted to sell the airport. To encourage the sale to the club members, there was to be no down payment and National Enterprises would hold the mortgage. Monthly payments would be the same as the present rent. Josh and Jack made the Aero Club an offer they could not refuse. As time passed, the economy turned downward and the aero club failed. The Club simply shut down and walked away from the deal. Josh and Jack did not want the airport back and sold it to B&M Flyers, another aspiring flight school. In time, it also failed, and once again Jack and Josh looked for a buyer. The Ray Kruser family now owns the airport property.

The Krusers reside in one of the homes on the residential portion of the airport. Although Mrs. Kruser's husband, Ray, passed away in 1996, Mrs. Kruser, her son Roger, and daughter Carol, have kept the business alive by subletting part of the operation to a flying school. Jim Strong Aviation has operated a flying school from the airport since 1992. Jim has a lifetime of flying experience that includes piloting corporate jets worldwide. His credentials are impressive, and he is succeeding where others have failed.

Josh and Jack agree it was a fun time. The airport provided them with many opportunities to enjoy flying and make many new lifelong friends. They have no regrets, they just do not want to do it again.

✈ Henderson Airfield, 1983 -

Dave Henderson, owner of Mid-Eastern Aviation, with his dream airplane, a 1946, Piper Cub. *Photo supplied by Dave Henderson.*

David Henderson never wondered about what profession he would choose in life. His earliest recollection of flying was building a large airplane from scrap lumber at the early age of seven. Although the airplane was not flyable in the true sense of the word, Dave envisioned many flights of fancy that have lasted a lifetime.

While his father was stationed at Castle AFB in California in 1956, Dave had the opportunity to take a ride with one of his father's friends in a Piper *Cub.* They flew near Castle AFB, and it was the first time he ever saw the huge military planes from the heavenly perspective. Dave recalled, "That flight is forever etched in my memory and removed any shadow of doubt that I would dedicate my life to aviation. The Piper *Cub* and I had formed an eternal bond."

After graduating from high school he got married. This coincided with the military conflict in Vietnam which decided Dave's next encounter with aviation. His choice of service would be the U.S. Air Force. During his tour of duty, he was assigned as crew chief on the F-4 and later crew chief on Helicopters. This exposed him to a wide variety of military aircraft as a crew chief and electronic-technician.

After the Vietnam conflict, Dave returned to civilian life. His father was stationed at Dover AFB, and he moved with his wife and son to Delaware. His early flight in the yellow Piper *Cub* still haunted him. He had to learn to fly his own airplane. By this time, Piper *Cubs* had given way to the new airplanes which offered electric starters, two-way radios, and radio navigation. He first took flying lessons at Delaware Airpark in 1970, but the low-wing Piper *Cherokee* did not recapture his dream. He decided to try Chandelle Airport where the instructor used a high-wing Cessna for instruction. Although the Cessna filled his present needs, the yellow *Cub* still held his fascination. This was the airplane he wanted to fly. Before he could complete the Cessna course, his flight instructor, who was in the Air Force, was shipped off to another base. Reynolds Jones, at Smyrna Airfield, also used a Cessna for flight instruction, and Dave decided to complete his training there. John Reed, FAA-approved examiner at Laurel Airport, approved his flight training in 1975, and he was now free to fly on his own.

Starting in the late 1930s the Piper *Cub* was the most popular trainer in the world. Piper introduced the *Cub* with a price tag of $1295. By the end of 1939, the production increased and this price was lowered to $900. During the period of the 1930s through the 1950s, more pilots took their initial flight training in a *Cub* than in any other airplane. It was so popular, almost every small airplane was called a Piper Cub by the non-flying public. Dave Henderson had to have one.

Dave's search for a *Cub* started in earnest in 1983. Most of the *Cubs* by this time had begun to show the hard life they led by neglect. Many had been used as trainers. Others were found stored in barns with their wings off. Their fabric was a series of patches. The plexiglass was fogged and shattered. Dave was heartbroken. The airplanes he located did not portray the image of the *Cub* he had held for all these years. But he finally found the *Cub* he wanted. The fact that it was located on Stapleton Airport in Denver, Colorado, did nothing to stop him. He bought a ticket on a jet and flew out at 500 mph and returned home with the *Cub* at 80 mph. The *Cub* was in excellent condition, with the exception of a couple weeks of detailing to make it a show stopper. Dave and his wife flew it to air shows where crowds would gather around the airplane in awe. After several shows, Dave sold the *Cub* for a price he could not refuse.

PIPER J-3 CUB SPECIFICATIONS

Engine	Continental
Type	4-cylinder
Horsepower	65
Wingspan	35.2 ft
Length	22.3 ft
Weight (empty)	640 lbs
Gross weight	1,100 lbs
Height	6.8 ft
Number of seats	2
Maximum speed	92 mph
Cruise speed	80 mph
Range	250 miles

This quick profit caused him to consider restoring *Cubs* professionally. Dave said, "Although I never had any intention of going into the Piper *Cub* restoration business, the airplane created such a demand, I had no choice. Before he could take on such a project, he had to find a place to work. Dave was working in the computer industry to support the family, and the decision to change professions was not an easy one. He found his *Field of Dreams* between Canterbury and Frederica in lower Kent County. It was a small airfield, owned by Roger Yorde, who established it as a private airfield on his eighty-acre farm around 1947. Roger decided to retire from farming and move to Florida and warmer weather. The sale, in 1985, of the Yorde property made both parties happy. The single hangar in back of the house would be used to restore Piper *Cubs.* Dave said, "I went into the *Cub* restoration business in 1985. He applied for, and was granted, a commercial airfield license by the State Aeronautics Commission with little fanfare. He named the business "Mid-Eastern Aviation" with a subtitle, *"Piper Cub Repair & Restoration."*

Dave researched everything he could find on the Piper *Cub* before he purchased his next *Cub.* This one was bought from Willard Jackson in Milford. Willard had bought the L-4 *Cub* as a military version in Europe after the war and had it shipped home. It was completely disassembled and stored in the back of Willard's hangar on the Milford Airfield. This *Cub,* he decided, would be restored back to its original military specifications in every detail possible. Dave remembered clearly, "The knowledge I had was now put into actual hands-on work. The welding was done by Jack Kennedy, a former aircraft welder for Bellanca. Kennedy could make welding an art. There was ample help along the way. Pappy Brubaker and Joe Jenkins are two important helpers Dave remembered.

When the *Cub* was finished in 1987, it looked new. It caused such a sensation at air shows, orders began to pour in. Dave said, "Over the years, I guess we have restored about 125 Piper *Cubs,* two 1930s *Travel Air* biplanes and several homebuilts. These airplanes were restored for customers in the U.S., England, France, Thailand, Alaska, and other areas around the globe. Some were true restorations, down to the last detail. Others were upgraded by customer request to include more powerful engines, electrical systems, wing fuel tanks, etc. The largest backlog of orders was about eighteen airplanes. Customers waited a year, or more, before their names came to the top of the list. Many were airline pilots who wanted to relive the experience of the "true feeling of flying." Apparently, customers did not object to the long wait, for the finished product was indeed a work of art.

Walking into Dave's office on the airfield is an experience in itself. It is anything but tidy. But the visitor never notices, because the walls are filled with the signatures of virtually everyone who passed through. Many are written in foreign languages, confirming Dave's statement that, "They come from virtually every corner of the earth." Patty Wagstaff, National Aerobatic Champion; Prince Tang of Thailand and his entire entourage; Barry Valentine, FAA Administrator; Mike Pangia, FAA Chief Litigator; General David McCloud, Joint Chiefs of Staff; Duane Cole, National Aerobatic pilot, and many more have been visitors to Henderson Airfield.

The Thailand connection with Prince Tang created a bridge of friendship that spanned both continents. There is virtually no civil flying in Thailand. Prince Tang, however, has amassed a number of collectable aircraft (which includes two of Dave's *Cubs* and the

Eight Piper Cubs, *all in various stages of restoration are displayed on Henderson Airfield. Photo by Dave Henderson.*

The Thailand military exhibition parachute team thrilled the spectators with their precision jumping. Photo by Dave Henderson.

two *Travel Airs*) and opened a huge aviation museum in his country. Since Thailand never manufactured airplanes, the collection represents aircraft and memorabilia from around the world. Dave Henderson has played a major role in the museum's collection authenticity.

Prince Tang sent his entire museum maintenance crew to Delaware to begin a two-year school, headed by Dave Henderson, to learn how to care for the collection. In turn, Dave also spent considerable time in Thailand, overseeing maintenance on the actual aircraft. Dave, an accomplished parachutist and aerobatic pilot, was asked by the Prince to do a Piper *Cub* aerobatic show in Bangkok for the public before he returned to the U.S. "He did not have to ask a second time," Dave replied. The people of Thailand had only seen military aircraft do high-speed flybys, rolls and rapid climbs. This would be the first time the Thai people would see a small airplane do tailspins,

Two of Dave Henderson's restored military versions of the Cub *in a blistering 80 mph flyby. Photo by Dave Henderson.*

Many locals turned out for the Royal Thailand visit to Henderson Airfield. Photo by Dave Henderson.

loops, hammerhead stalls, rolls, etc. ... all at the astonishing speed of eighty miles per hour!

Dave returned the invitation by inviting the entire Thailand Embassy staff to his Airfield for a day of flying and parachute jumping. Representatives from the Indian, Spanish, and Indonesian Embassies also attended, as invited guests, to partake in a day of historic flying. The hangar was cleaned out and tables set up for the Princess and Prince. (The embassy staff even brought the Royal Embassy chairs, linen and place settings for the Prince and Princess.) It was an "Open House" affair and the little airfield was overflowing with aviation-minded people from far and wide.

Over the years Dave Henderson and his *Field of Dreams* have been written up in virtually every general aviation magazine published. He has earned a worldwide reputation of excellence in the arena of early Piper *Cubs.* He works with his wife, brother Robbie, and one other full-time employee painstakingly disassembling the airplanes and making a list of parts needed.

Dave said, "Finding the parts for *Cubs* today is the ultimate challenge. The FAA will not allow anyone to make a new airplane part unless the process and material used meets their approval. Unless you are a serious supplier of aviation parts and have an engineering staff to back up your products, having an FAA bureaucrat approve remanufacturing a part is virtually impossible. Fortunately, Dave and Robbie made an impression on the FAA administrator with their high standards of handcrafting original parts, and they are reproducing *Cub* parts for their restorations and for sale in the marketplace.

Dave has a bitter feeling toward some government regulations. "The small guy, with no interest in becoming a corporate giant, is crushed out of existence by government. The County government has allowed a home to be built at the end of my runway in defiance of all existing federal laws. How can I fight them with my resources? Even if I win, I lose. The government is immune from financial restitution for wrongdoing, even if they are found guilty! They have allowed people to live in a home that is directly in the flight path of landing airplanes. That is dangerous! If someone gets killed crashing into the house, the lawyers will sue the pilot! This drives customers away."

Dave was fighting mad. He single-handedly approached the Delaware Legislature and fought for what he felt was right for future airport owners throughout Delaware. Under most circumstances, a single person, without an organized lobby, would not have a chance to have a bill introduced, much less passed, in the legislature. What Dave Henderson did was to have a law passed in Delaware that forbids anyone from constructing anything in the flight path of airport property. This included the planting of trees that will, in time, obstruct the safety of a landing aircraft.

Dave Henderson is no quitter. After a hard day's work, he attends college at night. Dave said, "I am realistic in that there are just so many *Cubs* out there to restore, and I have to be prepared."

■ SUSSEX COUNTY

✈ REHOBOTH AIRFIELD 1936-1984

Where Pine and Brine Meet

"As you soar over the new airfield you will see the combined beauty of the forests and ocean strand that has caused the phenomenal growth of this happy haven for vacationists and permanent residents from far and near.

Aerial view of 145-acre Rehoboth airfield circa 1943. Note there are four runways. Photo from the author's collection.

Between white breakers of the Atlantic on the east and the silver ribbon of the Lewes-Rehoboth Canal on the west, from the sail-dotted Rehoboth and Indian River Bays on the south to the pine forests and dunes stretching by broad streets and wooded drives, it welcomes the summer throngs; from its wide boardwalk to the rolling golf greens, and sand traps, and tennis courts of the Country Club, its white cottages sparkle in the sun.

From Washington, Baltimore, Wilmington, Philadelphia and other parts of this and other countries come the people to enjoy Rehoboth's welcome. Its attractions and restfulness appeal to all—from the child to the most mature—from the businessman to the diplomat. Bring your plane to a happy landing at Rehoboth, by the sea."

This description of Rehoboth was reprinted from the official program for the grand opening of the Rehoboth Airfield planned for June 3-4, 1938. The event would include the arrival of internationally known pilot Clarence Chamberlin and other dignitaries of the day. Wilmington parachute jumper Mickey Efferson, who jumped at almost every airport opening in the area, would also thrill the crowds with his free fall from 8,000 feet. Local newspapers reported that it would be the most spectacular aviation event ever held in the east. Unfortunately, mother nature intervened and torrential rain washed out the well-publicized event. Undaunted, the committee, headed by Leon "Blub" Thompson, rescheduled the grand opening for July 16 and 17. Many local people, including Frances Roch, Florence Salin, and Robert Rumbold, took advantage of an airplane ride with Clarence Chamberlin in the giant multi-engine Curtiss *Condor.*

Although the Rehoboth Airfield was opened officially in 1938, there was considerable flying activity in the area prior to that date. John Beach, Sr. and Forrest Wenyon made a living at barnstorming using the Dodd farm as their home base. The Dodd farm is now the site of the Rehoboth Elementary School. John Beach's son, Jack, would fly along with his father and help his father collect the fares when he was just a kid. Jack recalled his father's telling him, "Son, when they see you climb out of the front seat, they feel if they don't go for a ride, they will be called a sissy." John Beach would grow up to fly with the U.S. Army Air Corps during WWI. He later became the publisher of the *Delaware State News* in Dover.

Although barnstormers used almost any farm field to land, they were also known to land on the beach and offer rides. Blub Thompson, a former WWI Navy pilot, had been flying from small local farm fields since 1929. By 1931, he established the Rehoboth Beach Flying Club on the Dodd farm to encourage others to participate. The club members collectively purchased a Taylor *Cub* for members to

(Left to Right) John Beach, Sr., Forrest Wenyon, and Seaton Maynadier pose with their OX-5, Challenger. *This photo was taken on the Dodd farm in 1929. Photo courtesy of Jack Beach.*

use. The Rehoboth Beach Flying Club, and specifically Blub Thompson, were not satisfied with the accommodations on the Dodd farm. It lacked a hangar, or operations building, and was too small to accommodate larger airplanes.

Blub was in charge of the Pennsylvania Railroad Station in Rehoboth and talked flying to anyone who would listen. Thompson, who also served the Department of Commerce as Airfield Site Advisor, became very vocal in pressing for the establishment of an adequate public-use airfield to serve the area. He made speeches at the Rotary Club, Chamber of Commerce, town meetings and anywhere else anyone would lend an ear. Under the influence of Blub Thompson, the town of Rehoboth approved the idea of an airfield, which included the purchase of the 145-acre Ebe Tunnell farm for $8,800, three

FLY IN THIS GIANT ALL-METAL TRI-MOTOR AIRLINER

14 PASSENGER
1660 HORSEPOWER

AIR TOUR
$1
2 P. M. Until Dark

THE BIG SHIP WILL BE AT REHOBOTH AIRPORT
To Offer the General Public An Opportunity To Fly in One of America's Largest and Safest Planes

Arriving Aug. 16th-Will Remain a Few Days

Diamond State Garage--Texaco Distributors

Accord you this chance to fly with Penny Rogers, chief pilot, and Kenneth Starnes, co-pilot, in this $56,000 tri-motored Ford plane with transport Government-licensed pilots. Here is your chance to ride one of America's largest planes with all Pullman car luxuries—lavatory, electric lights, dressing room, running water and heat. Plane weighs 10,000 pounds loaded; wing spread 76 feet; cruising speed 125 miles per hour; top speed, 150 miles per hour. Dizziness and discomforts are unknown in this type of plane; only a sense of confidence and security occurs during each straight and smooth flight. See Rehoboth Beach from the air.

THE CHOICE OF THE OPERATORS OF THIS HUGE AIR LINER
TEXACO PRODUCTS
AVAILABLE AT
Diamond State Garage, Wm. J. Holland, Proprietor
REHOBOTH BEACH, DELAWARE

An event publicized in the Delaware Coast News *was the August 16th, 1937, arrival of a huge Ford, tri-motor airplane capable of flying 14 passengers. Photo courtesy of the George Hindall collection.*

Governor McMullen accepts the invitation to the Dedication Ceremonies of Rehoboth Airfield of Ernest Thompson, son of Leon "Blub" Thompson, who flew Ernest to Dover for the presentation April 15, 1938. Photo by Herman C. Taylor, from the George Jones collection.

miles outside of town, in 1933. The purchase was made possible by the State of Delaware's Enabling Act passed in 1931, which allowed cities to purchase land in the county for an airfield. A commitment of ninety percent federal funding from the Civil Works Administration (CWA) was also a deciding factor. All the good intentions of building the public-use airfield, however, did not bear fruit. The CWA was not forthcoming and the Airfield site was sold to Mr. George Shockley around 1935. Undaunted by the city's withdrawal, Thompson and his supporters pressed on.

The group continued to encourage the civic leaders to think positively toward aviation as part of the transportation system of the future. Their attempts were in vain, and they shifted their focus to soliciting private investments. With community support lacking, Thompson discussed Rehoboth's lack of aviation facilities with Richard "Dick" Morgan, FBO at Air Service Inc., on Bellanca Airfield in New Castle. An introduction to R.R.M. Carpenter resulted in a positive agreement to find a solution to purchase the original 145-acre site from Mr. Shockley for an airfield. Although the Carpenters did not invest in the airfield, they were the inspiration behind the plan of action. This was accomplished by selling stock in the venture and incorporating, "Rehoboth Airport, Inc." on April 4, 1936. The airfield was laid out with three runways to allow pilots to take advantage of landing into the wind. The longest of the runways was 3800 feet. Texaco aviation fuel and oil were now available on the airfield for the first time. At the same time, Blub and his followers created "Rehoboth Air Service, Inc." as a separate corporation which would be the operating business (FBO) on the airfield.

The officers and directors of Rehoboth Airport, Inc. were listed in the official program as follows;

OFFICERS

Leon L. Thompson .... president
Dr. Preston E. West ....vice president
Milton K. Fried .... secretary
Allen Vessels ....treasurer

DIRECTORS

Meyer Ableman	Hubert O. Nobel
S. P. Darden	Forrest Snyder
Walter C. Feaster	Leon L. Thompson
Milton K. Fried	Wm. Thoroughgood
Harold W. Hocker	Allen Vessels

Dr. Preston E. West

Thompson was appointed "Magistrate" by Governor McMullen in 1937, and quickly earned the title of *The Flying Magistrate.* Blub Thompson was the master of ceremonies at the Grand Opening announcing the events to the crowd. In traditional fashion, stunt flying, parachute drops, airplane racing and rides were offered to entertain the public at the grand opening. Clarence Chamberlin arrived in the largest land airplane in America, the Curtiss *Condor,* to sell rides to the public. This nine-ton airplane, with a wingspan of 96 feet, was powered by two, 650-hp Curtiss *Conqueror* engines. The *Condor* carried 27 passengers, while the engines consumed 100 gallons of fuel and two gallons of oil an hour. The fuel tanks held 400 gallons of fuel. After the Rehoboth opening, Clarence Chamberlin flew the *Condor* to the Dover Airfield and then on to Bellanca Airfield in New Castle where he spent the week selling rides.

This 1933 Curtiss Condor *18 was flown by Clarence Chamberlin at the Rehoboth Airfield Grand Opening. Photo by Herman Taylor, from the George Jones collection.*

PARACHUTE JUMP

JULY 14th

Rehoboth Beach, Del.

By the boy who laughs at death---the one and only smiling

MICKEY EFFERSON

COLLECTION WILL BE TAKEN

Passenger flights at Silver Lake by Dick Wenyon of Del-Mar-Va Flying School

In the early days, pilots would fly over towns and throw out paper flyers like this to announce their thrill shows. Flyer courtesy of Jack Beach.

An estimated 100 airplanes and pilots from other states were expected to fly in to join in the two-day festivities. The *Delaware Coast News* reported that over 5,000 people attended the event. Newly re-elected Mayor J. Arthur Downing flew over Rehoboth in the lead ship followed by a squadron of fifty aircraft to open the Sunday ceremonies.

The stockholders approved engaging Bill Kennedy as chief pilot to oversee the daily operations of the airfield. In addition, John Stevenson of Baltimore was hired as an instructor-pilot. An additional boost to the airport services included scheduled flights between College Park Airfield, Maryland,

Mickey Efferson spills the air from his parachute after another successful performance. Photo courtesy of Jack Beach.

This photo of the Rehoboth Airfield was taken during the CAP days of occupation. The airplane is a Waco Taperwing. *Note the Army vehicle to the left. Photo courtesy of Jean Cochran.*

and Rehoboth, on Saturdays, Sundays, and Mondays during the summer season.

The buildings on the airfield included a clubhouse and two hangars for the club airplanes. Aviation fuel and oil were available for purchase. The airport prospered, especially during the summer months, when pilots would use the field to fly people to and from the ocean resort. Sightseeing rides were offered as well as flight instruction, which helped bring in revenue. When war was declared in December of 1941, the Rehoboth Flying Service ceased all flying activity and the airfield was taken over by the Civil Air Patrol (CAP).

Within three short years, WWII changed the airfield's activities from a place for families to fly to for a vacation by the sea, to a CAP military operation run by the Army Air Corps. Rehoboth Airfield would be the second operational CAP base in the United States. (Atlantic City was first, although Rehoboth made the first operational flight.) The beautiful white sandy beaches would now become covered with the oil from tankers sunk by German submarines. It would become the duty of the CAP to search out the subs and report their positions so military action could be taken. The war years were dismal years for Rehoboth by the sea.

An unusual flying incident occurred on the airfield at the beginning of WWII. An Army Air Corps four-engine B-17 bomber got lost while on a training mission out of Langley Army Air Base, Virginia. Low on fuel with daylight coming to an end, the pilot recognized Rehoboth Airfield. Fortunately, the ground was frozen when the pilot decided to land the 49,650-lb bomber on the grass runway. A successful landing was made without incident. Unfortunately, the B-17 required 100-octane fuel for the engines and none was available. So there the plane sat. To protect the airplane from curiosity seekers, the Army posted armed guards around the clock. The wheels of government turned slowly while the weather began to get warmer and the big tires settled deep into the soft earth. By spring, a lone twin-engine Army C-47 flew in with a crew and enough fuel to fly back to Langley. With full power from the B-17's four engines (and some help from an old Fordson tractor) the huge airplane began to move out of the ruts. After a short taxi to the end of the runway the engines were tested for performance before takeoff. Surprisingly, the giant bomber was airborne before it used up half the runway. The episode was probably the most exciting event the town of Rehoboth recorded during the war. At the time, the B-17 was the largest airplane in the Army Air Corps inventory.

The buildings were spruced up with a new coat of paint and a sign painted after Aircrafters. Inc. took possession in 1945. Photo from Jean Cochran.

As the war scaled down, Blub Thompson moved on to accept a job with the Department of Commerce, Aviation Branch. The war was not quite over when on June 25, 1945, the investors sold Rehoboth Air Service, Inc. to a newly formed corporation named Aircrafters, Inc. The stockholders authorized 500 shares of stock and announced the new board as follows;

Clarence Flatt, Rehoboth, president

James R. Young, Rehoboth, vice president

William M. Wright, Silver Spring, Md., secretary-treasurer

The original wooden buildings left behind by the CAP and Rehoboth Air Service were the only infrastructures on the 145-acre airfield. Clarence Flatt would serve as FBO and Ed Chamberlin was hired as the chief pilot. Hal Wallin was hired the following year as flight instructor to fly sightseeing tours and tow banners.

Clarence Flatt's wife, Carolyn, wrote a front page newspaper column in *The Public Press* (formerly *The Delaware Coast News*), every week about the airfield activities. Written under the title, "Airporter," the articles portrayed the business and pleasure side of flying on Rehoboth Airfield. New student pilots were recognized, as well as those who obtained their pilots license. Group fly-outs and fly-ins were also well covered. The *Baltimore Sun* newspaper was reported to be flown into Rehoboth and several other airfields on the Peninsula daily in 1946. In one publicized "Fly to Rehoboth" event, over 400 airplanes

AOPA selected Rehoboth for their 1947, 1948, 1949, convention site. Note the makeshift control tower attached to the windsock pole. Photo from the Hagley Museum & Library.

This aerial view shows the huge turnout for the AOPA convention. Photo by Del Ankers.

flew in for a picnic. Carolyn, a student pilot herself, also wrote about her experiences in learning to fly. When the night runway lighting was installed on June 6, 1946, Carolyn flew the first flight with her husband. She wrote, "Flying at night is just like Christmas in every town, it is so beautiful!" Needless to say, Carolyn Flatt's public relations effort was a major influence on the positive attitude the public perceived about flying and Aircrafters, Inc.

Privately owned, public-use airfields can rarely be classified as moneymakers. This is especially true when there are multiple investors who are looking for a return on their investments. Rehoboth Airfield, although well managed, needed funds desperately to improve the infrastructure. The buildings were holdovers from WWII and left much to be desired in terms of modern conveniences. Hangars, with sagging doors and cracks in the siding, were less than ideal to service airplanes. New funds were necessary when it was announced the Aircraft Owners and Pilots Association (AOPA) had selected Rehoboth for their annual 1947 convention. It was estimated that 1000 airplanes would fly in for the two-day event.

Rehoboth Airfield gained national recognition and made aviation history September 1947, when AOPA held their first Round-Up convention at Rehoboth Beach. Max Karant, editor of *Flying* magazine and William T. Piper, president of Piper Aircraft, were among the many who attended the event. Prosperity in aviation was at its peak, and over 500 privately owned airplanes from all parts of the continent would be flying in for the meeting. Farmers, who grew crops around the unused portions of the airfield, were asked to harvest them to make room for parking. A makeshift control tower, which looked much like a ship's crow's-nest, held the air traffic controller and spotter. Together they guided airplanes in for landing by flashing a red or green handheld light.

Carolyn Flatt's wonderful articles on the happenings at Rehoboth came to an abrupt end in early 1949, coinciding with Clarence Flatt's leaving the board. Eleanor Wallin remembered that he was an accomplished artist and thought he resigned to pursue that career. In any event, a shift occurred in early 1949 when several investors sold out and a new board was elected. The new board authorized 8000 shares of

This specially designed Cushman motor scooter was used to transport pilots and passengers to and from remote parking areas. Photo courtesy of Les Ozier.

stock to be sold to upgrade the buildings. The board as filed with the Secretary of State was now defined as follows;

Edmund Chamberlin, Lewes, president

Tom Johnson, Lewes, vice president

William Wright, Silver Spring, secretary/treasurer

Lawrence Boaman, Rehoboth; Tom Johnson, Lewes; Edmund Chamberlin, Lewes; Carmen Naples, William Wright, and Emil Paroulek, of Silver Spring, Maryland, were listed as directors.

Edmund "Ed" Chamberlin was hired as FBO to replace Flatt and also serve as president of Aircrafters. Danny Cochran and Hal Wallin would be flight instructors, pilots, and mechanics. Of the 8000 shares authorized by the board, 5342 were issued. This brought in the necessary funds to improve the facilities for the upcoming AOPA convention in 1949.

After three successful events, the AOPA membership felt the national organization should hold the event at various locations throughout the country. It now became their policy to accept offers to hold their *Round-Up* convention from other communities.

In a surprise move on June 6, 1947, the CAA announced all small scheduled air carriers cease flights *immediately.* This action included passenger *and* cargo carriers. The CAA decided these scheduled carriers must apply for an airline certificate to be legal. Once filed, it could take the CAA a year or more to approve the request. Imagine, if you will, the government announcing all trucks to cease using the highways unless they held a commercial permit. The CAA action put many small operators out of business overnight. Rehoboth's scheduled weekend flights from Washington ceased immediately.

It had taken almost two years for the CAA to approve Rehoboth as one of their stops for passenger, freight, and airmail pickup. A seemingly minor infraction were the telephone and electric wires at the end of the west runway. At a public hearing, the utility companies decided they would bear one-half the cost, estimated at $2,000. The Rehoboth town council felt it was not its responsibility since the airfield was not on city property. The County also refused to accept responsibility. The lines were not on airport property, therefore the airport felt it was not responsible. The State of Delaware Aeronautics Commission claimed they did not have any funds, in spite of collecting an aviation fuel tax. Ironically, this went on for over a year before the issue was resolved by the Lewes Chamber of Commerce, Reho-

Edmund Chamberlin FBO for Aircrafters Photo courtesy of Eleanor Wallin

both Chamber of Commerce, the Lions Clubs, the Kiwanis Clubs, and the Rehoboth Art League.

Politicians came out in full force for the inaugural flight held on June 5, 1949. The list of politicos was indeed impressive: Governor Carvel; the mayors of Lewes and Rehoboth, Shaffer and Maull; Senators Williams and Frear; Representative Boggs; and Hugh Sharp, Delaware Aviation Commission. They cut a yellow ribbon strung across the entrance to All American's gleaming Douglas DC-3, followed by speeches of how wonderful this service was going to be for the community. (Where they were when the power lines needed to be removed is an unsolved mystery.)

All American's route would only be operational during the summer months. It claimed that an astounding 11,231 persons had flown the first season. In spite of these figures, All American ceased operations in 1951, although the airfield remained open.

The year 1951 was a tragic and turbulent one for Aircrafters and Rehoboth Airfield. Ed Chamberlin was flying a charter to Washington, D.C., when he apparently ran into some unknown trouble and crashed into the Chesapeake Bay. The two passengers escaped through the door before the water-laden *Stinson* sank. The pressure of the water held the door firmly shut and Ed could not escape.

Danny Cochran was elected president of Aircrafters, Inc. to replace Edmund Chamberlin. During the same year, vice president Tom Johnson bought controlling interest in Aircrafters by buying out several stockholders. Tom Johnson was the president of Tohickon Hosiery Mills, in Lewes, and also served as a State Senator. Tom had learned to fly at Rehoboth after the war. Danny Cochran was involved with airplanes all his life. As a kid he built models by the score, some of which are still hanging from the ceiling of their home in Rehoboth. When he was old enough to fly he took flying lessons and aviation mechanics: his wife, Jean, recalled that he always had an inner desire to work on ways to make something better.

Danny Cochran was moved up to replace Ed Chamberlin as FBO for Aircrafters. Photo courtesy of Eleanor Wallin.

David Kenton was named chief flight instructor at Aircrafters after Hal Wallin assumed the position of president and FBO. Photo from Eleanor Wallin.

Effective August 1954, Danny Cochran resigned as Aircrafters' president and FBO to accept a job with All American Engineering at Georgetown, although he was still listed as a director as late as 1962. Hal Wallin was now named Aircrafters president and FBO. David Kenton was hired to take Hal Wallin's place as chief flight instructor. Hal and his wife, Eleanor, would become a legend after serving Rehoboth Beach Airport and Aircrafters for almost three decades. Although the husband and wife team only owned one share of stock in Aircrafters, Inc., one could easily assume they were the owners by their commitment to customer service.

Like most pilots, Hal Wallin had also constructed model airplanes as a kid. The son of an interior decorator in Philadelphia, Hal entered his models in flying contests inside the Gimbel Building on the thirteenth floor. In 1930, he won a free ride on Ludington Airways' *Stinson* Tri-motor to New York and back. That flight hooked Hal on flying for life.

A Navy pilot's lack of shooting skills ventilated Hal's tow plane causing him to resign from the CAP. Photo courtesy of Eleanor Wallin.

Hal soloed in 1933, and received his private pilot rating from the Civil Aeronautics Board (CAB), in 1937. He added a commercial rating in 1940, and an instructors rating in 1941.

When the war broke out Hal moved to Riverhead, Long Island, New York, to fly submarine patrol for the CAP. He started a game of hopscotch with the CAP by transferring to Maryland, where he flew over the White House where the Army could practice trying to catch him in their searchlights. Then, he transferred to the Connecticut CAP, where he towed aerial gunnery targets for the Navy. When a Navy pilot shot his airplane full of holes, Hal felt this duty was a bit more than he bargained for and immediately resigned from the CAP.

Hal then applied for a job at the Philadelphia Navy Yard, where he was taught welding. By April of 1945, the Draft Board caught up with Hal and he was inducted into the U.S. Army Air Corps. He was sent to Japan to serve with the occupational forces in December of 1945. (Hal returned home in September of 1946.) He decided to try Rehoboth Beach for a couple of months' vacation with his wife Eleanor and daughter Norma before making any long-term commitment. A look at the local airfield landed him

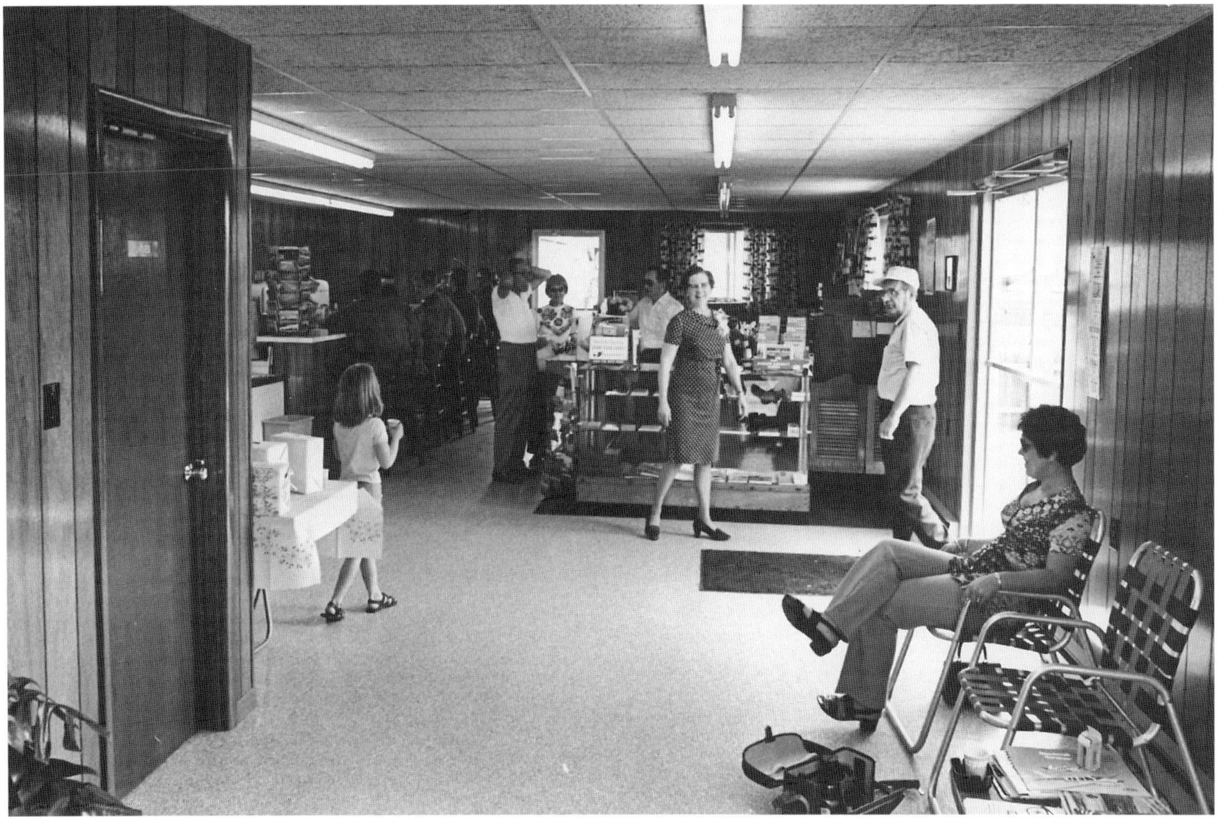

The interior of the new Operations Building was modern and attractive. Clean rest rooms, luncheon area, lounge and business offices were available to visitors. Photo courtesy of Eleanor Wallin.

a job as flight instructor for Aircrafters, Inc., and he decided to stay.

About 1964, Tom Johnson sold his controlling interest in Aircrafters. Richard "Kip" du Pont, along with two others were now the majority stockholders. This shift created the most drastic shift in the board of directors history. Hal Wallin retained the presidency, Eleanor Wallin became vice president, and Kip du Pont became secretary-treasurer. These same three, along with Joseph Wollin and C.R. Naples, served as directors.

On the weekend of May 9-10, 1970, a completely new 24' x 60' Operations Building was dedicated. The new all metal, air-conditioned building included a pilots' lounge, rest rooms and restaurant. Pilots from all over the east flew in for the dedication ceremonies which were opened by Mayor Lester F. Johnson. Atlantic Aviation and Summit Aviation displayed new Beechcraft and Cessna airplanes for the event.

In a newspaper interview by Nancy Lynch for the *Evening Journal* in 1978, Hal Wallin recalled the many notables and celebrities who used Rehoboth Airfield. Roger Mudd, reporter; John Sirica, Judge, District Court; Arthur Godfrey, entertainer; Harry Reasoner, reporter; Marvin Kalb, reporter; Maurice Udall, United Nations; Generals Spatz and Vandenburg and the governors of our neighboring states all passed through their portals. Other notables and many Washington D.C. government officials and their guests flew to Rehoboth to enjoy the Ocean. The number grew to the point that Rehoboth earned the nickname, "The Nation's Summer Capital."

Hal and Eleanor Wallin grew weary after thirty-one years of service and decided to give proper notice to retire. In the summer of 1977, they set a date as the next annual board meeting, which would be held April 1978. Unfortunately, Hal suffered a massive heart attack in November 1977. By January of 1978, Hal had passed away. With dedicated FBO people like the Wallins at the helm, the airfield ran the way an airfield should. They would be impossible to replace. They had become a legend in airfield management. The Wallins treated the airfield and its people as if it were their very own.

The board replaced Hal Wallin in January 1978, with George Garrison, who had leased the main hangar to service airplanes under the name "Garrison Aviation" since 1970. George Hindall, who held a long list of aviation credentials, was hired as a mechanic/pilot. George Garrison, by his own admission, had a monumental task to accomplish trying to fill the vacancy the Wallins' created. One thing is certain; Garrison was dedicated to aviation and loved Rehoboth Airfield. Unfortunately, the airfield needed a seasoned manager with a business/marketing background to make the parent company Aircrafters profitable.

By 1981, there were twenty-five investors in Aircrafters. Apparently none were happy with the operation because of the mounting deficit. The aviation industry entered a major recession due to product liability claims. Piper Aircraft went bankrupt and Cessna Aircraft ceased building single-engine aircraft. This decline, coupled with poor FBO management, created red ink on Aircrafters' financial statements. The board now consisted of Albert Morris (Greenville, Delaware), president, Eleanor Wallin, vice president, and Kip du Pont, secretary-treasurer. These three and David Clark, Monmouth, New Jersey, were the directors. Aircrafters, Inc. appeared to be in serious financial trouble.

Albert Morris, in June 1983, contacted Appraisal-Consultants, Inc. of Wilmington, with a request to provide an appraisal for the now 35.44-acre site. This acreage is confirmed by the Deed Record no. 356-477. Apparently, former owners of the airfield sold off much of the original 145-acreage over the years to raise cash to improve the facility and pay off debts. The longest runway was listed originally as 3,800 feet. The 1983 aviation directory listed it as 2,700 feet long. The appraisal for the property was submitted to the board on August 29, 1983. John Fortner and Joseph N. Nelson, certified appraisers, submitted along with the $530,000 appraisal, a detailed report outlining possible use for the property in the event the board decided to sell. Due to the expected growth in the area, Fortner and Nelson suggested the property be rezoned for a housing development.

The increased flying activity had ruined the grass runways to the point that a hard surface would be required to accommodate the heavier and faster airplanes. Pilots revved their engines and threw up clouds of dust which drifted off with the wind. Apparently, the board was reluctant, or incapable, of providing funds necessary to make the major improvements.

Historically, privately owned, public-use airfields are heavy on infrastructure requirements and short on gross income and profits. There are no landing fees and a token parking fee if you leave the airport. People fly to airports to spend their money in town. It is for these reasons that the Federal Aviation Agency (FAA) has imposed a tax on aviation fuel so the money can be used to improve non-revenue-producing areas of airport infrastructure. These include hard-surface runways, taxiway and parking ramps, but this money was available only to city, county, or state-owned facilities. The stockholders knew they had reached a point where vast amounts of private money would have to be raised to pave the runways, taxiway, and ramps. They would also have to seek out professional management, with marketing expertise to replace Garrison.

The board of directors decided to sell Rehoboth Airfield to George Garrison and Burke Flickinger, a Rehoboth land developer June 1, 1984. The funding for the purchase was made possible by Wilmington Savings and Loan Society; (WSFS) accepting a promissory note repayable in two years. The name of the FBO operation was now changed from "Aircrafters" to, "Garrison Aviation, Inc."

Garrison seemed to have been unable to meet his share of the financial obligation and Flickinger threatened to sell off the property. This is borne out by a Flickinger proposal to the zoning commission to rezone the property as residential. Garrison showed up for the hearing and protested the change. Since he was a legal partner in the property, the zoning commission tabled the request until the owners could come to an agreement.

Garrison, in a valiant effort to raise his share of the obligation, contacted Robert C. Wolhar, esq., Georgetown, to draw up a limited partnership agreement. This agreement, dated November 2, 1984, would offer 173 shares of Garrison's 340 shares at $5,000 per share and would thus raise enough cash to buy out Flickinger and start anew with some reserves. Much to Garrison's dismay, the shares did not sell. On April 15, 1985, a revised Limited Partnership was drawn up which offered 160 shares at $5,000 per share. Other enticements were offered to Limited Partners which included free hangar space as long as the purchaser of five shares owned an aircraft. The hangar could not be sublet or used for other purposes. Limited partners with four shares would be granted use of the land to build their own hangar. In the event a hangar was made available for sale, the Limited Partnership would have the first option to buy. Three shares would allow a free tie-down space for the owner year-round. Two shares would allow seasonal tie-down and reserved parking for one automobile. One share would give the owner the satisfaction of contributing to saving Rehoboth Airfield from developers. Any profits realized from the operation would be distributed on a per share basis.

This offering, obviously targeted at airplane owners, also failed to gain the anticipated support it was designed to produce. Time was running out. WSFS was threatening to call in the loan for delinquency in payments. In a final effort to save the airfield, an organization was formed named, "Friends of Rehoboth's Airport" (FRA). Supporters included Garrison, George Hindall, Charles Lloyd, John McTighe, John Scott, Waddy Owen, Donald Derrickson, Bill du Pont, reverend John Dean and Patty Cook. Arrangements were made by the FRA, headed by John McTighe, to investigate the possibility of the Rehoboth municipality purchase of the airfield through the sale of municipal bonds. McTighe and Lloyd served on the Rehoboth city council and found support with mayor John Hughes.

A community officials' dinner was planned for August 28, 1985, to bring all the political parties together to discuss the situation at hand. State Representative Jay Windgate, Rehoboth Commissioners Sam Cooper, Walter Brittingham, Walter Lehman, Billie Shields and Jim Horty were among the invited guests. In addition to these elected officials, Lewes Mayor Alfred Stango, and Councilmen George Smith, John Rhodes, Richard Gott and John Vessels, Jr. attended. Completing the list of invited officials were the Rehoboth Planning Commission Chairperson Mary Burton, secretary Edna Hottell, and the president of the Rehoboth Chamber of Commerce Dennard Quillin.

After the dinner, a short film was shown, titled "Airports mean Business." This was followed by a talk by the retired Director of the Maryland Aviation Administration, and Tildon Montant, manager of the Ocean City Maryland Airport. The mood of this meeting was overwhelmingly positive. There was, for the first time in many months, a feeling the Rehoboth Airfield would be saved. Press releases were written in an effort to gain community support. A planned 50th-year celebration of Rehoboth Airfield was announced for 1986.

In the end it was the FAA that struck the final blow. The FAA, after a lengthy survey of the facility, deemed the airfield too small to invest taxpayer dollars. Multi-engine aircraft could not land on the short 2,700-foot runway safely, especially when wet. Housing had also been built in what the FAA determined mandatory clear zones for approaches. This decision cooled all the political support. In desperation, the FRA suggested looking into a new site. The costs of site approval, public hearings, environmental approval, rezoning, FAA approvals, etc., were greeted with disapproval.

Burke Flickinger once again approached the county to rezone the property for residential development. This time he was unopposed.

A letter was circulated by George Garrison in June 1986, announcing, "the Rehoboth Airfield would be permanently closed on September 30, 1986." The three-month lead time is an FAA requirement to advise the flying public of the closure and to make the appropriate change on aeronautical charts. Leon "Blub" Thompson's *Field of Dreams* came to an end after fifty years of continuous operation.

✈ Bridgeville Airfield 1939-1958

Jim Scott was hooked on flying before he ever had the opportunity to leave the ground. Building and flying rubberband-powered models was his first introduction to aeronautics. Starting around 1928, he would spend hours cutting and gluing sticks of balsa

Jim Scott and his rubberband-powered biplane about 1928. Photo from Jim Scott.

wood to form the framework, then attach the fabric and dope it taut. To achieve longer flights, he would launch the models from the attic window of the house. By 1937, Jim joined the Bridgeville Flying Club and took flying lessons from Charlie Ricards at Tred Avon Airfield near Easton, Maryland. He soloed in a short three hours and twenty-four minutes of instruction in an Aeronca C. The Aeronca C was powered by a 36-hp Aeronca engine. To save weight, the designer did not install doors. The sides were scalloped out to allow entry. Cruising speed was about 50 mph and it landed at about 30 mph. A quantum leap in creature comfort was when he moved up to the 1937, Aeronca K which had two doors and an improved 40-hp engine.

The Bridgeville population in the 1930s was probably about 500 people. The land was flat, and farming was the main source of income for the area. The well-kept homes were mostly of wooden construction with elegant front porches and well-shrubbed yards. It was a typical farm town with a main street and railroad running directly through town. Cannon Foods, a local cannery, provided employment stability to many of the non-farming residents. The

Jim Scott's Aeronca K, with its two-cylinder, air-cooled engine developed a mere 40 hp. Photo from James Scott, II.

cannery was located next to the railroad to allow for ease of shipping to markets throughout the country. Main street (Rt. 404) consisted of a gas station, general store, barber shop, clothing store, fire house, and a grocery store which made up most of the business district. Passing through in 1998, an observer would marvel at how little anything has changed.

For James P. Scott, Bridgeville was home. The family history is one of merchants and businessmen. In the early years, the Scotts owned the town general store, a grocery store and later an upscale clothing store. Flying was for fun.

After learning to fly at Easton, James Scott purchased an Aeronca K in 1938 and brought it to a small farm east of town. Here he rented just enough farmland for a runway and built a hangar for his airplane. The airfield, located on what is now the northeast corner of Route 13 and 404, was created as his private *Field of Dreams* and he had no thoughts of further involvement. Later on, when the war broke out, his flying came to a halt for the duration. The propeller was removed from the airplane and the hangar was locked. On occasion, when time and weather permitted, he would join other pilots in driving to Pennsylvania, where there were no restrictions, and rent an airplane to fly.

After the war, Jim and his friend, Harvey B. Spicer, decided to open a flying school to take advantage of the GI Bill and teach others how to fly. On February

James Scott and his Fairchild 24W around 1948. Photo from James Scott.

A 1946 Piper, Cub *and a 1946 Taylorcraft were used for flight instruction and rentals. Photos from James Scott.*

The now nationally famous trial attorney, F. Lee Bailey, sold this 1949 Globe Swift *for two Botany 500 suits and $2,000. Photo from James Scott.*

27, 1946, they named their Business *Sussex Aero Industries, Inc.* Starting off with a couple of Piper *Cubs,* other airplanes were added as needed. The duo expanded the airport to include four hangars to store seven airplanes. At the peak of activity, four flight instructors were teaching on the airfield. In addition to five Piper *Cubs,* there was a war surplus Fairchild PT-19 *Cornell,* a Vultee BT-13, a Taylorcraft and a Stinson 108. These were a few of the airplanes that Jim remembered on the airfield. Jim Scott was also appointed to the first Delaware Aeronautics Commission in 1946. "It was sort of an honorary thing, we never had any money to do anything," Jim recalled.

Jim Scott's first student was his son, James, II, who soloed at the age of thirteen. His son would go on to make flying his career in the armed forces and retire with the rank of Colonel. After retirement, he worked for Flight Safety at New Castle County Airport.

Of all the airplanes Jim owned (too many to recall), one stood out in his memory. He purchased a Globe *Swift* from F. Lee Bailey for $2,000 and two Botany

The last remaining hangar, with its metal roof, is the sole remaining evidence of the Bridgeville Airfield and Jim Scott's Field of Dreams. *1998 photo by the author.*

500 suits. There would be no time for Lee to fly while in law school, and the money and suits could be put to better use. Jim recalled Lee saying, after he tried on the suits, "I am going to be the best dressed law student in college." (A suit and tie were required to enter college during this era.) Jim Scott, on the other hand, would have one of the most beautiful (and fastest) airplanes on Delmarva.

Aviation manufacturers had the vision of a private airplane in almost every family. The airplane technology had blossomed from the war and airplanes were no longer considered accidents waiting for a place to happen. The GI Bill ran out for veterans, and flying lessons would now have to be paid for by the student. Jim recalled, "Flying stopped after the government program ran out like you would shut off a running spigot." Jim bought out his partner, Harvey B. Spicer, for $2,500 and tried to keep the airfield open. Slowly, he began to sell off all the idle airplanes until they were all gone, except his beloved *Swift.* Officially the corporation came to an end on March 18, 1958.

The airfield closing did not dampen Jim's flying. He continued to fly for enjoyment until he was forced to stop for medical reasons at age 65 after flying more than 4,000 hours. During his lifetime he acquired a student license, private pilot, commercial pilot, instructor, multi-engine and seaplane ratings. When he wasn't flying, he was selling upscale clothing from his Bridgeville store that he owned for twenty-nine years.

✈ Laurel Airfield 1946

Randall and Irene Dickerson with their two daughters, Donna and Diane (standing). Photo courtesy of Diane Dickerson Boyce.

Howard H. Dickerson and his wife Mary, owned what is known as the *Blue Hen Farm* for as long as relatives can remember. The farm, located east of Laurel, lay in the heart of Delmarva's rich farmlands.

The Dickerson's raised two children, Gladys and Randall. Randall worked on the farm and went to local schools where he met and married his childhood sweetheart, Irene. After high school, Randall entered Lehigh University in Pennsylvania to study chemical engineering. During college, Randall became deeply involved in experimenting in ways of extracting oil from soybeans. Family members remember he was quite successful in his research, which he tested on the family farm. Upon graduation from college in 1945, Randall and his family decided to return to Laurel and start a crop-dusting business along with his soybean research from the family farm.

Irene and Randall Dickerson had the ideal place to start an aerial agricultural-spraying business. The farmlands were flat as a board and the soil was sandy, which would absorb rainwater and grow healthy crops. There was also a large farm house on the property which they could use for their residence. Randall's father and mother agreed to transfer 52.8 acres of the homestead to Randall to start his business. None of the family members could recall where Randall learned to fly. In six short months the family was struck with tragedy, when Randall was killed flying his agri-plane.

John Reed, FBO Laurel Airfield

Irene rented the entire airfield to John Reed, who was a U.S. Army Air Corps veteran from Pennsylvania. Like many others, the war changed John's life forever. No one could remember who or where John got started in flying. John was a flight instructor in Florida before flying supplies over the *Hump* to aid China in their conflict against Japan during WWII.

Before going off to war, John had attended Pennsylvania State University, majoring in law. He was also on the varsity track team and was considered for the Olympics. Somehow, his interest in law waned, and he transferred to the Wharton School of Business where he completed his college education. After the war, he was offered a job flying for Trans World Airlines (TWA). However, flying passengers for an airline was not the kind of flying John had in mind and he turned the offer down.

Laurel Airfield operations office, circa 1946. Photo courtesy of Jim Vreeland.

When John heard that the airfield at Laurel was available for lease, he promptly contacted Irene Dickerson to look over the property. Although the terms of the lease are not known, John Reed leased the airfield for his entire life. In the lease agreement, John and his wife, Marseille, also occupied the farm house which was conveniently located on the airfield.

John Reed had found his *Field of Dreams* and taught many people from the area how to fly. John chose the traditional Piper *Cub* for instruction and student rental because of its low operating costs, low mainte-

This later photo of the airfield was probably taken in 1947-48. Photo from the Delaware Public Archives.

nance and extraordinary flying characteristics. To offer more advanced flying beyond the Piper *Cub,* John purchased a military surplus North American AT6 *Texan.* John also used the *Cubs* for aerial power-line inspection flights for Delmarva Power & Light Company, banner towing, sightseeing, and aerial photography. In another move to increase income, John passed the FAA qualification test to become an instrument instructor and a flight examiner. The latter allowed John to approve the student pilots, who were trained by other flight instructors, to obtain their pilots license, or move up in their ratings. Many pilot log books (including the author's) have John Reed's signature of approval entered.

Jim Vreeland, a local pilot who worked for John early in his flying career recalled;

> I was as close to John Reed as anybody. John Reed became a legend in aviation to everyone who knew him. He was a quiet, soft-spoken man, whose size and maturity commanded authority and respect. If a stranger landed on the airfield, he always went outside to greet him personally and ask him to come inside for a free cup of coffee. It did not matter if he was going to buy anything or just wanted to visit.

This 1963 aerial photo shows the farm house nestled in the trees and expanded hangars. Photo courtesy of Jim Vreeland.

Although he quit instructing in his later years, John was still actively flying his Piper *Cub* inspecting power lines before he died in 1980.

After John Reed passed away, the Dickerson family was faced with the choice of closing the airfield or leasing to another FBO. Fortunately for Delaware, they chose to keep the airfield alive. John Reed, Jr. assumed the responsibility of running the airport for a year after his father's death. Although John had taught his son to fly, John Jr. did not seek pilot ratings beyond his private pilot license. His interest in flying remained simply that of flying for pleasure. Several years after his father's death, John Jr. searched through his father's records copying the registration numbers of Piper *Cubs* his dad owned. With this information he located the owners and asked if they were willing to sell. John Jr. found the yellow *Cub* he wanted and completely restored it to new condition. He proudly flies his family *Cub* on nice days to recapture the times he flew with his father.

After about a year, John Jr. turned the airport management over to Mr. Cannon, a businessman from Bridgeville. Mr. Cannon, although committed to his own business (Cannon Foods), added some additional hangars and kept the airfield open. Business pressures caused Cannon to decide to turn the operation to someone else.

Ed Young

The choice was Edward (Ed) Young, who oversees the operation today. Ed Young is well-suited for the job. He is a quiet, sincere person who has spent his life in the world of small privately owned airplanes. Ed's other responsibility, beyond overseeing the airfield, is one that he has had since 1970. Ed has his own separate business of servicing and licensing airplanes on the Laurel Airfield. With approximately forty resident airplanes on the airfield and many customers from outlying areas, idle time is not abundant. Although the airfield structures do not portray the image of the DuPont Airport, it does have a loyal group of flying supporters, which includes a skydiving group.

John and Wanda Griffith operate a successful flight school from the airfield. They are very marketing-oriented and constantly seek ways to introduce flying to the public. Their roster presently lists forty-one students who are learning to fly their Cessna 152 and 172. The four-seat, Cessna 172 is available for rent and advanced training.

John Reed's wife, Marcella, still resides in the old farm house with as many cats as there are airplanes on the airfield. Marcella was a delight to interview and, although not a flyer, thoroughly understood her husband's commitment to his profession. She recalled that they were "set up" by friends for a dinner date in Rehoboth in 1958. A year later they were married and moved into the rented farm house on the airfield. About five years before John passed away from cancer, they bought the house. It is now hers and is no longer a part of the airfield property.

Diane Boyce, Irene Dickerson's daughter, has taken the responsibility of overseeing the family-owned airfield. Her mother is failing and being cared for in a nursing home. Diane lives near Dover with her husband Douglas. Although Diane and the family are not aviation-minded, they continue to hold a deep feeling toward the airfield's people and heritage. As far as she is concerned, her parents *Field of Dreams* will live on in memory. That decision sits well with the people who love Laurel Airfield.

✈ Private Airfields

In the early days of flying, landing on someone's private-landing airfield could cause an hour's delay from the owner while you exchanged flying tales and shared a cup of coffee. Today, it is a different matter. Most private airfields are marked, "Landing by prior permission only." It is highly recommended that pilots obtain permission to land or face the consequences. The owner has the right to impound your airplane for trespassing. This list was submitted by Harry Van Den Heuvel, Delaware Aeronautics Office, April 1998, and may be incomplete. No further attempt was made to research additional privately owned airfields in Delaware.

New Castle County

Evergreen Acres, R.D. #3, Middletown, William A. Wollerton, Jr.
Shallcross Airfield, Rt. 301, Middletown, Fred M. Shallcross
Reed Airfield, R.D. #2, Middletown, David H. Reed
Red Lion Airfield, Bear, Elmer Segers
*Duffy Airfield, Townsend, Mr. Duffy
Smitty's Airfield, Newark – (near Christiana Hospital and Delaware Park)

Kent County

Ewing Airfield, N. Garrison's Golf, Smyrna, Charles M. Ewing
Lasch Airfield, Behind Elementary School, Hartley, Erich Lasch
Lasch Airfield, Hartley, Harry C. Lasch
*Johnson Airfield, Magnolia, Al Johnson (aerial applicators)
Sheridan Airfield, Carter Road, Smyrna, John J. Sheridan
*Mervine Airfield, Greenwood, John M. Mervine
Flying Jem Airfield, Wyoming, James H. Durrill
*Evans Airfield, Cheswold, Edward F. Evans
Brown Airfield, Felton, Herman C. Brown, Esq.
Hobbs Airfield, Harrington, George E. Hobbs
*Holden Airfield, Little Creek, Wayne Holden
Bombay Hook Airfield, U.S. Govt., Bombay Hook Refuge

Sussex County

Drummond Airfield, R.D. #1, Milford, William D. Drummond
Pepper Airfield, Route #2, Laurel, Horace Pepper
Whaley Airfield, Route #2, Laurel, George R. Whaley
*Hudson Airfield, R.D. #1, Milton, Joseph R. Hudson
Parker Airfield, Lincoln-Waples Pond, Lincoln, Edwin E. Parker
DuPont Airfield, Seaford, Emile du Pont
*Mervine Airfield, Greenwood (Now Corman, Aerial Applicators)
*Joseph Airport, Georgetown, Melvin L. Joseph
Thomas Airfield, Seaford, John Thomas
*Townsend Airfield Dagsboro, Delaware, Sherman Townsend
Lyons Airfield, Greenwood, John R. Lyons
Sugar Hill Farms Airfield, Greenwood, John R. Lyons
White House Farm Airfield, Millsboro, Sam H. Showell, Sr.
Sussex Airfield, Frankford, D.W. Long
De Vries Airfield, Lewes, Grace Brittingham
*Del-Ag-Aero, Rt. 78, Reliance, Barry Pezey (Aerial Applicators)
Delaware National Guard, Bethany

**Still in use*

Aeronautical Dictionary

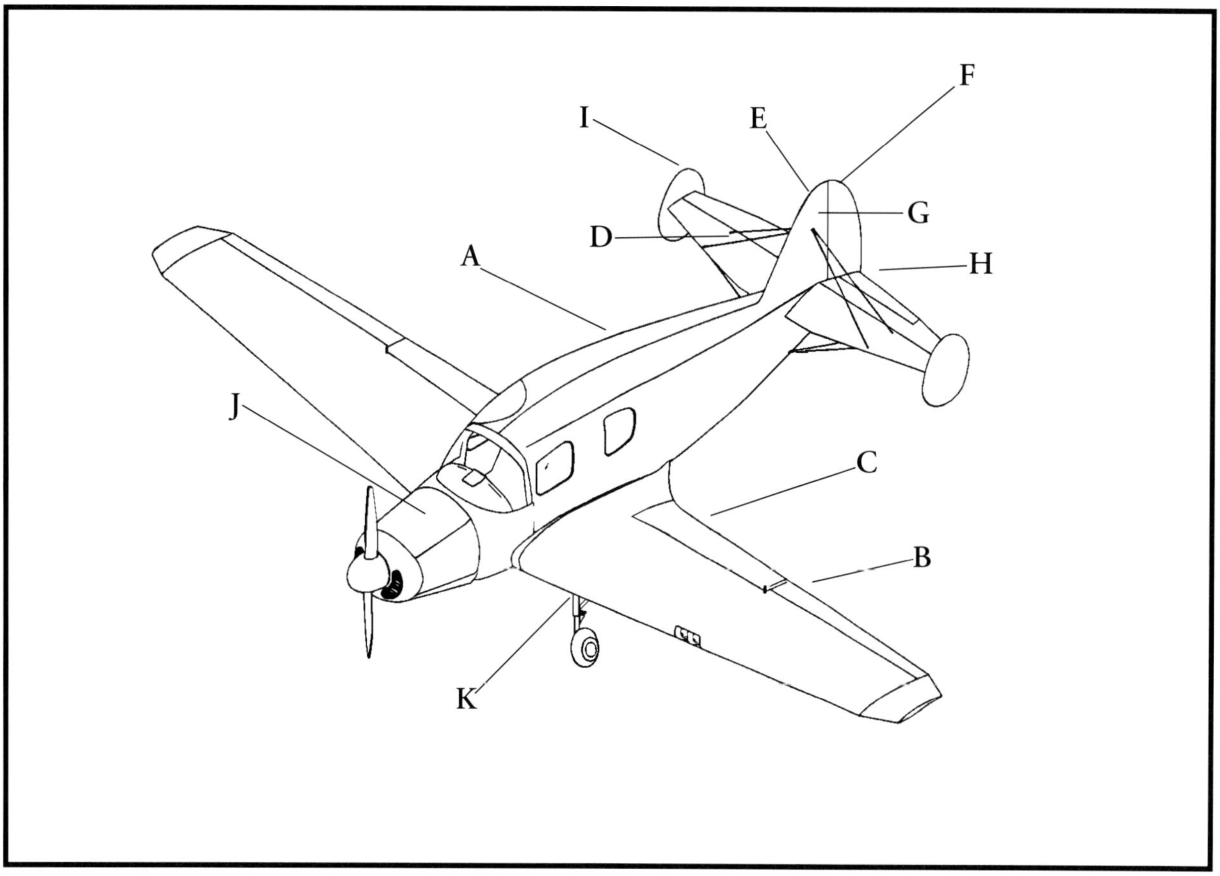

A	-	Fuselage	G	-	Bracing Wires
B	-	Aileron	H	-	Elevator
C	-	Wing Flap	I	-	Stabilizer Fins (Bellanca only)
D	-	Horizontal Stabilizer	J	-	Air-cooled Engine
E	-	Vertical Stabilizer	K	-	Retractable Landing Gear
F	-	Rudder			

ace • A combat pilot with five or more victories.

aileron • A hinged section of a trailing-edge area of a wing, controllable by the pilot for the purpose of raising or lowering the wing during flight.

airfoil • A surface shaped for movement through the air in order to obtain a desired reaction. (Wing, rudder, stabilizer, etc.)

air-speed indicator • An instrument for showing the relative speed of an aircraft's movement through the air.

altimeter • An instrument, similar to an aneroid barometer, graduated to indicate the altitude at which an airplane is flying above sea level or above a set reference level.

bank • The position of an airplane when its wings are tilted laterally to prevent skidding while turning.

bank indicator *(Spirit of St. Louis)* • A steel ball in one instrument, and an air bubble in another instrument, which showed when the wings were level in straight flight, or properly inclined in a turn.

barnstorm • To travel around the country, carrying fare-paying passengers on local flights (usually from farm fields or prairies).

biplane • An airplane with two main supporting surfaces (wings), placed approximately one above the other.

blackout • Temporary loss of vision and dulling of senses caused by a steep bank or a quick pull-out from a dive during which blood is forced downward from the head.

breakaway • An acrobatic stunt used in flying exhibitions, in which the wingwalker slips off the tip of a lower wing and swings down under the airplane at the end of a rope ladder or cable attached to the landing gear.

breather • A vent in the crankcase of an engine, which allows gases to escape and keeps crankcase pressure approximately equal to atmospheric.

cabane strut • A strut in the system of trussing used for supporting the overhang of an upper wing, located above the outer-bay wing struts.

cannibalize • To remove a part from one airplane for the purpose of using it on another.

canopy *(parachute)* • The main cloth (umbrella-shaped when open) portion of a parachute.

Canuck • A Canadian modification of the Curtiss JN4 (Jenny). More lightly built and with slightly higher performance.

carburetor • An apparatus used on internal combustion engines, in which passing air is charged with gasoline.

ceiling *(absolute)* • The maximum height above sea level at which an airplane can maintain horizontal flight under standard atmospheric conditions.

ceiling *(weather)* • The bottom of a cloud layer. Also the distance between the ground and the bottom of a cloud layer.

center of gravity • The point about which an aircraft will balance, irrespective of its attitude.

chandelle • An abrupt climbing turn which makes use of the momentum of the airplane to gain altitude.

chord • The longitudinal dimension of an airfoil section (wing width).

cirrus • A thin, white, featherlike cloud which forms at altitudes of 20,000 to 40,000 feet.

cockpit • A portion of an airplane in which one or more members of the crew, or sometimes passengers, are seated.

compass heading • The compass reading to be followed in maintaining the desired direction of flight. The compass heading includes compensation for variation, deviation, and wind drift.

compass rose • A circle, graduated in degrees, printed on charts to simplify measuring courses and directions.

contact • The command given when it is desired to have the engine switch turned to the "on" position.

contact flying • Flying in which the pilot maintains visual contact with the ground or water.

course, compass • The compass reading to be followed in maintaining the desired direction of flight when there is no wind draft. The compass course includes compensation for variation of deviation. When there is no wind drift, the compass course is the same as the compass heading.

course, magnetic • The direction of flight in relation to magnetic north, measured clockwise in degrees between true north and the desired path of the airplane over the ground.

course, true • The direction of flight in relation to true north, measured clockwise in degrees between magnetic north and the desired path of the airplane over the ground.

cowling • A removable metal covering on the nose of a fuselage or nacelle, used to house the engine and to reduce its air resistance.

crab • To head an airplane at an angle into the cross wind in order to keep from drifting to the leeward of the desired course.

crankshaft • The steel shaft driven by the pistons acting through the connecting rods, at one end of which the propeller is attached in the case of direct-drive aircraft engines.

cumulo-nimbus • A mountainous, billowy storm cloud, accompanied by rain and often by violent air currents.

cumulus • A billowy white cloud, varying in size from small to mountainous.

cylinder *(engine)* • The chamber in which the combustion takes place and the piston moves back and forth.

dead stick • A propeller which has stopped turning during flight. Often used to describe the condition when a propeller is still turning but no longer exerting useful thrust because of engine trouble.

deviation *(compass)* • The angular difference between the compass indication and the magnetic heading of the plane. Caused by magnetism in iron and steel parts, and by magnetic fields created by electric circuits.

direct drive engine • An engine in which the crankshaft is attached directly (not through gears) to the propeller.

dope • A cellulose acetate or cellulose nitrate material used to shrink and strengthen the cloth surfaces of aircraft and to make them impervious to air and water.

drag the field • To fly over a field at very low altitude in order to observe its surface, usually with the intent of landing.

drift • The angular movement of an aircraft in relation to the ground, to one side or the other of its heading, caused by a cross wind.

drift angle • The angle at which an airplane must be headed into the wind in order to avoid being drifted off the desired course.

earth-inductor compass • A compass which depends for its indications on the current generated in a coil revolving in the earth's magnetic field.

elevator • An airfoil, usually hinged to the back of the stabilizer, which is controllable by the pilot for the purpose of inclining the longitudinal axis of the airplane in relation to the horizon (climbing, gliding, etc.).

empennage • The tail surfaces of an airplane, including stabilizing and control surfaces.

fin • The vertical stabilizing surface (Jennies, Standards, CHs) above the fuselage, forward of the rudder.

flying speed • Any normal speed above stalling speed.

flying wire • A wire or cable running angularly between upper and lower wings (Jennies, Standards, DHs), which transmits the lift on the outer portion of the wings in toward the fuselage.

fuselage • The approximately streamline body of an airplane.

GI Bill • After WWII, Congress passed a law that funded GIs for educational programs it felt were interrupted by the war.

great circle • Any portion of the line cut on the surface of the earth by a plane surface which passes through the center of the earth. The shortest surface distance between any two points on a hemisphere.

ground loop • An uncontrollable turn while an airplane is rolling over the ground at a speed too low for the rudder to be effective.

ground school • A school which gives instruction in various subjects to students undergoing flight training.

ground speed • The speed of an aircraft in relation to the ground. It is affected by wind velocity.

hedge-hopping • Flying so low over the ground that it is necessary to pull up to clear trees and other obstructions.

Hispano-Suiza *(Model A)* • An eight-cylinder, direct-drive, vee-type, water-cooled engine, rated at 150 hp at 1450 rpm. Dry weight 445 lbs.

hp *(horsepower)* • A unit of power equal to a rate of 33,000 foot-pounds of work per minute.

Immelmann • A half-roll on top of a half-loop, reversing the direction in which the airplane was flying before the maneuver was started.

inner bay • The area and structure of the wings between (Jennies, Standards, DHs) the fuselage and the nearest pair of wing struts.

inner-bay strut • One of either pair (Jennies, Standards, DHs) of the upper-to-lower wing struts nearest the fuselage.

isogonic lines • Imaginary lines on the earth's surface, at all points on which the magnetic variations is the same.

Jenny *(Curtiss JN4-D)* • A two-place, open tandem-cockpit, tractor biplane, powered by an OX-5 engine. Used for training in World War I. Max. speed 75 mph. Landing speed 45 mph. (See photograph.)

Keystone Pathfinder *(Commercial model)* • A trimotored, fabric-covered, cabin biplane, powered by Wright Whirlwind tractor-engines. Max. speed 120 mph. Landing speed 56 mph. Built at Bristol, Pennsylvania.

landfall • The sighting of land after a sea voyage.

landing wire • A wire or cable running angularly between upper and lower wings (Jennies, Standards, DHs) which opposes the flying wire, and resists forces in the opposite direction to that of normal flight.

lean out *(the mixture)* • To reduce the amount of gasoline which is mixed with the air passing through the carburetor.

Liberty • A twelve-cylinder, direct-drive, vee-type, water-cooled engine, developing 400 hp at 1700 rpm. Dry weight 844 lbs. Designed and built for use in World War I.

line *("on the line")* • The area of a flying field on which airplanes in active use are parked between flights.

liquid compass • A compass in which the magnets and cards are immersed in liquid in order to damp out oscillation.

lock stitch • To attach the cloth covering to the ribs of an airfoil by means of heavy thread passed, with a long needle, around the rib.

longeron • One of the main longitudinal members forming the framework of the fuselage.

magnetic storm • A major disturbance of the earth's magnetic conditions.

magneto • A small dynamo, geared to an internal combustion engine for the purpose of generating and distributing high-tension current for ignition.

mixture control • The control, operated by the pilot, which regulates the ratio of gasoline to air before the mixture is drawn into the engine cylinders.

monoplane • An airplane with only one main supporting surface (wing).

mph • Miles per hour.

nacelle, engine • The approximately streamlined enclosure of an engine.

navigation lights • The colored lights which show the position and direction of movement of an airplane at night. (Green at right wing tip; red at left wingtip; yellow on tail.)

NC-4 • A biplane flying boat, specially modified for transatlantic flight; fabric-covered wings; wood hull; powered by four 400 hp Liberty engines (three tractors and one pusher). Crew of five men. Max. speed 85 mph. Landing speed 62 mph. Range, economical — 1690 statute miles.

north, magnetic • The direction indicated by the needle of a compass unaffected by local magnetic attraction.

North Magnetic Pole • A point in northern Canada where a freely suspended magnet would point vertically (latitude about 71° north; longitude about 96° west).

north true • The direction indicated by the needle of a compass unaffected by local magnetic attraction.

obstruction lights • Red lights placed on poles, towers, buildings, and other obstructions in the vicinity of an airport to warn pilots of objects which must be cleared in landing and taking off.

one-eighty • A training maneuver in which the cadet closes his throttle as he flies above a field, executes a 180-degree turn with his plane, and lands without the further use of power.

Oriole *(Curtiss)* • An open tandem-cockpit, tractor biplane with fabric-covered wings and wood monocoque fuselage. One pilot, two passengers. Powered by one Curtiss C-6 hp engine. Max. speed 98 mph. Landing speed 50 mph.

ornithopter • A flying machine with wings that flap; designed in an attempt to copy the flight of a bird.

outer bay • The area and structure of the wings (Jennies, Standards, DHs) between the inner and outer pairs of struts.

outer-bay struts • One of either pair (Jennies, Standards, DHs) of the upper-to-lower wing struts nearest the wing tip.

OX-5 *(Curtiss)* • An eight-cylinder, direct-drive, vee-type, water-cooled engine, rated at 90 hp at 1400 rpm. Dry weight 390 lbs.

OX-6 *(Curtiss)* • An eight-cylinder, direct-drive, vee-type, water-cooled engine, rated at 100 hp at 1400 rpm. Dry weight 410 lbs.

pancake • To hit the ground hard with an airplane after leveling off too high for a proper landing. The wings are in a stalled but approximately horizontal position.